Hydrodynamics of Oil and Gas

Hydrodynamics of Oil and Gas

Ian Lerche

University of South Carolina
Columbia, South Carolina

René O. Thomsen

SAGA Petroleum a.s.
Sandvika, Norway

Plenum Press • New York and London

Library of Congress Cataloging-in-Publication Data

On file

ISBN 0-306-44872-6

A Division of Plenum Publishing Corporation
233 Spring Street, New York, N. Y. 10013

Printed in the United States of America

Preface

There has long been interest in the flow of fluids through permeable aquifers. Stratigraphic trapping of oil and gas by permeability changes in an aquifer and the amounts of hydrocarbons so trapped are major concerns to the oil industry. The variations of aquifer width and geometry and of the positions in an aquifer where hydrocarbons can be trapped by hydrodynamic forces are intimately intertwined in determining the shape, and thus the volume, of hydrocarbons.

Perhaps the seminal work in this area is reflected by King Hubbert's massive review paper "Entrapment of Petroleum under Hydrodynamic Conditions" (*Am. Assoc. Pet. Geol. Bull.* **37**(8), 1954–2026, 1953), in which a wide variety of effects, such as capillarity, buoyancy, surface tension, and salinity of water, are incorporated as basic factors influencing the positioning and shaping of hydrocarbon masses in hydrodynamically active aquifers. In those days, while the basic physics could readily be appreciated, development of a detailed quantitative understanding of the interplay of the various factors in controlling or modulating hydrodynamic shapes was severely limited by computer abilities. Indeed, Hubbert actually constructed and photographed physical models, using alcohol and water, to illustrate basic concepts. It is difficult to obtain an appreciation of the behavior of flow geometries from such experiments when all factors are permitted to vary simultaneously. Such an understanding is rapidly becoming of paramount necessity as the petroleum industry struggles to diminish the risk factor in exploring for oil in the last decade of this century.

The point is that, when found (mostly serendipitously), stratigraphic reservoirs are, on a one-to-one basis, as prolific as the more easily identified structural reservoirs. However, using current technology it is extremely difficult to identify stratigraphic trapping of hydrocarbons without actually drilling to see if hydrocarbons are present, and drilling is an expensive operation. In addition, the positioning of hydrocarbons in a structural trap is, of itself, also modified by hydrodynamic driving forces, as is the access of hydrocarbons to the putative structural trap along a connecting thoroughfare. Thus, structure, stratigraphy, and permeability, by virtue of their intertwined evolution as a sedimentary basin evolves, are all connected in their impact on the migration and hydrodynamic trapping of hydrocarbons.

The advent of high-speed computers capable of handling a wide variety of nonlinear problems has meant that a reevaluation of the hydrodynamic effects on flow patterns can be handled very quickly in a workman-like fashion in a wide variety of situations. Thus, a generic "feel" for the resolution, sensitivity, uniqueness, and precision of understanding of particular regimes can be achieved rather readily.

The purpose of this book is to provide a compendium of behaviors under a variety of conditions so that an appreciation of the interesting patterns of behavior is collected in one place.

The understanding of the structure of hydrodynamic flow patterns and their influence on hydrocarbon behaviors is far from being fully developed. The impact of time-dependent effects (e.g., permeability changes by compaction, fracturing, cementation, or dissolution) on the fundamental properties of the aquifer and the basic response patterns to variable hydrodynamic flow or to oil and/or gas masses being added to (or subtracted from) the total mass budget in an aquifer, etc., have hardly been explored. Equally, solubility effects, temperature and pressure dependence of surface tension, miscibility, hydrocarbon kinetic evolution from oil to gas while in the aquifer, and a wide variety of related phenomena (e.g., episodic fracturing and resealing of capped aquifers) would seem to be worthy of a more detailed treatment than has so far been their lot.

Aware of these limitations, and also of our own abilities, we consider it appropriate to record here the variety of steady-state shapes that are permissible, and the conditions under which they obtain, so that an appreciation can be gleaned of the complexity of behaviors available.

This book is intended both for practioners in the oil industry who are actively involved in the search for subtle stratigraphic traps and for those entering the field, either as students or as full-fledged professionals.

We have tried to develop systematically the subject of hydrodynamic shaping starting with the conditions and assumptions of Hubbert. These conditions are then systematically weakened and generalized throughout the book as more involved patterns of development, more closely mimicking reality, are discussed. At each level of new development in each chapter we provide as an underpinning a short theoretical development before presenting case histories and examples. The reader interested in obtaining a quick overview of the importance of particular factors can skim through the pictorial representations without first being forced to examine a detailed quantitative development. The reader more interested in the progressive development of the subject *per se* can, on the other hand, read the quantitative development sections and bypass the case histories. The reader interested in an integrated development should, of course, read from cover to cover.

We will have succeeded in our purpose if others can develop the quantitative aspects of hydrodynamic impact to levels significantly beyond the level herein. Suggestions for such developments are recorded in the last chapter.

The results we report would not have been possible without a considerable amount of help and suggestions from many sources. Discussions and collaboration with Zhiyong He on the early phases of this work were crucial, as was the insistence of Mike Zeitlin, Christian Hermanrud, and other industrial contacts that the hydrodynamic impact on oil accumulations needed to be reviewed. We are grateful for scientific and financial support from the Industrial Associates of the Basin Analysis Group at the University of South Carolina. These Associates are Texaco, Oryx Energy, Phillips Petroleum, Conoco, BP, Chevron, Saga Petroleum, Statoil, Unocal, Marathon, and Arco. Support from the Danish Research Academy and the Danish Research Council is also gratefully acknowledged.

We thank our secretary, Donna Black, who had the task of converting our scrawl into type, and Yuan-Yuan Li, who had the thankless task of producing a great fraction of the figures using the Hydrodynamics Code given in the Appendix.

This venture would not have been possible without the support of friends and families who were infinitely forbearing during the long months it took to write this book.

Ian Lerche
Rene O. Thomsen

Columbia, South Carolina

Contents

1

Introduction

1.1. General Considerations

Two dominant factors determine that hydrodynamic forces are of paramount importance in influencing the location, shape, and amount of hydrocarbons in permeable aquifers and reservoirs. On the one hand, oil has a specific gravity of around 0.8–0.9 g/cm^3, while subsurface waters range in specific gravity from around 1 g/cm^3 to about 1.3 g/cm^3 for hypersaline saturation conditions. Thus, there is a vertical buoyancy drive on oil of around 0.1 to 0.5 g/cm^3, depending on conditions. For gas, with a density at production of around 0.1–0.3 g/cm^3, the equivalent buoyancy drive is about 0.7–1.2 g/cm^3. Thus, hydrocarbons have an innate tendency to seek out structural highs if flow conditions and flow pathways exist.

On the other hand, the deposition, and later compaction, of sediments in a basin provides a major water flow throughout the basin's geologic evolution. For example, sediments deposited with 50% porosity and then compacted to a typical reservoir porosity of, say, 10% have decreased their depositional thickness by a factor of 1.8 with the loss of 80% of pore water. Since the pore water originally made up 50% of the sediment thickness, this fact implies that the water lost corresponds to 40% of the rock thickness. For a 15,000-ft column of rock, all at 10% porosity, the implication is that 12,000 columnar feet of water have been "lost." Clearly, the subsurface flow of water in sedimentary basins is a large factor dominating the evolution of the basin on a geologic time scale.

Thus, as hydrocarbons are produced in a basin, they are subject to both an intrinsic buoyancy drive and a significant hydrodynamic drive. In order to assess the pathways of hydrocarbon migration and accumulation, it is essential that the interaction of buoyant hydrocarbons with hydrodynamic forces be examined under as wide a swath of conditions as possible. In this way we can build up an understanding of the overall picture of evolution of hydrocarbons from source to reservoir and of the subsurface conditions which aid or deter the migration.

While buoyancy and flow predominantly control the evolution of hydrocarbon migration, they have to be encapsuled within the geometric and physical frameworks that govern subsurface phenomena. Thus, the influence of permeability—its spatial and temporal variations—plays a marked role in discriminating viable flow paths from nonviable modes of transport. The variable width and variable shape of aquifer pathways, both spatially and temporally, also control the allowed locations of hydrocarbon accumulations. The variability of hydrodynamic flow and the ability of water to take gas and oil into solution, and so convect hydrocarbons toward unanticipated pathways and accumulation sites, represent another aspect requiring investigation. In addition, the leakage of an aquifer, the variability of grain sizes influencing capillary pressure effects, the precipitation (or dissolution) of minerals decreasing (increasing) the permeability, the temperature dependence of solubilities and surface tensions, and the increase in oil density with time, lowering the buoyancy by solution of the light miscible components of the oil, all have roles to play in influencing the behavior of hydrocarbon transport.

Addition of oil to an already existing hydrocarbon accumulation changes the hydrodynamic forces acting on the oil, thereby influencing the accumulation amount and also impacting the sites of allowed accumulation.

Some of these factors are undoubtedly of minor importance in relation to others that have a more pervasively rugged control on the hydrocarbon behavior. Investigation of dynamic ranges of conditions under which particular factors make their presence felt is one of the major concerns throughout this book.

1.2. Historical Perspective

Hubbert (1953) has noted that the basic acceptance of hydrodynamic flow as a dominant player in the siting of hydrocarbon accumulations had a long and checkered career for the first 80 to 100 years after oil was discovered by Drake in 1859 at Titusville, Pennsylvania. While there had been sporadic attempts to include the influence of the flow of subsurface water prior to the early 1950s, these attempts often waxed and waned with the polemic vigor of the proponents and opponents of the idea. By the early 1950s two major factors firmly entrenched the idea of hydrodynamics as a major influencing factor. First, Tiratsoo (1952) published a text in which hydraulic activity was regarded as the *sine qua non* in ascertaining hydrocarbon migration pathways and accumulation sites in the petroleum geology arena. Second, Hubbert (1953) published a massive review paper detailing 12 years' involvement (Hubbert, 1940) with the basic problem and exemplifying, by both case histories and laboratory experiments, the basic patterns of behavior impressed on mobile hydrocarbons as a consequence of hydrodynamic flow. In addition, Hubbert (1953) also spelled out a variety of basic factors influencing hydrocarbon flow and provided a fundamental quantitative development for examining the relevant factors, as well as estimates of the size and scale of various factors. A short, but fascinating, account of the early years of development of the hydrodynamic concept is also contained in Hubbert's review paper.

Since that time, there has been a steady and significant improvement in our understanding of hydrodynamic control on hydrocarbon motion. The ability to identify from field observations the likelihood that hydrodynamic action is, or was, active has been examined by Dahlberg (1982), using Hubbert's (1953) paper as a basis from which to build practical rules of thumb. Also, Berg (1975) and Davis (1987) have more recently reinterpreted the basic physics of hydraulic impact on hydrocarbon evolution in interesting manners while remaining within the basic framework of the overriding assumptions set down by Hubbert (1953) governing thin oil stringers. However, as we shall see in Chapter 2, even within the restrictive assumptions of the thin oil stringer approximation, there are degrees of indeterminacy in assessments of the length of an oil stringer.

1.3. Fundamental Rules

1.3.1. Darcy's Law

Darcy (1856) showed experimentally that the velocity, v, of a fluid flowing through a stationary porous medium is proportional to the applied pressure gradient, ∇p, and inversely proportional to the viscosity, μ, of the fluid, so that

$$v = -\frac{\kappa}{\mu}\nabla p. \tag{1.1}$$

The proportionality factor, κ, is the permeability of the porous medium. Note that Eq. (1.1) is an equation of motion for the fluid, thereby replacing Newton's equations of motions.

Darcy's law refers specifically to the applied pressure gradient. In terms of subsurface fluid pressures with a total fluid pressure, p_{tot}, and a fluid pressure p_e in excess of hydrostatic, we have

$$\nabla p_{\text{total}} = \nabla p_e + \rho_w g \hat{z} \tag{1.2}$$

where ρ_w is water density, and $\hat{z}$ is a unit vector in the direction of gravity, of strength g. The relevant pressure in Eq. (1.1) is the excess pressure, p_e.

Darcy's law is often written in terms of an equivalent hydraulic head, H:

$$v = -\frac{\kappa \rho g}{\mu} \nabla H \equiv -K \nabla H \tag{1.3}$$

where $K \equiv \kappa \rho g/\mu$ is the hydraulic conductivity. The form (1.3) is often used in applications to groundwater flow. Throughout this book, we use the excess pressure formalism because this formalism makes it easier to visualize some important concepts.

Darcy's law operates only under conditions where the spacing, b, between solid matrix particles is much smaller than any other scale of concern in the fluid flow (Scheidegger, 1960; de Wiest, 1969; Bear, 1972). In addition the permeability, κ, depends on both the lithology type and the porosity, ϕ, or void ratio, $V\ [\equiv \phi/(1\text{-}\phi)]$, of the rock in question. For instance, Dutta (1987) has shown that for Gulf Coast shales and clays a good representation is

$$\kappa = k_0 V^m \tag{1.4}$$

with $m \approx 5$ and $k_0 \approx 3.4\ \mu\text{D}$ ($1\ \text{D} \cong 10^{-8}\ \text{cm}^2$) (Fig. 1.1).

Empirically, it appears that a power law representation of permeability and void ratio of the form (1.4) describes a considerable variety of rock types, but the parameters k_0 and m vary significantly, with m typically in the range 3 ± 3 and k_0 varying from microdarcies for shales and clays to darcies for clean sands and gravels. As we shall see later, this massive change in permeability (by factors of order 10^6) between shales and sands causes some interesting behaviors.

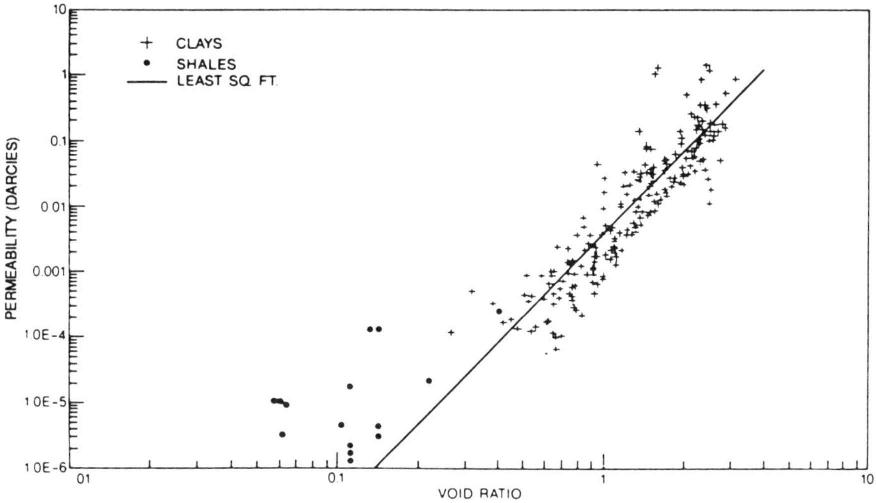

Figure 1.1. Permeability versus void ratio, $\phi/(1 - \phi)$, of U.S. Gulf Coast clays and shales (after Dutta, 1987).

1.3.2. Surface Tension and Capillary Pressure

1.3.2.1. Surface Tension

At the interface between two dissimilar fluids (oil and water, say), there is a surface tension keeping the two fluids in separate phases (Laplace, 1789).

For a common radius of contact, r, between the two fluids, the equivalent discontinuity in surface pressure is

$$\Delta p_{\text{S.T.}} = 2T/r \tag{1.5}$$

where T is the surface tension. For oil and water the surface tension at 20°C is around 35 dyn/cm and decreases as the temperature increases. The surface tension is composition specific, being higher for thick heavy crude oils and also higher for hypersaline waters replacing fresh water.

1.3.2.2. Capillary Pressure

If a volume of oil is injected into a water-saturated rock, there is a pressure difference p_c across the interface between the oil and the water. If the rock is preferentially wet by water with respect to oil, p_c will be positive so that the pressure inside the oil will be greater by p_c than that inside the water; if the rock is preferentially wet by oil, p_c will be negative. Except in rare cases, the rocks are preferentially wet by water (p_c positive) (Fig. 1.2).

The oil volume will adjust so that capillary energy, p_c/ρ_o, will be a minimum, where ρ_o is the density of oil. For constant ρ_o, the configuration must be such that p_c is at its minimum value. The pressure p_c between two nonmiscible fluids whose interfacial tension is T is related to the principal radii of curvature, r_1 and r_2, of the interface by Plateau's equation (Shorthose, 1936):

$$p_c = T\left(\frac{1}{r_1} + \frac{1}{r_2}\right) \tag{1.6}$$

with the higher pressure on the side of the center of sharpest curvature. In a water-wet rock, p_c is then at a minimum when the oil occupies the largest accessible voids.

If the rock is wet by oil, the capillary pressure p_c would be negative, and for p_c/ρ_o to be a minimum would require that oil be in the smallest accessible voids (Fig. 1.3).

The pressure p_c is the minimum possible value of the capillary pressure, the initial displacement pressure, obtained when oil is injected into a water-saturated rock, as given by the equation

$$p_c = \frac{CT\cos\alpha}{d} \tag{1.7}$$

where C is a dimensionless factor of proportionality, α is the contact angle in the water phase that the oil–water interface makes with the solid boundary, and d is the mean grain diameter (Hubbert, 1953).

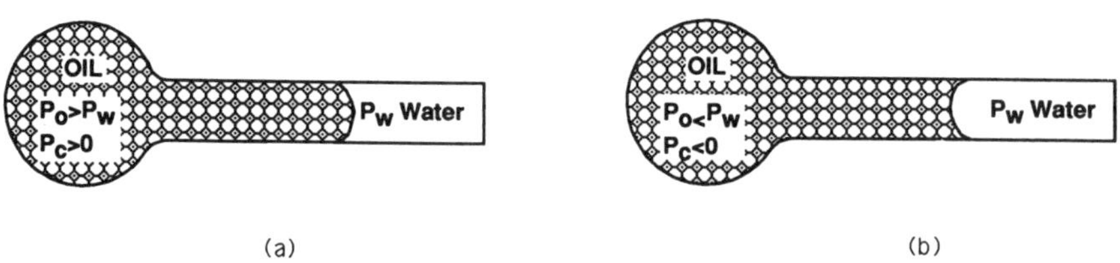

Figure 1.2. Diagrams illustrating variation of capillary pressure with wettability. (a) In preferentially water-wet container, capillary pressure in oil is positive. (b) In preferentially oil-wet container, capillary pressure in oil is negative.

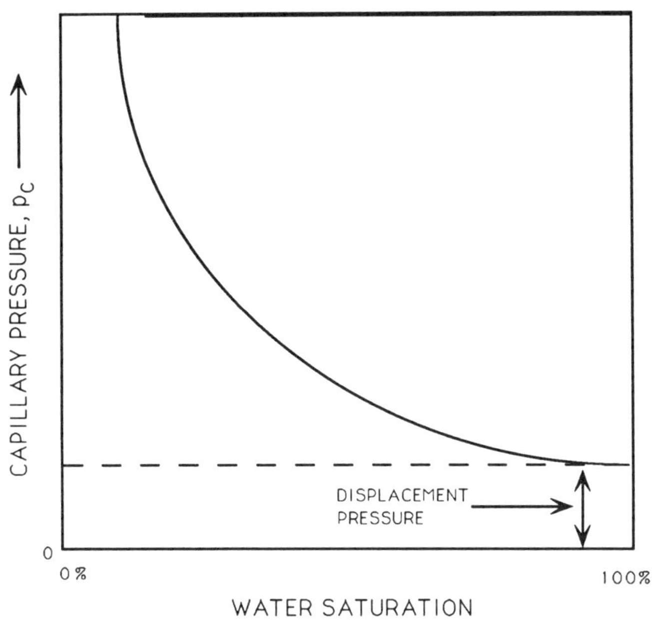

Figure 1.3. Capillary displacement pressure of oil or gas in water-saturated sand.

In order of magnitude, $T \cos \alpha$ may be taken to be about 25 dyn/cm, and the factor C, as determined from the work of Purcell (1949), has a value of about 16 (Hubbert, 1953). Thus, p_c is about $(400/d)$ dyn/cm^2. The magnitude of the capillary displacement pressures in sediments of various degrees of coarseness is shown in Table 1.1.

The force due to capillarity for oil is given by

$$-\frac{\text{grad } p_c}{\rho_o} = \frac{CT \cos \alpha}{\rho_o} \frac{\text{grad } d}{d^2} \tag{1.8}$$

Table 1.1. Capillary Displacement Pressure of Water against Oil in Sediments of Various Grain Sizes

Sediment	Grain diameter, d (mm)	Capillary pressure, P_c (atm)
Clay	Less than 2^{-10}[a]	Greater than 1
Silt	2^{-10} to 2^{-4}	Between 1 and 2^{-4}
Sand	2^{-4} to 2	Between 2^{-4} and 2^{-11}
Granules	2 to 2^2	Between 2^{-11} and 2^{-12}

[a]The value of 2^{-10} mm for clay particles is a maximum; much finer clays are known. In a clay with a particle size of 10^{-4} mm, for example, P_c would be about 40 atm.

Under capillary forces alone, oil will tend to move in the direction of increasing grain size. Hubbert (1953) provides a special example:

> If we let the increase in grain diameter be proportional to the grain diameter itself by some fixed factor β, we should then have
>
> $$E_0 \equiv -\nabla p_c/\rho_o = \frac{CT \cos\alpha}{\rho_o}\frac{\beta d}{d_2} = \frac{C'}{d} \tag{1.9}$$
>
> where
>
> $$\beta = (1/d) \text{ grad } d$$
>
> and
>
> $$C' = \frac{\beta CT \cos\alpha}{\rho_o}$$
>
> If we arbitrarily set $\beta = 1/500$ per centimeter, corresponding with a rate of change of grain size that would double the grain diameter in about 350 cm, the factor C' assumes the value of approximately 1 cm^2/sec^2. Then for a shale with particle diameter 10^{-5} cm, E_0 would have the value of 10^5 dynes per gram, or 100 times gravity; for a fine sand with $d = 10^{-2}$ cm this would reduce to 100 dynes per gram, or 0.1 of gravity.

Between a sandstone and a shale, discontinuities occur, and these require that we deal with a *finite* difference

$$\Delta p_c = CT \cos\alpha \left[\frac{1}{d_2} - \frac{1}{d_1}\right] \tag{1.10}$$

where Δp_c is the difference in the capillary displacement pressures in the two media, and d_2 and d_1 are their respective mean grain diameters.

Table 1.1 shows that the capillary pressure of oil in a shale is on the order of tens of atmospheres, while in a sand it drops to the order of tenths of an atmosphere. Hence, oil would be expelled from the shale into the sand by an unbalanced pressure of the order of tens of atmospheres.

A shale–sand interface then appears as a surface of unidirectional flow to oil (or gas). Across such a boundary the oil flows from the shale to the sand. Unless a pressure is applied to oil in sand that is greater than the capillary pressure opposing the flow of oil into shale, no flow of oil into shale occurs. For a single-phase fluid, this difference of capillary pressure does not exist, and the fluid can flow equally in either direction. A sand–shale boundary can be regarded as an impermeable barrier to oil trapped in sand, but water can flow in either direction. Note also that in a monolithologic aquifer, in which the grain size varies with position along the aquifer, there will be a capillary pressure forcing the oil to the largest pores, or preventing it from crossing less porous (lower permeability) regions. These factors lead to sealing conditions and stratigraphic trapping conditions by a shift from coarse-grained to fine-grained material or by preferential cementation cutting down the pore sizes.

1.3.3. Buoyancy Pressure

By Archimedes' principle, displacement of water, of density ρ_w, by oil, of density ρ_o, causes a vertical buoyancy pressure gradient,

$$\nabla \text{buoyancy} = -(\rho_w - \rho_o)g\hat{z} \tag{1.11}$$

pushing the oil upward. If no other forces act, then, under Darcy's law, the vertical oil speed is

$$\begin{aligned} v_o &= -\frac{\kappa_o}{\mu_o} \nabla \text{buoyancy} \\ &= \frac{\kappa_o g}{\mu_o} (\rho_w - \rho_o) \end{aligned}$$

where κ_o is the permeability of oil, and μ_o is the oil's viscosity.

For a permeability of 1 D, a viscosity of 1 cP (cP = centipoise), and a density difference of $\rho_w - \rho_o = 0.1$ g/cm^3, the vertical oil flow speed is 3×10^5 m/yr. This flow is exceptionally fast compared to geologic evolution speeds, such as sediment deposition rates (of order 10^{-4}–10^{+3} m/Myr), implying that oil will adjust virtually instantaneously (on a geologic time scale) to prevailing conditions even if the conditions, in turn, change with time. Thus, the hydrodynamic flow of oil and/or gas can be considered to be approximately in a steady state at each instant of time, adjusting quickly to changes in conditions. This behavior is the dominant reason that we can consider steady-state situations throughout this monograph.

1.4. Quantitative Procedures

The dominant concern throughout will be to combine the effects of capillary pressure, surface tension, buoyancy, and Darcy's law into a single equation describing the flow of water around a self-consistent shape, in such a way that the shape is in force balance equilibrium at all times with prevailing conditions. The volumetric contents of the self-consistent shape will be taken to be a fluid phase distinct from the water phase so that we can examine the impact of hydrodynamic conditions on the position and shape of a hydrocarbon body in an aquifer.

To set the stage for further development within a familiar context, in the next chapter we examine the simple case of a thin oil stringer—a situation which has formed the cornerstone of nearly all developments since its detailed exposition by Hubbert (1953).

2

Hydrodynamics of Thin Oil Stringers

Over half a century has passed since Hubbert (1940) discussed the role of hydrodynamic forces in controlling both the position, relative to structural and stratigraphic variations, and the length and shape of a thin oil stringer. In a review in 1953, Hubbert commented on the impact of hydrodynamics on oil accumulations. In particular, he stated, "If many such accumulations are not to be overlooked, we must supplement our customary knowledge of structure and stratigraphy with the three-dimensional ground-water hydrology of every petroliferous basin."

In the 1953 review, Hubbert analyzed a wide variety of phenomena in terms of their impact on the entrapment of hydrocarbon accumulations under hydrodynamic and hydrostatic conditions. For instance, the subject of groundwater potential is discussed on pp. 1958–1960; the field of force for a fluid between two slant, parallel, surfaces on pp. 1961–1964; the field of flow on pp. 1964–1965; the relation between force and flow on pp. 1965–1969; the effect of inhomogeneities on pp. 1969–1973; migration and entrapment of oil *and* gas on pp. 1974–1975; the effect of capillary pressure on pp. 1975–1979; the migration in reservoir beds on pp. 1979–1992; the order of magnitude of effects on pp. 1992–1993; the impact of changes from fresh water to salt water on pp. 1993–1997; types of hydrodynamic traps and variations in trap position with flow state changes on pp. 1997–2005; and experimental confirmation both in the laboratory and from geologic evidence on pp. 2006–2025. Hubbert's 1953 review has formed the basis over the last half century for major developments in primary hydrocarbon migration from shales to carrier beds, the recognition of shales as good seals, the creation of stratigraphic traps, and so on, as detailed in both reservoir engineering and petroleum geology texts (e.g., Hobson, 1954; Amyx *et al.*, 1960; Nind, 1981; Dahlberg, 1982).

The effect of capillary pressure as a dominant control in such hydrodynamic conditions was clearly recognized by Hubbert (1953), and the understanding of its role has developed further since then (e.g., Berg, 1975). While rearrangements (Davis, 1987) of Hubbert's equations to examine sensitivity of the capillary pressure dominance to parameters related to fluid type (water, oil, gas) and geologic conditions (permeability, lithology, geometry) are of interest, nevertheless the basic rules of operation were set down long ago by Hubbert (1953).

Perhaps the most widely used part of Hubbert's (1953) paper is the equation in the abstract which relates the angle of inclination of a thin oil stringer to the potentiometric gradient. Hubbert (1953) writes:

> Under hydrodynamic conditions accumulations of oil or gas will invariably exhibit inclined oil– or gas–water interfaces with the angle of inclination given by
>
> $$\tan\theta = \frac{dz}{dx} = \frac{\rho_w}{\rho_w - \rho_o}\frac{dh}{dx} \tag{2.1}$$
>
> where dz/dx is the slope of the interface, ρ_w the density of the water, and ρ_o that of the oil (or gas), and dh/dx the component of slope of the potentiometric surface in the horizontal direction, x.

Quite often, the length, Δz, of an oil stringer is estimated from the difference, Δh, in measurements at the same depth of potentiometric head at two locations, using

$$\Delta z = \frac{\rho_w}{\rho_w - \rho_o}\Delta h \tag{2.2}$$

Unfortunately, neither Eq. (2.1) nor Eq. (2.2) contains any reference to capillary pressure—in accord with Hubbert's (1953) explicit derivation from a general result [p. 1980; "as obtained from equation (29) with the capillary term omitted"]. In an attempt to obtain an expression which combined the effects of capillarity and hydrodynamic drive on an oil stringer, Berg (1975) returned to Hubbert's equation (29) and obtained a correction factor, which we shall discuss a little later in this chapter. A less general rederivation of the capillary effect discussed by Hubbert (1953) is presented in Davis (1987).

The purpose of this chapter is to provide a discussion of the appropriate formulas for describing the length of an oil stringer and to spell out the conditions of derivation and application of the relevant results. We do this for two main reasons: (i) to examine the conditions under which the popular Hubbert equation (Eq. 2.1) can be safely used to describe an oil stringer, and (ii) to examine the general result and exhibit its limiting forms under different conditions.

2.1. Hydrodynamic Conditions

In order to set the stage for later development, in this section we consider several illustrative cases of hydrodynamic and capillary forces acting on an oil stringer. These cases illuminate very sharply the consequences of the underlying direct and indirect assumptions made in deriving the appropriate tilt, shape, and length formulas describing oil stringers.

2.1.1. Parallel Oil Stringer, No Capillary Effect

Consider the situation sketched in Fig. 2.1 in which a homogeneous stringer of oil is in a uniform aquifer in which there is a steady down-dip uniform flow of water. We will make the following assumptions:

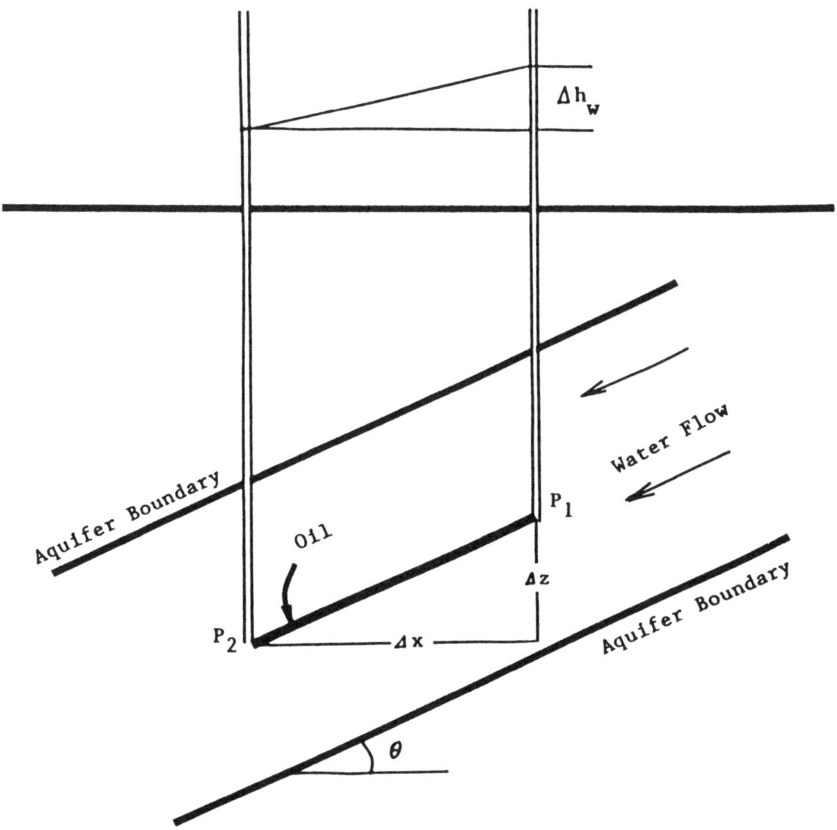

Figure 2.1. Diagram of the geometry of an oil stringer in a confined aquifer.

1. The stringer is aligned parallel to the direction of flow and is at rest.
2. The flow lines are uninfluenced by the presence of the stringer.
3. Capillary pressure within the stringer and at the stringer–water surface is negligible.
4. The stringer is sufficiently thin that variations across the stringer and effects at the ends of the stringer can be ignored.
5. Boundary effects at the edges of the aquifer are negligible.

Then there is a pressure difference $\Delta p(\equiv p_2 - p_1)$ between the upper and lower ends of the stringer because of the difference in the height of the water column. This pressure difference is given by

$$\Delta p = \rho_w g \Delta z_o - \rho_w g \Delta h_w \tag{2.3}$$

where g is the acceleration due to gravity, ρ_w is the (constant) water density, Δz_o is the vertical difference in height between the top and bottom of the stringer, and Δh_w is the difference in water head between the top and bottom of the stringer.

This pressure difference, Δp, must be balanced by the pressure due to the weight of the stringer, of density ρ_o, i.e.,

$$\Delta p_o \equiv \rho_o g \Delta z_o \,(= \Delta p) \tag{2.4}$$

Substituting from Eq. (2.3) in Eq. (2.4), we obtain

$$\Delta z_o = \frac{\rho_w}{(\rho_w - \rho_o)} \Delta h_w \tag{2.5}$$

which is immediately recognizable as corresponding to Eq. (2.2). Expressing this relation in terms of gradients, we can write

$$\Delta z_o = \frac{\rho_w}{(\rho_w - \rho_o)} \left(\frac{\Delta h_w}{\Delta x} \right) \Delta x \tag{2.6}$$

which, taking limits (thin stringer approximation), gives

$$\tan \theta \equiv \frac{dz_o}{dx} = \frac{\rho_w}{(\rho_w - \rho_o)} \frac{dh_w}{dx} \tag{2.7}$$

where θ is the dip angle of the aquifer, by assumption 1. Equation (2.7) is just Hubbert's formula in the abstract of his 1953 paper. Here, Δz_o is supposed to represent the height of an oil stringer trapped by down-dip flow, while Δx is the lateral width over which the column stringer exists.

However, there are clearly several problems with this interpretation of the Hubbert formula. Suppose, first, that we want to estimate the maximum column height, Δz_o, of trapped oil (gas), a problem of considerable interest to the oil industry in estimating reserves in an oil field.

Then, given a fixed measurement of the lateral potentiometric gradient, dh_w/dx, it follows that as $\Delta x \to \infty$, so too does Δz_o. Thus, we have the problem that we cannot deduce the length of an oil stringer because if the oil stringer is indeed parallel to the flow, as assumed, then since

$$\frac{dh_w}{dx} = \frac{\rho_w - \rho_o}{\rho_w} \tan\theta \tag{2.8}$$

and since, further, θ, ρ_w, and ρ_o are all constant, then the lateral potentiometric gradient would also have to be constant! This constancy is unlikely in real situations. We would have to actually measure the down-dip and up-dip ends of the oil column in order to ascertain its length, and that is precisely what we are trying to avoid being forced to do—we would prefer to predict the stringer's maximum length. Furthermore, even if we did measure a constant value for dh_w/dx, we still cannot predict Δz_o without measuring Δx. What the Hubbert formula really shows is that Δx is proportional to Δz_o under the conditions and assumptions of its derivation. This point was also made in a different manner by Dahlberg (1982), who commented that the shape of the oil accumulation does in fact alter the flow lines—contrary to assumption 2—and this alteration was also discussed and demonstrated by Hubbert (1953, Figs. 32 and 33), and even as early as a century ago by Lamb (1879).

Let us leave this problem to one side for the moment and concentrate on the question of motion of an oil stringer.

2.1.2. Motion, Capillary Pressure, and Flushing

Consider the same oil stringer under the same assumptions as in the previous subsection. Can we predict its direction of migration even if we cannot predict its length? Clearly, the answer is affirmative for when dh_w/dx is less than, equal to, or greater than $[(\rho_w - \rho_o)/\rho_w] \tan\theta$, the oil stringer migration is up dip, stationary, or down dip, respectively.

It is also clear that as long as we demand that the Hubbert formula is precisely satisfied, then we have "frozen" the location of the oil stringer, no matter how long it is. Because we want to investigate the effect of capillary pressure forces on the oil stringer behavior, it is appropriate to investigate the oil stringer problem *de novo*. Suppose then that we retain the same geometry as sketched in Fig. 2.1 but relax assumption 3 (capillary pressure negligible) while retaining the other four assumptions.

Then the pressure difference caused by water flow is not equal to the pressure difference in the oil column, and, in order to hold the oil stringer at rest, the difference in pressures must be balanced by a capillary pressure difference, p_c, between the stringer and its surrounding milieu (Shorthose, 1936); i.e.,

$$p_c = \Delta p - \Delta p_o \tag{2.9}$$

Inserting the expressions for Δp from Eq. (2.3) and Δp_o from Eq. (2.4) into Eq. (2.9), we can rearrange matters to read

$$\Delta z_o = \frac{p_c \tan \theta}{g\rho_w[\tan \theta(\rho_w - \rho_o)/\rho_w - dh_w/dx]} \tag{2.10}$$

Inspection of Eq. (2.10) shows that as the capillary pressure is ignored, the classical Hubbert formula results by setting the denominator of Eq. (2.10) to zero at the same time. Berg (1975) appeared to be hinting at the need for such an equation but did not arrive at the required form given by Eq. (2.10).

Two different ways of deriving Eq. (2.10), presented below, also shed light on the balance of forces involved in a thin oil stringer.

2.1.2.1. An Argument Based on Buoyancy Pressure

Suppose the buoyancy pressure of the oil stringer is balanced by the excess pressure difference, Δp_{excess}, caused by the flow and by the capillary pressure difference. Then

$$g\Delta z_o(\rho_w - \rho_o) = \Delta p_{\text{excess}} + p_c \tag{2.11}$$

However, the excess pressure difference is no more than $g\rho_w\Delta h_w$, and this in turn can be written

$$\Delta p_{\text{excess}} = g\rho_w\Delta h_w = g\rho_w \frac{\Delta h_w}{\Delta x}\left(\frac{\Delta z}{\tan \theta}\right) \tag{2.12}$$

Inserting Eq. (2.12) into Eq. (2.11) and rearranging, we obtain

$$\Delta z_o = \frac{p_c \tan\theta}{g\rho_w}\left[\frac{(\rho_w - \rho_o)\tan\theta}{\rho_w} - \frac{dh_w}{dx}\right]^{-1} \qquad (2.13)$$

2.1.2.2. An Argument Based on the Potential Method

The water head, h_w, is related to both the height, z, and the fluid pressure, P, in excess of hydrostatic through

$$h_w = z + (P/\rho_w g) \qquad (2.14a)$$

That is,

$$P = \rho_w g(h_w - z) \qquad (2.14b)$$

Along the flow direction, the fluid pressure gradient is

$$\frac{dP}{dL} = \rho_w g\left(\frac{dh_w}{dL} - \frac{dz}{dL}\right) \qquad (2.15)$$

where dL is an increment in distance along the flow line, and $dL = dx \sec\theta$, or $dL = dz/\sin\theta$.

Then the force on a unit mass of oil along the flow direction is

$$F_o = \frac{-dh_w}{dL} = -\left(\frac{gdz}{dL} + \rho_o^{-1}\frac{dP}{dL}\right) \qquad (2.16)$$

The total force (per unit cross section) along the length of the oil stringer is

$$\begin{aligned} F_t &= F_o\rho_o L \\ &= \Delta zg\left[(\rho_w - \rho_o) - \frac{dh_w}{dx}\frac{\rho_w}{\tan\theta}\right] \end{aligned} \qquad (2.17)$$

If this force is again balanced by the capillary pressure difference, p_c, we obtain, as before,

$$\Delta z = \frac{p_c \tan\theta}{g\rho_w[\tan\theta(\rho_w - \rho_o)/\rho_w - dh_w/dx]} \qquad (2.18)$$

2.2. Self-Consistent Oil Stringers

By whatever means, we see that as long as assumptions 1, 2, and 5 are in force, it appears superficially as though the inclusion of capillary pressure allows a determination of the length of an oil stringer.

And yet all is not well even with the formula given by Eq. (2.18). First, we note that if the water density is lower than the density of oil, perhaps as a result of gas charging of the water for instance, then Δz is negative. The correct inference is that no stable state then exists so the oil stringer will be flushed down dip.

We also note, as commented on previously, that if the denominator of Eq. (2.18) is zero, so that we recover Hubbert's (1953) formula, we have a critical condition under which the oil stringer can be of arbitrary length, which condition is unimpacted by capillary pressure. Only if $\Delta z > 0$ [denominator of Eq. (2.18) is positive] do we then have a finite length for the oil stringer. These difficulties in determining the length of an oil stringer are not peculiarly beholden to the presence or absence of capillarity, although they are influenced by p_c. Rather, they stem directly from the more fundamental fact that any *finite*-width oil stringer (no matter how thin) *does* change the flow lines around the stringer, and it is these distortions of the flow by the oil stringer that provide constraint forces limiting the length of a stringer.

Indeed, this point was clearly recognized by Hubbert (1953), who, through his equation (49), gave the equation for the shape of the bounding curve between an oil region and a water domain (in the absence of capillarity) in the form

$$\frac{dz}{dx} = \frac{\partial h_w/\partial x}{[(\rho_w - \rho_o)/\rho_w] - (\partial h_w/\partial z)]} \tag{2.19}$$

Note the presence of the *partial* derivatives of hydraulic head in Eq. (2.19), implying that h_w is a function not only of lateral position, x, but also of vertical position, z. Thus, if one knew the functional shape of $h_w(x,z)$, then one could solve Eq. (2.19) for the bounding flow line. Unfortunately, the variation of hydraulic head with vertical and horizontal position is also determined by the *shape* of the oil stringer, because the flow of fluid past the oil stringer is influenced by the stringer and, in turn, influences the oil shape itself. This fundamental point seems to be underappreciated despite its recognition over a century ago by Lamb (1879).

Physical arguments also show that the problem can be resolved only by allowing for flow distortion around an oil stringer. Consider a long stringer whose width makes up a fraction f of the aquifer width. Then if the stringer is indeed parallel to the flow, it follows that the flow speed must increase by an amount $1/(1 - f)$ in transiting past the stringer. For a very wide stringer ($f \rightarrow 1$), the flow speed becomes very large and is indeed seriously influenced by the stringer. Further, the assumption of the stringer "paralleling" the flow is not valid as Hubbert's equation (49) [our Eq. (2.19)] clearly shows.

The overriding controls on distortions to the fluid flow are (a) the presence of the aquifer boundaries and (b) the presence of an oil stringer whose shape is to be found.

The essence of this problem was solved over a hundred years ago by Lamb (1879). The argument goes as follows. The vertical force on a buoyant oil stringer is intrinsically upward. Hence, the only stable equilibrium is when the oil hugs the top surface of the aquifer, where the buoyancy pressure is balanced by the resistance of the boundary. This result is clearly seen in the figures of Hubbert's (1953) paper.

Thus, we conclude that:

(i) The shape and location of a thin oil stringer are *not* determinable by elementary means, as Hubbert (1953) noted. However, the stability with respect to up-dip or down-dip hydrodynamic motion of such a stringer *is* determinable by Hubbert's (1953) arguments.

(ii) The shape and stability of a *finite*-thickness oil stringer are determinable by arguments presented in detail by Lamb (1879) over 100 years ago.

2.3. Salinity and Gas Effects

Within the framework of the thin oil stringer approximation, the effects of saline waters, and of gas replacing oil as the buoyant phase, are most easily accommodated by increasing the water density, ρ_w, in approximate proportion to the salinity; i.e., $\rho_w = \rho_f(1 + S)$, where ρ_f is the density of fresh water, and S is the salinity (in parts per million). Effects of temperature, T, can readily be included by allowing for the thermal expansion coefficient, α, in ρ_f, i.e., by writing $\rho_f(T) = \rho_f(T_s)[1 - \alpha(T - T_s)]$, where T_s is a reference temperature. Gas buoyancy is accommodated by replacing ρ_o by ρ_g, with a gas law connection replacing the (virtually) incompressible density of oil. Thus, at a subsurface depth z where the gas pressure is P_g and the temperature is T, both of which vary with depth, the gas density is described by the perfect gas law,

$$\rho_g = mP_g/kT$$

where m is the mass of a gas molecule and k is Boltzmann's constant, provided the pressure–temperature conditions are far removed from critical point behavior (Nind, 1981), that is, provided the mean free path, λ, between gas molecule collisions is much larger than $N^{-1/3}$, where $N = \rho_g/m$. Should λ be much smaller than $N^{-1/3}$, then the gas is best described as being quasi fluid. The two end-member cases of a perfect gas and a perfect fluid bracket the transitional cases for gas transport.

We shall return to these factors, and the impact of hydrodynamic forces on shaping mobile masses of gas and/or saline water, in later chapters.

3

Generalization of Hubbert's Conditions

3.1. Fixed Angle Aquifer

The overriding controls on distortions to the "thin stringer" fluid flow are (a) the presence of aquifer boundaries and (b) the presence of a buoyant stringer whose shape is to be found. Now, the vertical force on a buoyant stringer is intrinsically upward. Hence, the only stable equilibrium is when the stringer hugs the top surface of the aquifer, where the buoyancy pressure is balanced by the resistance of the boundary, as seen in the figures of Hubbert's (1953) paper dealing with physical experiments.

Under these conditions, the shape of the stringer can be examined if a steady-state condition is assumed. Consider an ideal aquifer that has a uniform thickness, H, and through which there is a down-dip flow of water (Fig. 3.1). The aquifer is taken to make a *fixed* angle θ with respect to the horizontal to z, the direction of gravity; that is, $\theta = 0$ represents a horizontal aquifer, and $\theta = \pi/2$ represents a vertical aquifer. The flux of water, Q, through the aquifer is given by

$$Q = \frac{\kappa \ell}{\mu}(H - h)\frac{dP_{ex}}{d\ell} \tag{3.1}$$

where $\kappa(\ell)$ is the permeability of the aquifer as a function of the distance, ℓ, along the aquifer from a reference point, μ is the viscosity of water, $dP_{ex}/d\ell$ is the water pressure gradient, H is the width of the aquifer, and h is the width of the oil stringer.

If the flux, Q, is to be conserved along the aquifer, the excess water pressure due to water flow at any point along the oil–water interface is

$$P_{ex} = Q\mu \int_0^{\ell} \frac{d\ell'}{\kappa(\ell')[H - h(\ell')]} \tag{3.2}$$

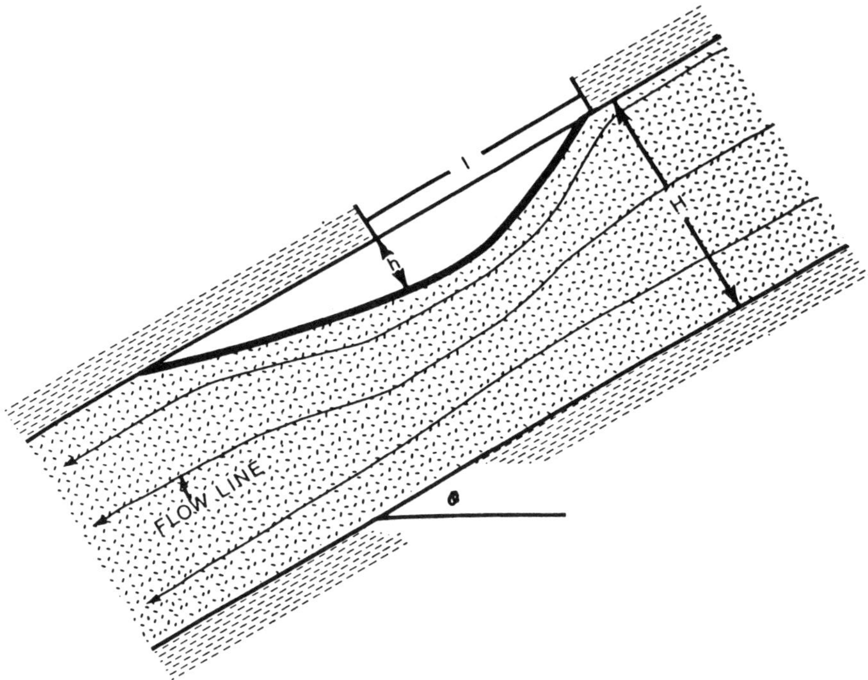

Figure 3.1. Diagram of the geometry and location of an oil stringer in a confined aquifer.

Here, the reference pressure level is set to zero at the *upper* end of the oil stringer for convenience. Thus, the total water pressure at any point is given by

$$P_w = \rho_w g(\ell \sin\theta + h\cos\theta) - Q\mu \int_0^{\ell} \frac{d\ell'}{\kappa(\ell')[H - h(\ell')]} \qquad (3.3)$$

while the oil pressure is simply a function of the weight:

$$P_o = \rho_o g(\ell \sin\theta + h\cos\theta) \qquad (3.4)$$

where ρ_w and ρ_o are the densities of water and oil, respectively, and g is the acceleration due to gravity. The pressure resulting from interfacial tension is a function of the curvature of the oil–water interface:

$$P_s = \frac{2T(d^2h/d\ell^2)}{\left[1 + \left(\frac{dh}{d\ell}\right)^2\right]^{3/2}} \qquad (3.5)$$

where T is the interfacial tension between oil and water.

Balancing these pressures and rearranging matters, we have

$$(\rho_o - \rho_w)g(\ell \sin\theta + h\cos\theta) + Q\mu \int_0^{\ell} \frac{d\ell'}{\kappa(\ell')[H - h(\ell')]} = \frac{2T(d^2h/d\ell^2)}{\left[1 + \left(\frac{dh}{d\ell}\right)^2\right]^{3/2}} \tag{3.6}$$

Equation (3.6) gives the shape of the oil stringer in terms of the oil's width h as a function of ℓ along the aquifer under steady-state conditions. Equation (3.6) can be simplified by taking a Dupuit approximation (Turcotte and Schubert, 1982; $|dh/d\ell| << 1$), assuming that the width h varies slowly along ℓ; i.e.,

$$(\rho_o - \rho_w)g(\ell \sin\theta + h\cos\theta) + Q\mu \int_0^{\ell} \frac{d\ell'}{\kappa(\ell')[h - h(\ell')]} = 2T\left(\frac{d^2h}{d\ell^2}\right) \tag{3.7}$$

Equation (3.7) can be solved numerically for a given mass amount of oil. To understand the influence of various factors on the shape of the oil stringer, several synthetic cases are simulated using Eq. (3.7). In each case one of the parameters is varied with the others being held constant at the "default" values in Table 3.1. A linear variation of permeability, $\kappa = 100(1 + \beta\ell)$, is applied along the aquifer to simulate gradual facies change, where permeability at the upper end ($\ell = 0$) of the oil stringer is 100 mD. With ℓ expressed in centimeters, a β of 0.0001 is used based on the assumption that over a kilometer scale the permeability changes from 100 mD, representative of a barrier facies of lagoonal origin, to 1100 mD,

Table 3.1. Default Parameters Used in the Simulations

Permeability [$\kappa = 100(1 + \beta l)$] (mD)	$\beta = 0.0001\ \text{cm}^{-1}$
Mass of oil (cm^2)	200,000
Flow rate of water (cm/yr)	0.2
Tilt angle of aquifer	30°
Density of water (g/cm^3)	1.0
Density of oil (g/cm^3)	0.9
Viscosity of water (cP)	1.0
Interfacial tension (dyn/cm)	35
Width of aquifer (cm)	1000
Acceleration due to gravity (cm/s^2)	980

representative of a good reservoir sand of littoral marine bar type. A flow rate of 0.2 cm/yr is somewhere between the rate of flow caused by compaction and that due to typical meteoric recharge. Interfacial tension between oil and water is a function of temperature and pressure. At pressures of about 100 psi or more, medium-gravity oils have interfacial tensions of 30–35 dyn/cm against their connate waters. Gases usually have a higher interfacial tension (Berg, 1975).

The role of permeability variation in trapping oil is well understood, although it may be further complicated by the presence of a down-dip flow of water. Figure 3.2 clearly shows that with a higher permeability gradient, the width of the oil stringer increases and the maximum thickness is further up dip. With a uniform permeability throughout the aquifer ($\beta = 0$), the oil stringer still exists, because of the flow and the steady-state assumption, but with a fixed amount of oil supply the "reservoir" would be more economical if the permeability increased down dip.

With increasing amounts of oil trapped in the aquifer, the width of an oil stringer increases more rapidly compared to its length (Fig. 3.3).

A dramatic change of the shape is observed when different flow rates are applied to the system (Fig. 3.4), and the changes in length and width are somewhat proportional. The maximum width is obtained at a flow rate of about 0.8 cm/yr, while the stringer is found to be unstable when the flow rate is even higher.

The exact limits at which the stringer becomes unstable differ when parameters are varied from the default values in Table 3.1, and the precise

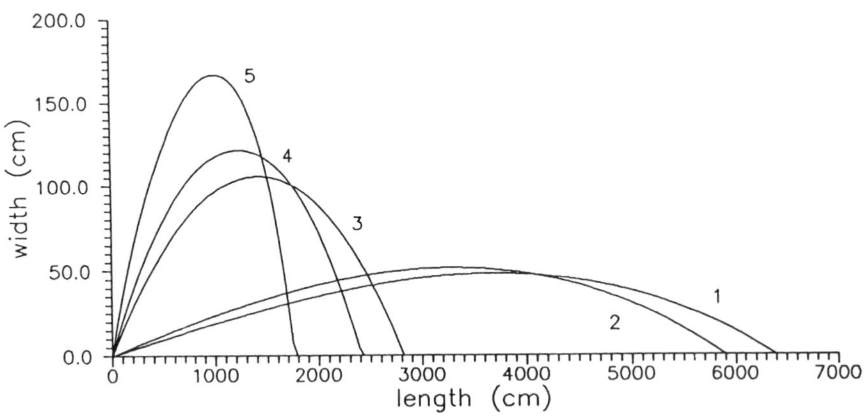

Figure 3.2. Shape of oil stringer changes with different gradients of permeability variation, simulated with the linear equation $\kappa = \kappa_o (1 + \beta \ell)$, where $\kappa_o = 100$ mD at the upper end of the stringer. Beta: 1, 0.0/cm; 2, 0.001/cm; 3, 0.0005/cm; 4, 0.001/cm; 5, 0.0005/cm.

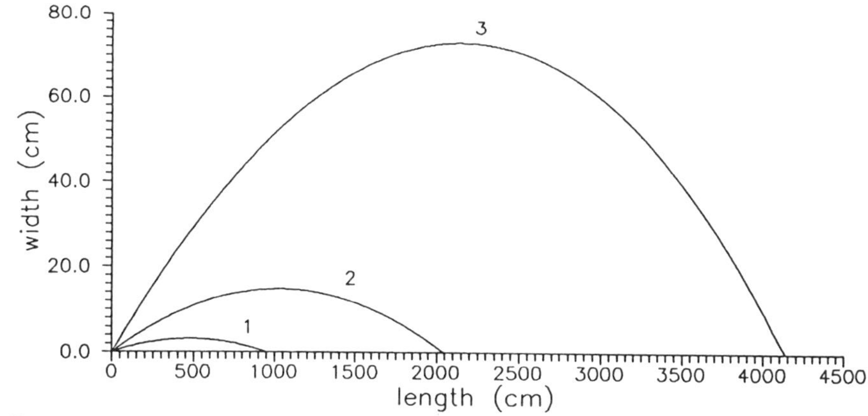

Figure 3.3. Oil stringer shape with different mass amounts of oil available. Mass: 1, 2000 cm^2; 2, 2000 cm^2; 3, 200000 cm^2.

limits are hard to find due to numerical difficulties at the point where no solution exists. The influence of the aquifer inclination angle is also an important factor (Fig. 3.5). Similarly to the variation of flow rate, there is no steady-state solution when the angle is too high or too low. It is important to note that the results are of greater importance to oil explorationists in relation to the way that the thin-stringer equations of Chapter 2 can be used. A longer oil stringer does not necessarily mean a larger amount of accumulation as can be seen from Figs. 3.2, 3.4, and 3.5. The relationship between the width and length of the oil stringer is determined not only by the amount of oil trapped but also by the permeability varia-

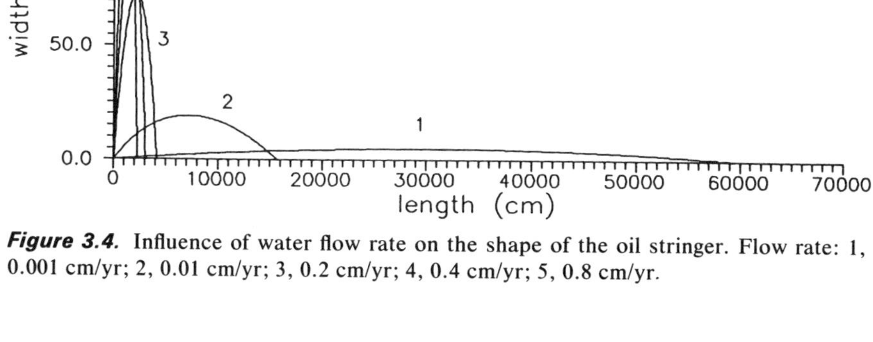

Figure 3.4. Influence of water flow rate on the shape of the oil stringer. Flow rate: 1, 0.001 cm/yr; 2, 0.01 cm/yr; 3, 0.2 cm/yr; 4, 0.4 cm/yr; 5, 0.8 cm/yr.

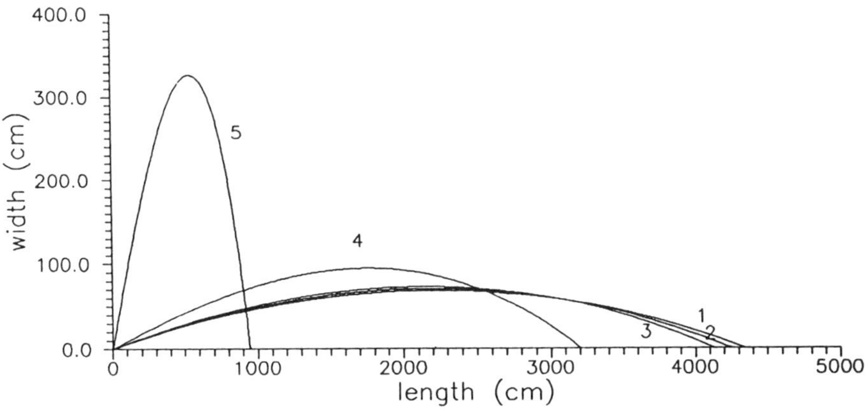

Figure 3.5. Oil stringer shape with different tilt angles of the aquifer. Tilt angle: 1, 7.5 degrees; 2, 20.0 degrees; 3, 30.0 degrees; 4, 60.0 degrees; 5, 85.0 degrees.

tion, flow intensity, aquifer inclination, interfacial tension, and other factors.

We conclude that the shape and location of a thin oil stringer are *not* determinable by elementary means, as Hubbert (1953) noted. However, the stability with respect to up-dip or down-dip hydrodynamic motion of such a stringer *is* determinable by Hubbert's (1953) arguments.

The shape and stability of a *finite*-thickness oil stringer are determinable on the basis of arguments presented in detail by Lamb over 100 years ago. Under steady-state conditions, the shape and stability of such an oil stringer may be predicted using Eqs. (3.6) and (3.7) if estimates of reservoir properties and hydrodynamic conditions are provided.

3.2. An Analytic Solution

The advantages of having available an analytic solution to the nonlinear integro-differential equation (Eq. 3.7) describing the shape of a steady-state profile are fairly obvious. First, such a solution can be investigated rather simply for its sensitivity to changes in parameters describing the flow; second, when it comes to more complex situations, which can often only be solved numerically, analytic solutions can be used to test the accuracy and reliability of numerical codes. We have found this second aspect to be of great benefit in assessing step sizes to be taken and in uncovering subtle "bugs" in the codes describing profiles that might otherwise have gone unnoticed until they created major difficulties.

Suppose then that we treat a constant-angle, fixed-permeability, fixed-width aquifer in which the maximum height of the buoyant accumulation is much smaller than the aquifer width, i.e., $|h| << H$. Then the factor $(H - h)^{-1}$ in the integral of Eq. (3.7) can be expanded, and, to first order in h/H, we can then write

$$\frac{d^2h}{d\ell^2} = \frac{Q\mu}{2T\kappa H}\left(\ell + H^{-1}\int_0^\ell h d\ell\right) - \frac{\Delta\rho g}{2T}(\ell \sin\theta + h\cos\theta) \quad (3.8)$$

where $\Delta\rho = \rho_w - \rho_o$.

If we measure h in units of aquifer width H with

$$h = yH \quad (3.9a)$$

and measure length, ℓ, along the aquifer similarly,

$$\ell = xH \quad (3.9b)$$

then Eq. (3.8) takes on the form

$$\frac{d^2y}{dx^2} = (a - b\sin\theta)x - yb\cos\theta + a\int_0^x y\,dx \quad (3.10)$$

where $a = \dfrac{Q\mu H}{2T\kappa}$ and $b = \dfrac{\Delta\rho g H^2}{2T}$. When differentiated once with respect to x, Eq. (3.10) yields

$$\frac{d^3y}{dx^3} = (a - b\sin\theta) - b\cos\theta\,\frac{dy}{dx} + ay \quad (3.11)$$

which we recognize as a third-order differential equation with constant coefficients. The general solution is given by

$$y = -[(1 - b\sin\theta/a)] + \sum_{i=1}^{3} y_i \exp(\sigma_i x) \quad (3.12)$$

where the three coefficients y_1, y_2, and y_3 have to be determined by boundary conditions, and where the σ_i ($i = 1, 2, 3$) are the three roots of

$$\sigma^3 + (b\cos\theta)\sigma - a = 0 \quad (3.13)$$

given by

$$\sigma_i = u_i - \frac{1}{3}(b \cos \theta)/u_i \tag{3.14a}$$

with

$$u_1 = [a + (a^2 + \frac{4}{27} b^3 \cos^3 \theta)^{1/2}]^{1/3} \tag{3.14b}$$

and with $u_2 = u_1 \exp(2\pi i/3)$ and $u_3 = u_1 \exp(-2\pi i/3)$.

The boundary conditions follow from the requirement that $h = 0$ on $\ell = 0$ (correspondingly, $y = 0$ on $x = 0$), which, from Eq. (3.10), also then forces the requirement that $d^2y/dx^2 = 0$ on $x = 0$. The third boundary condition is that at the end of the oil profile (occurring at $x = L$) we again require $y = 0$.

Thus, we require

$$1 - [(b \sin \theta)/a] = \sum_{i=1}^{3} y_i \qquad (y = 0 \text{ on } x = 0) \tag{3.15a}$$

$$\sum_{i=1}^{3} y_i \sigma_i^2 = 0 \qquad (d^2y/dx^2 = 0 \text{ on } x = 0) \tag{3.15b}$$

$$(1 - [(b \sin \theta)/a]) = \sum_{i=1}^{3} y_i \exp(\sigma_i L) \qquad (y = 0 \text{ on } x = L) \tag{3.15c}$$

The linear equations (3.15) can be solved algebraically for the unknown coefficients y_1, y_2, and y_3.

Pragmatically, for ease of obtaining numerical plots of the analytic profile, it is easier to specify a given slope on $x = 0$, i.e.,

$$dy/dx = \tan \phi \qquad \text{on } x = 0 \tag{3.15d}$$

Then, in place of Eq. (3.15c), we write

$$\tan \phi = \sum_{i=1}^{3} y_i \sigma_i \tag{3.15e}$$

which, together with Eqs. (3.15a) and (3.15b), provides three equations for determining y_1, y_2, and y_3. The variation of $y(x)$ with increasing x is then followed using Eq. (3.12) until y again crosses zero, which defines the length L of the oil profile. The areal mass, M, is then given by

$$M = H^2 \int_0^L y \, dx$$

That is,

$$M/H^2 = -[1 - b \sin \theta/a]L + \sum_{i=1}^{3} \sigma_i^{-1} y_i[\exp(\sigma_i L) - 1] \quad (3.16)$$

Thus, by varying the initial angle θ for the oil stringer, the shape, length L, and total mass M of the stringer can be examined arithmetically.

3.3. Variable Aquifer Width and Aquifer Angles

In deriving Eq. (3.7), we note that the assumptions of constant angle for the aquifer and constant width are not fundamental. The reason is that Darcy's equation deals with pressure *gradients* driving flow so that at each location, ℓ, along the aquifer, where the aquifer angle is $\theta(\ell)$, we have, for example, the differential water pressure gradient along the aquifer of

$$\nabla P_w = \rho_w g\left(\sin \theta + \frac{dh}{d\ell} \cos \theta\right) - \frac{Q\mu}{\kappa(\ell)[H(\ell) - h(\ell)]} \quad (3.17)$$

Tracing through the corresponding derivation for flow balance, we can write the equivalent of Eq. (3.8) in the form

$$2T \frac{d^2h}{d\ell^2} = Q\mu \int_0^{\ell} \frac{d\ell'}{\kappa(\ell')[H(\ell') - h(\ell')]} - g \int_0^{\ell} \Delta\rho(\ell')[\sin \theta(\ell') + \frac{\partial h(\ell')}{\partial \ell'} \cos \theta(\ell')] \, d\ell' \quad (3.18)$$

where the density difference $\Delta\rho(\ell')$ [$\equiv \rho_w(\ell') - \rho_o(\ell')$] may also vary along the length of the aquifer, perhaps because of thermal or salinity gradients in the water or perhaps because of compositional variations in the buoyant oil phase, and so on.

It is the equation of the form of Eq. (3.18) that will be used considerably throughout this monograph.

As well as a direct interfacial tension at the oil–water interface, it is also possible to include directly a component due to any capillary pressure caused by rock–oil and water–rock differences. This inclusion is easily handled by adding to the left-hand side of Eq. (3.18) a term

$$p_c \frac{dh}{d\ell} \tag{3.19}$$

where p_c is the capillary pressure as a consequence of rock effects.

3.4. Scaling Effects

In order to provide accurate and stable numerical solutions to the shape equation (3.18), the step size, $\Delta\ell$, for numerical integration must be chosen small enough to preserve precision. In order to determine how small is "small enough," we first consider Eq. (3.18) written in the following units, commonly used in assessing reservoir characteristics: μ, centipoise; H, meters; κ, millidarcies; T, dynes per centimeter; v, meters per year; ℓ, meters; g, multiples of 980 cm/s²; and $\Delta\rho$, grams per cubic centimeter. With these units, it follows that Eq. (3.18) takes the form

$$\begin{aligned}\frac{d^2h}{d\ell^2} &= \frac{3}{2} \times 10^8 \frac{H_0 v \mu}{T} \int_0^\ell \frac{d\ell'}{\kappa(\ell')[H(\ell') - h(\ell')]} \\ &\quad + 4.9 \times 10^6 \frac{g}{T} \int_0^\ell \Delta\rho(\ell')\left[\sin\theta(\ell') + \frac{dh(\ell')}{d\ell'}\cos\theta(\ell')\right] d\ell'\end{aligned} \tag{3.20}$$

where we have written the flux $Q = H_0 v$ with H_0 the aquifer width at $\ell = 0$, the beginning point of the oil accumulation, and v the speed of the water flow at that point.

Consider a Taylor series expansion of the oil shape width $h(\ell)$ around $\ell = 0$ starting with a slope $dh/d\ell = \tan\phi$. Then,

$$h(\ell) \approx \Delta\ell \tan\phi + \frac{1}{2}\left(\frac{d^2h}{d\ell^2}\right)_{\ell=0} (\Delta\ell)^2 + \frac{1}{6}\left(\frac{d^3h}{d\ell^3}\right)_{\ell=0} (\Delta\ell)^3 + \cdots \tag{3.21}$$

Now $d^2h/d\ell^2 = 0$ on $\ell = 0$. Thus, from Eqs. (3.20) and (3.21), we have

$$h(\ell) \approx \Delta\ell \tan\phi + \frac{1}{6}(\Delta\ell)^3 \left[\frac{3}{2} \times 10^8 \frac{H_0\mu v}{T\kappa_0 H_0} + 4.9 \right.$$
$$\left. \times 10^6 \frac{g}{T} \Delta\rho(0) \times (\sin\theta_0 + \tan\phi\cos\theta_0)\right] \tag{3.22}$$

where the subscripts "0" refer to evaluation on $\ell = 0$.

In order that the term in $(\Delta\ell)^3$ in Eq. (3.22) be small in magnitude compared to the term $\Delta\ell \tan\phi$, so that a Taylor series provides the basis for a valid numerical scheme, an estimate of the largest step size allowed is

$$\Delta\ell \leq (6|\tan\phi|)^{1/2} \left[\frac{3}{2} \times 10^8 \frac{\mu v}{T\kappa_0} + 4.9 \times 10^6 \frac{g}{T} |\Delta\rho(0)| \, (1 \right.$$
$$\left. + |\tan\phi|)\right]^{-1/2} \equiv \Delta\ell_{max} \tag{3.23}$$

To obtain a rough estimate of $\Delta\ell_{max}$, note that for $\mu \approx 1$ cP, $T \approx 30$ dyn/cm, $g = 1$, $K_0 \approx 1$ D, $|\Delta\rho| \approx 1$, $\tan\phi \approx 1$, and $v \approx 30$ m/yr, we have

$$\Delta\ell_{max} \approx 6^{1/2} \left[\frac{3}{2} \times 10^8 + \frac{(4.9 \times 10^5)(2)}{3}\right]^{-1/2}$$
$$\approx 10^{-4}\ \mathrm{m} \approx 10^{-2}\mathrm{cm} \tag{3.24}$$

Any step size larger than $\Delta\ell_{max}$ will rather quickly cause numerical instability of the equation solutions being sought.

3.5. Permeability Variations along the Aquifer

Because of facies changes, grain size sorting within a facies, or cementation/dissolution effects, it is quite likely that the permeability along an aquifer will vary.

We model the variation of permeability with aquifer position by an equation of the form

$$\kappa(\ell) = \frac{1}{2}(\kappa_a + \kappa_b) + \frac{1}{2}(\kappa_a - \kappa_b) \tanh\left(\frac{\ell - \ell_0}{L_s}\right) \tag{3.25}$$

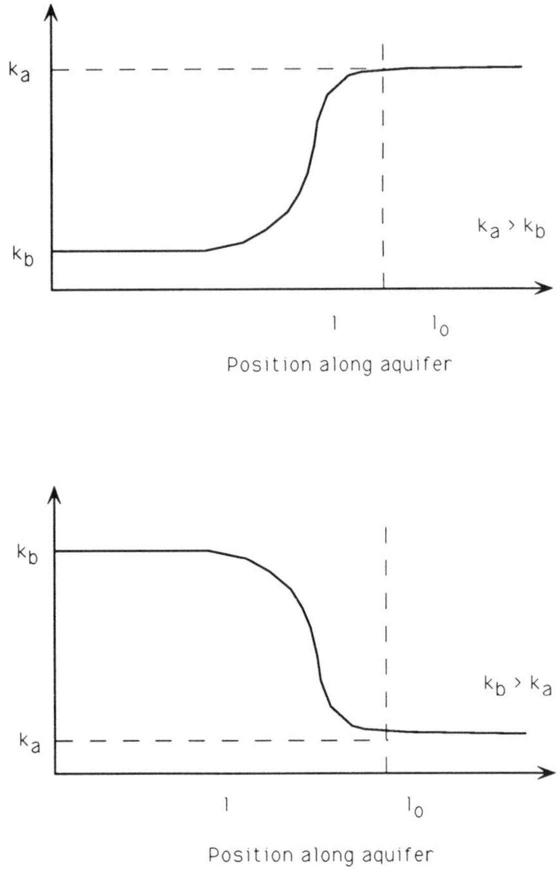

Figure 3.6. Sketch of permeability variations along the aquifer from high to low, and from low to high.

so that the permeability changes from the value κ_a at $\ell - \ell_0 >> L_s$ to κ_b at $\ell - \ell_0 << L_s$ and does so over a scale length L_s centered at a position $\ell = \ell_0$ relative to the start position of the oil stringer, as shown in Fig. 3.6. Here κ_a, κ_b, ℓ_0, and L_s are freely specifiable by the user.

3.6. Slow Variations and the Born Approximation

For a situation in which κ, H, θ, and $\Delta\rho$ all vary slowly along the length of the aquifer, it is possible to borrow techniques from physics (Morse and Feshbach, 1953) which provide approximate analytic representations (the Born approximation) of solutions to the integro-differential equation (3.18). The sense of the argument proceeds as follows. First, differentiate Eq. (3.18) with respect to ℓ, obtaining

$$\frac{d^3h}{d\ell^3} = \frac{Q\mu}{2T\kappa(\ell)[H(\ell) - h(\ell)]} - g\,\frac{\Delta\rho(\ell)}{2T}\left(\sin\theta + \frac{dh}{d\ell}\cos\theta\right) \quad (3.26)$$

Then rewrite Eq. (3.26) in the form

$$\begin{aligned}\frac{d^3h}{d\ell^3} &- \frac{\mu\nu h}{2T\kappa_0} + g\,\frac{\Delta\rho(0)}{2T}\cos\theta_0\,\frac{dh}{d\ell} - \frac{\nu\mu}{2T\kappa_0} \\ &+ \frac{g\Delta\rho(0)\sin\theta_0}{2T} = \frac{H_0\nu\mu}{2T}\left\{\frac{1}{\kappa(\ell)}[H(\ell) \right. \\ &\left. - h(\ell)]^{-1} - \frac{1}{\kappa_0 H_0}\left(1 + \frac{h}{H_0}\right)\right\} \\ &- \frac{g}{2T}\,[\Delta\rho(\ell)\left(\sin\theta + \cos\theta\,\frac{dh}{d\ell}\right) \\ &- \Delta\rho(0)\left(\sin\theta + \frac{dh}{d\ell}\cos\theta_0\right)]\end{aligned} \quad (3.27)$$

Let the right-hand side of Eq. (3.27) be denoted by a source term $S(\ell)$. The factors H_0, κ_0, θ_0, and $\Delta\rho(0)$ are scaling values for the slowly varying components H, κ, θ, and $\Delta\rho$. Thus, if there were no variation at all with position along the aquifer, then the right-hand side of Eq. (3.27) would take on the value $\nu\mu/2T\kappa_0 \times h^2/(H_0 - h)$ so that, to order h/H, i.e., for a thin stringer, the right-hand side could be ignored. Hence, basic use can be made of the thin-stringer form to provide a Born approximation solution to Eq. (3.27).

Using the roots σ_i of the characteristic equation (3.13), we can write the Born approximation to Eq. (3.27) as

$$\begin{aligned}\frac{h(\ell)}{H_0} &= H_0^2 \int_0^{\ell} S(\ell')\,d\ell' \sum_{i=1}^{3} \exp[\sigma_i H_0^{-1}(\ell - \ell')](3\sigma_i^2 + 1)^{-1}\,d\ell \\ &+ \sum_{i=1}^{3} y_i \exp(\sigma_i \ell/H_0)\end{aligned} \quad (3.28)$$

where, as previously, the coefficients y_1, y_2, and y_3 are to be found by satisfying the three boundary conditions given by Eq. (3.15).

Iterations of the Born approximate solution (Eq. 3.28) can be made, but, by and large, we have not found such iterative procedures to be stable beyond the first Born approximation. We have, however, found that the Born approximation provides a good rough approximate behavior in a wide variety of cases, and it can be used as a first approximate shape for more general numerical investigations in which the full nonlinear integro-differential equation is used to construct solutions.

3.7. Numerical Considerations

The general functional form of Eq. (3.18) is

$$\frac{d^2h}{d\ell^2} = -\frac{P_c}{2T}\frac{dh}{d\ell} + f\left(h, \frac{dh}{d\ell}, \ell\right) \equiv G\left(h, \frac{dh}{d\ell}, \ell\right) \tag{3.29}$$

where we have added the extra capillarity term for the sake of completeness.

If Eq. (3.29) is integrated formally once, we obtain

$$\frac{dh}{d\ell} = \tan\phi + \int_0^\ell G\left[h(\ell'), \frac{dh(\ell')}{d\ell'}, \ell'\right] d\ell' \tag{3.30}$$

where $\tan\phi$ is the chosen slope of $dh/d\ell$ on $\ell = 0$.
Set

$$y(\ell) = dh(\ell)/d\ell, \tag{3.31}$$

so that

$$h(\ell) = \int_0^\ell y(\ell')d\ell' \tag{3.32a}$$

and

$$y(\ell) = \tan\phi + \int_0^\ell G[h(\ell'), y(\ell'), \ell']\, d\ell' \tag{3.32b}$$

Then the iteration scheme

$$y_n = y_{n-1} + \Delta\ell G(h_{n-1}, y_{n-1}, \ell_{n-1}) \tag{3.33a}$$

$$h_n = h_{n-1} + \Delta\ell y_{n-1} \tag{3.33b}$$

yields accurate values for h_n at the nth position $\ell_n = n\Delta\ell$ along the aquifer as long as $\Delta\ell$ is chosen small enough [$\leq\Delta\ell_{max}$; see Eq. (3.23)].

The initial starting conditions on $n = 0$ are

$$h_o = 0 \tag{3.34a}$$

$$y_o = \tan\phi \tag{3.34b}$$

The iteration scheme proceeds forward until at some value of $n = N$, the value for h_n again crosses zero. Then the stringer calculation is terminated. Note that all variables may vary freely along the aquifer including permeability, aquifer width, aquifer angle, density contrast, and so on, at the user's choice.

3.8. Conversion to Fractional Mass as Independent Coordinate

While the fundamental structure of Eq. (3.29), governing the thickness variation of an oil stringer with respect to length along the aquifer, is of considerable utility in numerical investigations, nevertheless there are advantages to having a formulation that uses the mass directly. First, from a computational viewpoint, there is the advantage that the mass of oil contained is then calculated at the same time as the oil width variation. Second, the appearance of a very wide, but quickly varying, oil stringer (or, indeed, of a very thin and slowly varying stringer) means that the length step size has to be adjusted continually to allow for shaping variations. To an extent, a formulation based on fractional mass alleviates some, but not all, of these numerical problems.

Return then to the generic equation structure,

$$\begin{aligned}\frac{d^2h}{d\ell^2} &= a\int_0^{\ell}\frac{d\ell'}{[H(\ell') - h(\ell')]\kappa(\ell')} + b\,\frac{dh}{d\ell} \\ &+ \int_0^{\ell}[ch(\ell')\sin\theta\,(\ell') \\ &+ f\cos\theta(\ell')]d\ell' \\ &\equiv S(h,\ell)\end{aligned} \tag{3.35}$$

where b, c, f, θ, H, and κ can all be dependent on length, ℓ.

Introduce the areal coordinate

$$dA = h\,d\ell \tag{3.36}$$

which, for constant density, ρ_o, of the hydrocarbons is directly equivalent to a differential mass coordinate:

$$dM = \rho_o h \, d\ell = \rho_o \, dA \tag{3.37}$$

Note also that

$$\frac{d^2h}{d\ell^2} = h \frac{d}{dA}\left(\frac{hdh}{dA}\right) \equiv \frac{h}{2} \frac{d^2}{dA^2} (h^2) \tag{3.38}$$

Equation (3.35) can then be written

$$\frac{h}{2} \frac{d^2}{dA^2} (h^2) = S(h, \ell) \tag{3.39}$$

with solution

$$h^2 = 2 \int_0^A dA' \, (A - A')S[h(A'), \ell(A')]/h(A') + 2A \tan \phi_0 \tag{3.40}$$

where $\tan \phi_0$ is the tangent of the initial slope angle of the oil stringer on $\ell = 0$ ($A = 0$) where $h = 0$. It follows from Eq. (3.40) that

$$h = \{2A \tan \phi_0 + 2 \int_0^A dA' \, (A - A')S[h(A'), \ell(A')]/h(A')\}^{1/2} \tag{3.41}$$

with

$$\ell(A) = \int_0^A dA'/h(A') \tag{3.42}$$

and where the source term $S(h, \ell)$ on the right-hand side of Eq. (3.39) can be written as

$$S(h, \ell) = a \int_0^A \frac{dA'}{\kappa(A')h(A')[H(A') - h(A')]} + \frac{bhdh}{dA} + \int_0^A [ch(A') \sin \theta(A') + f \cos \theta(A')] \, h(A')^{-1} \, dA' \tag{3.43}$$

Numerically, there is a difficulty caused at the leading edge (near $\ell = 0$ where $h = 0$) of the oil stringer by the conversion to mass as a basic coordinate. The difficulty stems from the division by h in Eqs. (3.41) and

(3.43), which tends to make their evaluation numerically unstable in the vicinity of $h = 0$ ($\ell = 0$). However, the advantage of using equal differential area as a basic coordinate is so great that it is useful to devise a strategy to handle the "leading edge" problem in combination with the differential area procedure. A suitable arrangement is as follows. Evaluate the shape of the oil stringer from its leading edge ($h = 0$ at $\ell = 0$) using differential *length* as a basic coordinate. Carry out this evaluation until a point $\ell = \ell_*$ is reached on which point $h = h_*$ and the area is $A = A_*$, with

$$A_* = \int_0^{\ell_*} h(\ell')\, d\ell' \tag{3.44}$$

Then rewrite Eq. (3.41) in the form

$$\begin{aligned} h = \{h_*^2 + [2(A - A_*)\tan\phi_0 + 2\int_{A_*}^{A} dA'\,(A - A')^x \\ S[h(A'),\, \ell(A')]/h(A')]\}^{1/2} \qquad A \geqslant A_* \end{aligned} \tag{3.45}$$

Thus, a small differential step length, $\Delta\ell$, can be used near the leading edge of the oil stringer, and then conversion to a differential area unit, ΔA, can be used [in conjunction with Eq. (3.45)] once A_* is reached. This combined procedure has the advantage of accurately determining the leading edge shape and, at the same time, of accurately evaluating the bulk of the shape of the oil stringer without being restricted to extremely short step lengths.

One final potential source of numerical instability can arise in using the differential mass formalism near the far end of the oil stringer where $h \to 0$ [so that the integrands of integrals on the right-hand side of Eq. (3.45) again are being divided by a very small quantity], leading to the possibility of unstable evaluation of the final end point of the oil stringer. However, in practice this potential problem does not arise for several reasons. First, note that the square root is taken on the right-hand side of Eq. (3.45). Thus, if the right-hand side becomes negative, the square root operation immediately forces a complex value for h, signaling that the point has been crossed where $h = 0$. It is then a simple matter to return to the previous mass point and use half the differential mass to carry through the calculation. If a complex square root is again obtained, then the differential mass can be halved once more until the null point ($h = 0$) is identified. If no complex square root is obtained, then the zero of h must occur in the second half-mass zone. The procedure is repeated with the mass in the zone being halved until the $h = 0$ point is identified.

4

Steady-State Oil Stringers in Capped Aquifers

4.1. Hubbert's Experiments

In a series of elegant experiments, Hubbert (1953) forcefully illustrated the evolution of "oil" accumulation shapes under hydrodynamic conditions. A sand box with a Plexiglas front had a rubber stratum embedded in the sand to simulate an impermeable upper boundary to a reservoir sand. The box was filled with water under pressure, and then colored alcohol was added. After settling to equilibrium, the alcohol was found to be at the structural high.

Hubbert (1953) then caused water to flow toward the structural nose and recorded the following observations:

> For slow rates of flow the "oil" accumulation became tilted downward in the direction of the flow, just so long as the tilt continued to be less than the dip angle the oil remained in the anticline. When, by increasing the rate of flow, the tilt was made to exceed the dip, the oil was migrated completely out of the anticline and down the dip until it was arrested by the steeper dip of the nose. . . . Here, for a fixed rate of flow, it remained in a completely stable equilibrium. At a still higher rate of flow the tilt was made to exceed the steeper dip of the terrace and the oil was migrated farther down dip out of that structure also.

Inspection of Hubbert's figures (Hubbert, 1953; Figs. 32 and 35) shows that (i) the aquifer is extremely wide compared to the "oil" stringer and (ii) the oil stringer is not an infinitely thin layer. Both effects require that a quantitative understanding of the shape of the stringer be dependent on the nonlinear self-consistent behavior of a finite-thickness oil stringer developed in Chapter 3. It seems to be appropriate to first ask whether the formal mathematical treatment of Chapter 3 can be used to account for the shapes found experimentally. If so, then other, more complex situations can be investigated numerically with a firm underpinning of concrete support based on the match of theory and observations. We address the Hubbert experimental problems on two fronts. First, we present a mathematical analysis for wide reservoirs, and then we provide computer simulations.

4.2. Mathematical Analysis

For an extremely wide aquifer ($H_0 >> h$) in which the variation of aquifer width ($|H(\ell) - H_0|$) is itself small compared to the fiducial width, H_0, at the upper end of the oil stringer, it follows that the hydrodynamic pressure term

$$\frac{Q\mu}{2T}\int_0^\ell \frac{d\ell'}{\kappa(\ell')[H(\ell') - h(\ell')]}$$

can be written, approximately, as

$$\frac{v_0\mu}{2T}\int_0^\ell \frac{d\ell'}{\kappa(\ell')}\left(1 + \frac{[h(\ell') - (H(\ell') - H_0)]}{H_0}\right)$$

so that, to lowest order in variations, we have Eq. (3.18) expressible in the form

$$\frac{d^2h}{d\ell^2} = \frac{v_0\mu}{2T}\int_0^\ell \frac{d\ell'}{\kappa(\ell')} + \frac{\Delta\rho g}{2T}\int_0^\ell \left[\sin\theta(\ell') + \frac{\partial h(\ell')}{\partial \ell'}\cos\theta(\ell')\right] d\ell' \qquad (4.1)$$

For a constant-permeability aquifer, and for each region of the aquifer where the aquifer angle is constant, Eq. (4.1) reduces to

$$\frac{d^2h}{d\ell^2} = \left(\frac{v_0\mu}{2T\kappa} + \frac{\Delta\rho g\ell}{2T}\sin\theta\right)\ell + \left(\frac{\Delta\rho g}{2T}\cos\theta\right)h(\ell) \qquad (4.2)$$

Consider, for instance, the simple situation, as sketched in Fig. 4.1, in which only two fixed angles control the shape of the aquifer—the right-hand (left-hand) side angle being negative (positive), thereby representing an anticlinal shape. Start the oil stringer at $\ell = 0$ to the right of the anticlinal nose so that $\theta = \theta_1$. Then in the region where θ remains at $\theta = \theta_1$ we have the solution to Eq. (4.1) as

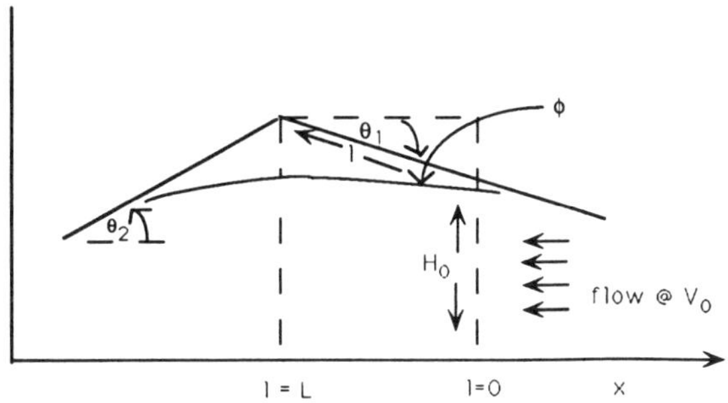

Figure 4.1. Sketch of the sharp-edged anticline geometry addressed quantitatively in Section 4.2.

$$h(\ell) = \ell\left(\tan\theta_1 - \frac{v_0\mu \sec\theta_1}{\kappa\Delta\rho g}\right) + h_1 \sinh\left[\left(\frac{\Delta\rho g\cos\theta_1}{2T}\right)^{1/2}\ell\right] \quad (4.3)$$

where h_1 is related to the angle ϕ at $\ell = 0$ through

$$\left.\frac{dh}{d\ell}\right|_{\ell=0} = \tan\phi = \tan\theta_1 - \frac{v_0\mu}{\kappa\Delta\rho g}\sec\theta_1 + h_1\left(\frac{\Delta\rho g}{2T}\cos\theta_1\right)^{1/2} \quad (4.4)$$

so that the solution given by Eq. (4.3) becomes

$$\begin{aligned} h(\ell) &= \ell\left(\tan\theta_1 - \frac{v_0\mu}{\kappa\Delta\rho g}\sec\theta_1\right)\left(1 - \frac{\sinh\left[\left(\frac{\Delta\rho g}{2T}\cos\theta_1\right)^{1/2}\ell\right]}{(\Delta\rho g\cos\theta_1/2T)^{1/2}\ell}\right) \\ &\quad + \tan\phi\left(\frac{2T}{\Delta\rho g\cos\theta_1}\right)^{1/2}\sinh\left[\left(\frac{\Delta\rho g\cos\theta_1}{2T}\right)^{1/2}\ell\right] \end{aligned} \quad (4.5)$$

Note that if $v_0\mu/\kappa\Delta\rho g < \sin\theta_1$, i.e., if $v_0 < \kappa\Delta\rho g\sin\theta_1/\mu \equiv v_*$, then $h(\ell)$ remains positive as ℓ increases. If, however, $v_0 > v_*$ then $h(\ell)$ will eventually *decrease* as ℓ increases if, further,

$$v_0 \geqslant v_* + \frac{\kappa\Delta\rho g\cos\theta_1}{\mu}\tan\phi \quad (4.6)$$

However, we note one further complication. If $\ell \geqslant L$, then the aquifer angle changes abruptly on $\ell = L$ from $\theta = \theta_1$ to $\theta = \theta_2$. For $\theta = \theta_2$, the general solution to Eq. (4.2) matching onto the solution from the region $\theta = \theta_1$, i.e., the solution with

$$h(L) = L \tan \theta_1 \left(1 - \frac{v_0}{v_*}\right) + h_1 \sinh \left[\left(\frac{\Delta\rho g \cos \theta_1}{2T}\right)^{1/2} L\right] \quad (4.7)$$

and

$$\frac{dh}{d\ell} = \tan \theta_1 \left(1 - \frac{v_0}{v_*}\right) + h_1 \left(\frac{\Delta\rho g \cos \theta_1}{2T}\right)^{1/2} \times \cosh \left[\left(\frac{\Delta\rho g \cos \theta_1}{2T}\right)^{1/2} L\right] \quad (4.8)$$

on $\ell = L$, is

$$h(\ell) = -\ell \left(\tan \theta_2 + \frac{v_0}{v_*} \sec \theta_2 \sin \theta_1\right) + h_2 \sinh \left[\left(\frac{\Delta\rho g \cos \theta_2}{2T}\right)^{1/2} \ell\right] + h_3 \cosh \left[\left(\frac{\Delta\rho g \cos \theta_2}{2T}\right)^{1/2} \ell\right] \quad (4.9)$$

with [from Eqs. (4.7) and (4.8)]

$$h_2 \sinh \left[\left(\frac{\Delta\rho g \cos \theta_2}{2T}\right)^{1/2} L\right] + h_3 \cosh \left[\left(\frac{\Delta\rho g \cos \theta_2}{2T}\right)^{1/2} L\right] = L \left[\tan \theta_1 \left(1 - \frac{v_0}{v_*}\right) + \tan \theta_2 + \frac{v_0}{v_*} \sec \theta_2 \sin \theta_1\right] + h_1 \sinh \left[\left(\frac{\Delta\rho g \cos \theta_1}{2T}\right)^{1/2} L\right] \quad (4.10)$$

and

$$h_2 \cosh\left[\left(\frac{\Delta\rho g \cos \theta_2}{2T}\right)^{1/2} L\right] + h_3 \sinh\left[\left(\frac{\Delta\rho g \cos \theta_2}{2T}\right)^{1/2} L\right] = \left(\frac{2T}{\Delta\rho g \cos \theta_2}\right)^{1/2} [\tan \theta_1\left(1 - \frac{v_0}{v_*}\right)$$

$$+ \tan \theta_2 + \frac{v_0}{v_*} \sec \theta_2 \sin \theta_1]$$

$$+ \left(\frac{\cos \theta_1}{\cos \theta_2}\right)^{1/2} h_1 \cosh \left[\left(\frac{\Delta\rho g \cos \theta_1}{2T}\right)^{1/2} L\right] \tag{4.11}$$

Thus, it is possible to compute h_2 and h_3 directly in terms of h_1 from Eqs. (4.10) and (4.11). Because h_1 is already given in terms of the initial angle, ϕ, of the oil stringer, and in terms of the parameters of the flow system, it follows that an exact analytic solution is available in this case. However, the determination of the shape of the oil stringer depends upon the total mass contained in the aquifer as well as upon the assumed initial starting value ($\ell = 0$) at a distance L to the right of the anticlinal apex. But, nowhere, as yet, in constructing the exact analytic solution have we made use of the requirement that $h(\ell) \geqslant 0$ or that the mass must be determined. We now address these concerns.

4.3. Areal Mass Limitations

A pragmatic procedure for calculating the position of the end of the oil stringer, at $\ell = L_M$, is as follows. The generic form of solution in $\ell < L$ is

$$h_1(\ell) = \ell\beta_1 + h_1 \sinh(\alpha_1\ell) \tag{4.12}$$

while in $\ell > L$ we have

$$h(\ell) = -\beta_2\ell + h_3 \sinh(\alpha_2\ell) + h_3 \cosh(\alpha_2\ell) \tag{4.13}$$

provided $h(\ell) > 0$. The areal mass, M, of oil is

$$M = \rho_o \int_0^{L_M} h(\ell)\, d\ell \tag{4.14}$$

Suppose first that $\ell < L$. Then the mass is

$$\frac{M}{\rho_o} = \frac{\beta_1}{2} L_M^2 + \frac{h_1}{\alpha_1} [\cosh(\alpha_1 L_M) - 1] \tag{4.15}$$

and, because $h(L_M) = 0$, we also have

$$L_M\beta_1 + h_1 \sinh(\alpha_1 L_M) = 0 \tag{4.16}$$

Eliminating h_1 gives

$$\frac{M}{\rho_0} = \frac{L_M\beta_1}{\alpha_1}\left[\frac{L_M}{2}\alpha_1 - \frac{\cosh(\alpha_1 L_M) - 1}{\sinh(\alpha_1 L_M)}\right] \tag{4.17}$$

For $\alpha_1 L_M << 1$, we have

$$\frac{M}{\rho_o} \cong \frac{L_M\beta_1}{2}\left[\frac{(\alpha_1 L_M)^3}{24}\right] \tag{4.18}$$

which is positive (negative) according as $\beta_1 > 0$ (< 0). If $\beta_1 < 0$, then $M/\rho_o < 0$, and no solution can exist. In that case, the solution must start to the left of $\ell = L$. Thus, $v_0 > v_*$ provides the critical velocity of flow forcing the oil stringer to the left side of the anticlinal "nose." For smaller values ($v_0 < v_*$), then $\beta_1 > 0$ and a stable solution exists to the right side of the anticline "nose." The solution may extend to the left of the anticline if the mass is large enough; that is, if $L_M \leqslant L$, the oil stringer is confined to the right side of the anticline. The relation between contained mass, M, length of oil stringer, L_M, and angle, ϕ, at the upper (right) side of the stringer is such that M is a highly nonlinear function of L_M; L_M is, in turn, a function of tan ϕ through $h(\ell = L_M) = 0$. Thus, by a process of elimination it is, in principle, possible to express ϕ and L_M in terms of the oil mass contained. In practice, because of the highly nonlinear nature of the system, it is easier to specify the angle ϕ, at the upper end of the oil stringer, calculate the first place where $h = 0$, thus providing L_M, and then compute the mass contained by direct integration of $h(\ell)$ from $\ell = 0$ to $\ell = L_M$.

Clearly, for systems made up of several sections, each of constant but different angles, the starting and ending positions of the stringers and the initial angle (or contained mass), as well as the shape of the boundary curve between the oil and following water, are more easily determined by computer modeling experiments.

However, we must first address the technical difficulty of what happens if the hydrodynamic forces are large enough to drive the "tail" of the stringer beyond the left end of the aquifer, as sketched in Figure 4.2a–c. In that case the generic pattern of evolution depends on precisely what is claimed to happen at and beyond the left side end of the aquifer.

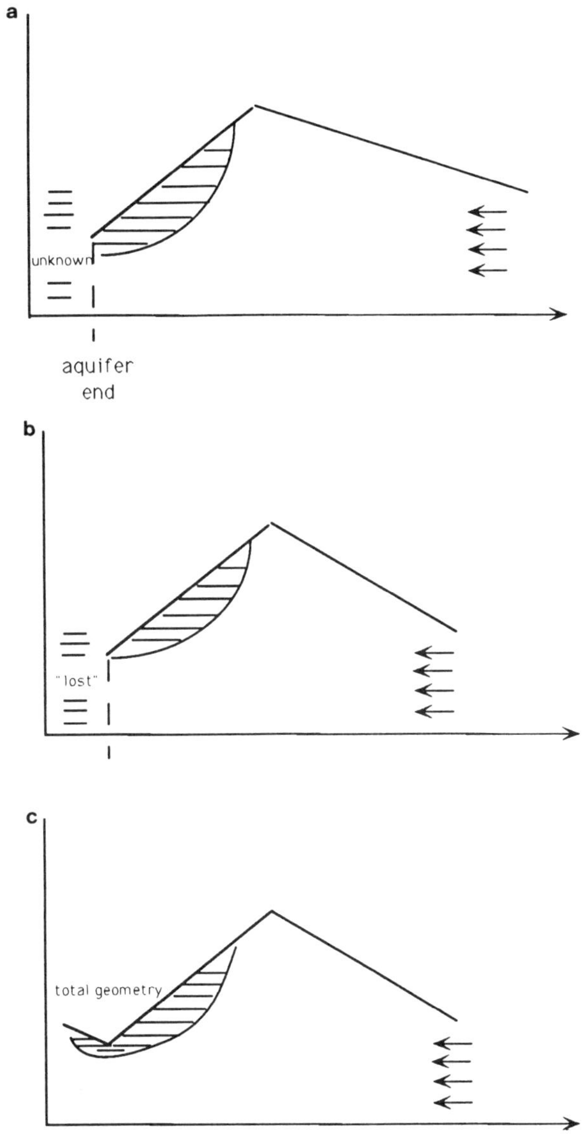

Figure 4.2. Illustration of possible "end conditions" applicable to a hydrocarbon stringer with hydrodynamic flow from the right: (a) An "unknown" situation, so that it is inappropriate to extend the stringer beyond the left edge of the aquifer top; (b) a "lost" situation in which any component of the stringer extending beyond the left edge of the aquifer is truly lost, that is, the aquifer ceases to seal; (c) schematic representation indicating how knowledge of the global geometry, beyond the area of apparent interest, can impact the shaping and amount of an entrapped hydrocarbon mass within the declared area of interest.

If the oil is just thought to be "lost" to the system, then the only steady-state solution must have $h = 0$ at the left aquifer end. In this case, as the upper end of the stringer moves closer to the lower end of the aquifer, the oil mass contained in the aquifer decreases (Fig. 4.2b). However, if it is recognized that the aquifer upper stratal boundary must extend beyond the aquifer "end," as shown in Fig. 4.2c, then one can track the oil shape once the *total* global shape of the aquifer is given. This problem will be taken up again later in the monograph.

The other fundamental reason why computer simulations are crucial is that the Hubbert experiments and the analytical procedure given above are both developed for extremely wide (effectively infinitely wide) aquifers in which variations of aquifer width have no effect on the flow. A problem of concern is to examine the changes in shape and position of the oil in the Hubbert experiments when the upper aquifer shape is kept fixed, but the lower boundary starts to encroach on the upper boundary, so that one can no longer ignore aquifer width variation effects. Such problems are best examined by numerical simulations.

4.4. Numerical Simulations

To examine the influence of flow conditions on the shaping of oil/gas masses, we start by illustrating the determination of shapes for the physical experiments conducted by Hubbert. The series of pictures given in Fig. 4.3a–c show the development of hydrodynamically controlled oil stringers as the flow speed systematically increases. The permeability of the formation underlying the upper boundary is set at the fixed value $\kappa = 1000$ mD. The lateral scale of the figures is 32 inches. These conditions allow for a close numerical mimicking of the physical shapes reported by Hubbert (1953). The flow velocity is systematically increased (with flow direction from right to left across the figures) from relatively low flow (Fig. 4.3a) to a flow speed of 300 cm/yr. As the flow speed increases, the sequence of figures shows a gradual displacement of the oil to the left, away from the anticlinal culmination. The dynamic steady-state shapes sketched should be compared with Figs. 32 and 35 of Hubbert's (1953) paper. The close geometrical correspondence is manifest, lending strong corroborative support to the veracity of the numerical implementation of the governing flow equations.

As we shall see in detail in later chapters, the behavior of a gas stringer under exactly the same conditions as above for the oil stringer has dominant differences due mainly to the higher buoyancy of the gas relative to that of the oil, with a secondary effect being due to the increased surface tension between gas and water relative to that between

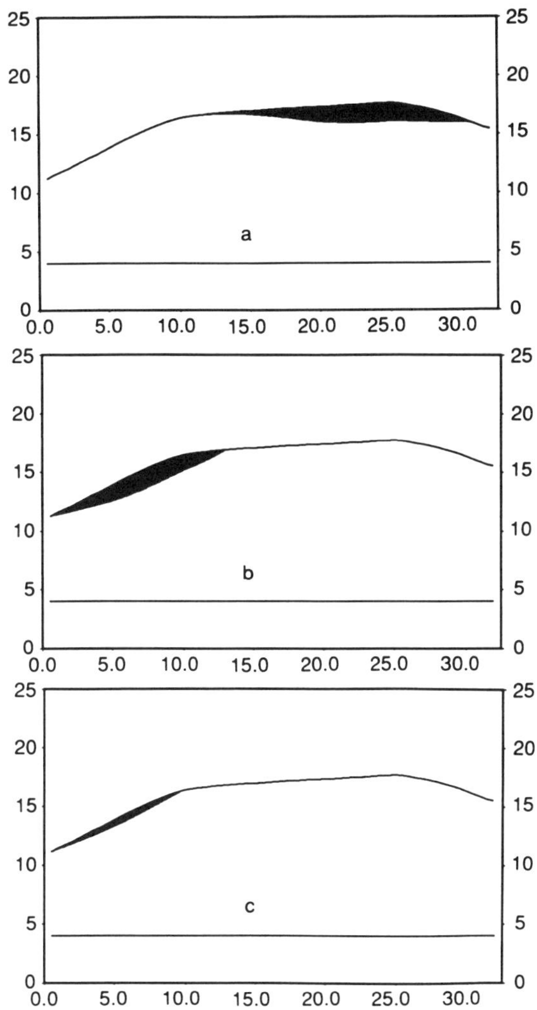

Figure 4.3. (a) Shape of an oil stringer under low flow conditions. Flow is from right to left. (b) Corresponding shape at higher flow conditions. (c) Shape at the highest flow conditions just prior to the left side of the oil stringer being "lost" over the edge of the aquifer sketched. Lateral and vertical scales are in inches to mimic the scale of Hubbert's experiments.

oil and water. Thus, the flow speed has to be increased to significantly higher values before the gas can be driven as far from the anticlinal culmination as is the oil.

As the oil (or gas) mass is decreased or increased, the flow pattern and the restoring forces on the shape change. For both oil and gas, a smaller areal mass gives rise to a more rapid motion away from the anticlinal culmination as the flow speed increases, while for an increased areal mass of oil or gas, a higher flow speed is required in both cases to drive the hydrocarbons down dip away from the culmination.

5

Finite-Width Aquifers

The examples in the previous chapter dealt with the influence of an upper, impermeable cap on an aquifer in controlling the shape and position of a buoyant mass under hydrodynamic forces. The effects of any lower boundary were considered negligible. This situation was specifically tailored so that Hubbert's physical experiments could be mimicked as closely as possible.

However, aquifers must have a finite width, and, as the mass of the buoyant stringer increases, the width of the stringer can become comparable to the aquifer width. In such cases, significant departures can be expected from the results obtained for the single-surface capped aquifers of the previous chapter.

In this chapter we look at the effects of wide and narrow aquifers in influencing hydrodynamic shapes. First, we start with the Hubbert aquifer shapes of the previous chapter, and, for the same total flux of water, the same mass of buoyant oil, and the same physical attribute parameters, we gradually move a basal aquifer surface toward the upper capping surface. The steady-state flow pattern at each move of the surface is obtained. In this way we build up a series of progressive "snapshots" of the accommodation of the system to a gradually encroaching boundary. Second, as the lower aquifer boundary encroaches, we then investigate the differences in positioning and shape as the buoyant fluid is changed from oil to gas. Third, we investigate the situation of the effect of an increasing flux of water on the positioning and shape of a buoyant mass as the lower boundary encroaches. Finally, we put together a sequence of behaviors as the mass of buoyant fluid (of different types) in the aquifer increases.

5.1. Encroaching Lower Aquifer Boundary

For the same conditions of water flux considered in examples mimicking the Hubbert experiments in the previous chapter, we have investigated the behavior of oil stringer shapes and positions in an aquifer as the base of the aquifer gradually approaches the upper boundary of the aquifer. Figures 5.1–5.4 illustrate the essential response for oil and gas under a fixed water flux.

On general arguments, note that as the lower boundary approaches the upper boundary, the flow speed past the oil stringer must increase because in the limit, as the lower boundary is infinitesimally close to the upper boundary, the flow speed would tend to an infinite value. It follows that at any finite distance separating upper and lower boundaries, there must be a distortion of the flow pattern relative to the case of an infinitely wide aquifer (the Hubbert problem). Thus, the hydrodynamic pressure pushing on the oil stringer must increase as the aquifer width narrows. In turn, this increase implies that the steady-state shapes of oil accumulations must be influenced by the width of the aquifer. The immediate task is to explore that influence.

Figures 5.1 and 5.2 illustrate the increasing distortions of oil stringer shape and position in the aquifer as the aquifer width narrows. Note that the stringer systematically increases its length away from the anticlinal culmination as the aquifer narrows, indicating a thinner and longer oil accumulation.

For a gas stringer, the effects of a narrowing aquifer width on the shape changes are less dramatic until the aquifer width is much narrower than in the corresponding oil case, as exhibited in Figs. 5.3 and 5.4. Basically, the reason why a lesser impact is felt by the gas stringer is that, because of its much higher buoyancy force, the gas is retained closer to the anticlinal apex under an aquifer of a given width than the less buoyant oil would be. Thus, the aquifer width has to be much narrower before the hydrodynamic forces are sufficiently elevated to overcome the excess buoyancy of the gas stringer.

5.2. Oil and Gas Shapes and Positions

Even within the framework of a fixed hydrodynamic water flux, significant variations in shape and position can still be brought about. For example, as the aquifer is buried more deeply in a sediment pile, the subsurface temperature increases. Viscosity and surface tension then both decrease so that the differences between gas and oil are minimized. In addition, the density of gas-phase material increases, roughly according to the perfect gas law, so that the differential buoyancy between gas and water is reduced. In addition, as sedimentary sequences are buried, they tend to compact, resulting in a decrease in the stratigraphic porosity and, most often, also a decrease in the aquifer permeability.

In this section of the chapter, we examine numerically the patterns of behavior of oil and gas stringers in both broad-width and narrow-width aquifers under conditions of decreasing or increasing surface tension, viscosity, permeability, and density. In order to illuminate the role that each individual factor plays, we consider variations in stringer shape and position in response to changes in one component at a time. Clearly, in a real geologic situation all variable factors act in concert so that the explicit roles played by individual components are masked.

5.2.1. Surface Tension Variations

As the surface tension decreases, the differential pressure difference between buoyancy and hydrodynamic forces must lessen, because it is only the surface tension (and, of course, capillary) forces that require a pressure difference. This effect is shown most readily in Fig. 5.5a–d for an oil mass and, to a lesser extent, in Fig. 5.6a–d for entrapped gas as the aquifer narrows. Clearly, the lesser response of a gas stringer is due to the intrinsically higher surface tension at a gas–water interface than at an oil–water contact. Note also from Figs. 5.5 and 5.6 that, as the aquifer width decreases, there is not much of a shift in either the position or the shaping of the hydrocarbon accumulation as the surface tension decreases.

5.2.2. Viscosity Variations

Under steady-state conditions, as the water viscosity increases, the hydrodynamic driving force on a hydrocarbon accumulation decreases. Thus, the hydrocarbon accumulation will tend to remain close to the hydrostatic position it originally occupied until, as the aquifer narrows enough, the increase in the speed of water flow past the accumulation is sufficient to forcibly move the stringer away from the anticlinal culmination. This effect occurs more easily for oil than for gas, as shown in the sequence of Figs. 5.7a–d and 5.8a–d, which compare and contrast oil and gas behaviors.

5.2.3. Permeability Variations

Direct inspection of the equations governing steady-state hydrocarbon stringer shapes shows that aquifer permeability, κ, and viscosity, μ, enter only in the proportionate ratio κ/μ. Thus, a decrease in viscosity can be compensated for by a corresponding decrease in permeability to maintain the ratio κ/μ at a fixed value.

Thus, Figs. 5.9a–d and 5.10a–d show the effects of permeability variations for oil and gas, respectively, although the same effects can be achieved with the substitution of fractional variations in viscosity for like variations in permeability.

5.2.4. Density Variations

Although oil is virtually incompressible, its density does change depending on the composition of the oil. Thus, tarry oil has a density of around 1.01 g/cm^3 whereas light Pennsylvania crude has a density of around 0.8 g/cm^3. In addition, uptake of gas (CH_4, CO_2, N_2, SO_2) by oil also lowers the density of the oil.

However, these variations in density are relatively small in the sense that the fractional buoyancy density differential, $(\rho_w - \rho_o)/\rho_o$, between water (ρ_w) and oil (ρ_o) roughly varies in the range -0.05 to $+0.2$. The highly compressible nature of gas, on the other hand, means that the gas (CH_4) density varies markedly from a value of around 7×10^{-4} g/cm^3 at a pressure of 1 atm and a temperature 273 K to around 0.1 g/cm^3 at a pressure of about 150 atm. For values of the gas density in excess of about 0.1 g/cm^3, the gas behaves more as a quasiliquid (condensate), and the perfect gas law is no longer valid. Thus, the fractional buoyancy differential $(\rho_w - \rho_{gas})/\rho_{gas}$ varies from around 10^3 to around 9—a much greater dynamic range than in the case of oil.

Figure 5.11 shows the effect of density variation on the hydrocarbon shaping and positioning in a thin aquifer for gas. The lighter the hydrocarbons, the higher is the buoyancy drive and so the closer is the positioning of the stringer to the anticlinal culmination. Correspondingly, the denser the gas, the more rapidly it moves away from the hydrostatic shape, but the buoyancy drive in gas is so large relative to that in oil that the gas holds much closer to the anticlinal culmination even for a significant change in density from 0.1 to 0.3 g/cm^3. These effects are clearly seen in Fig. 5.11a, b.

5.3. Increasing Water Flux

In a general sense, increasing the flux of water through an aquifer increases the hydrodynamic force attempting to dislodge the hydrocarbon stringer from its hydrostatic location. Thus, irrespective of the closeness of the lower and upper aquifer boundaries, there is an expectation that higher water flux conditions should show a proclivity for transporting a steady-state hydrocarbon stringer farther from the anticlinal culmination. For oil and gas stringers of fixed physical attributes, these effects are indeed seen, as illustrated in Figs. 5.12–5.17, in a ruggedly stable manner for all conditions of lower aquifer boundary position.

Note that as the lower boundary encroaches, the hydrodynamic forces increase for two reasons: first, because the water flow speed must increase in passing around a hydrocarbon stringer as the aquifer narrows and, second, because the higher the flux of water the greater is the hydrodynamic drive in the first place. Thus, we see from Figs. 5.12–5.17 that the departure of a hydrocarbon stringer (oil or gas) from its hydrodynamic position is not only larger the narrower the aquifer under given water flux conditions, but is also larger the higher the water flux under a fixed aquifer width condition.

5.4. Varying Masses of Oil and Gas

Up to this point, we have explored the effects on a hydrocarbon stringer due to changes in position of an encroaching lower aquifer boundary when the physical attributes of the hydrocarbon mass are changed or when the hydrodynamic effect of water flux is increased. A question of some interest is, what is the maximum areal mass of hydrocarbons that can be held in an aquifer under given conditions? The actual mass can, of course, be considerably less than the maximum.

For the case where the upper aquifer surface follows the classical Hubbert form, we have investigated this maximum-mass-limit problem for different physical attributes of the fluid components in the aquifer system and fixed and variable water flux speeds, for both oil and gas making up the hydrocarbon stringer. We ask that the leading edge of the stringer be started at each location along the top aquifer. Then we explore the maximum angle that the stringer can make with respect to the top aquifer boundary. If the angle is too steep, then hydrodynamic pressure will force a diminution; if the angle is too shallow, then buoyancy forces and surface tension forces allow the angle to increase. A balance is struck at some angle representing the maximum angle at which a hydrocarbon stringer can be obtained at that location in the aquifer. At particular locations (e.g., down dip to the right of the culmination in the Hubbert geometry), both the hydrodynamic force (directed flow from right to left) and the buoyancy force act together to push any hydrocarbon stringer toward the anticlinal culmination. There is no allowable stringer starting at such locations.

Once the maximum angle at each location has been identified (and the angle does depend on the buoyancy, and hence on whether oil or gas makes up the stringer), the calculations of shape are carried through, thereby producing the maximum areal extent of hydrocarbon stringer permitted at each starting location.

Representations of the variation of maximum areal mass along the aquifer are given for oil and gas in Figs. 5.5–5.10 for a very wide aquifer. As the width of the aquifer is decreased, the hydrodynamic force increases as water transits past the aquifer. In turn, this increase in force impacts the shape of the stringer and changes the maximum areal mass allowed at each location. Figures 5.12–5.17 exhibit this effect for narrower aquifer widths for stringers composed of both oil and gas. The effect of the narrowing aquifer in influencing the maximum allowable mass is, once again, more noticeable for oil than for gas owing to the lesser buoyancy of oil relative to that of gas.

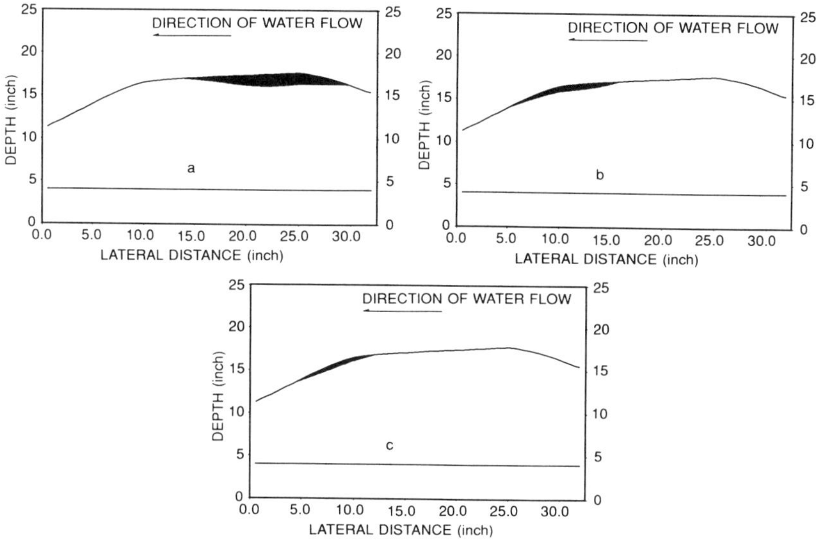

Figure 5.1. General patterns of behavior for oil flow with increasing flow speed in an aquifer of the Hubbert type with a base about 15 inches below the aquifer top. Fixed parameters are $T = 35$ dyn/cm, $\rho_{oil} = 0.8$ g/cm^3, and $\kappa = 1000$ mD. Flow speed: (a) $v = 1$ cm/yr; (b) $v = 100$ cm/yr; (c) $v = 200$ cm/yr.

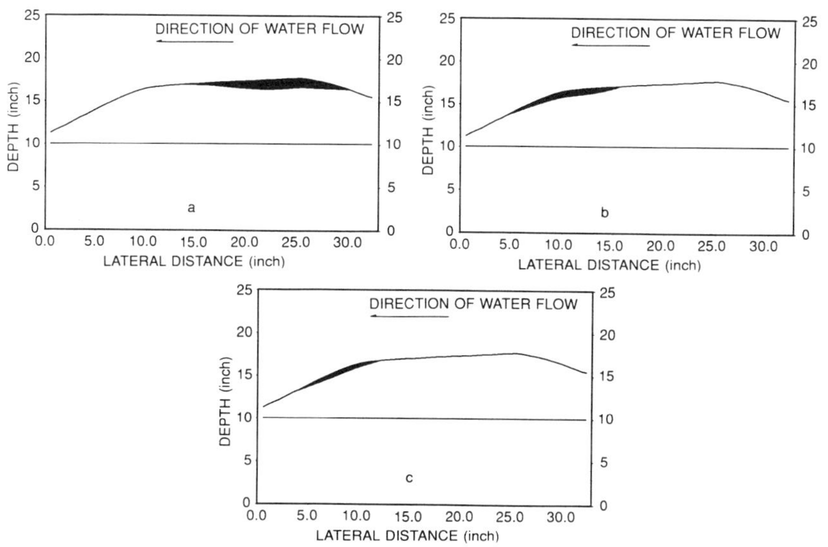

Figure 5.2. As for Fig. 5.1 except that the aquifer base has been moved to about 5 inches below the aquifer top.

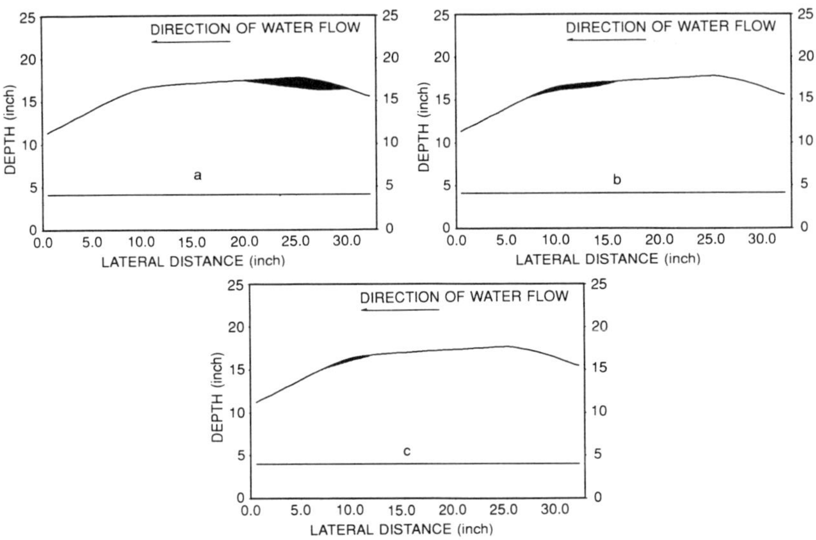

Figure 5.3. General patterns of behavior for gas flow with increasing flow speed in an aquifer of the Hubbert type with a base about 15 inches below the aquifer top. Fixed parameters are T = 25 dyn/cm, ρ_{gas} = 0.1 g/cm^3, and κ = 1000 mD. Flow speed: (a) v = 1 cm/yr; (b) v = 100 cm/yr; (c) v = 200 cm/yr.

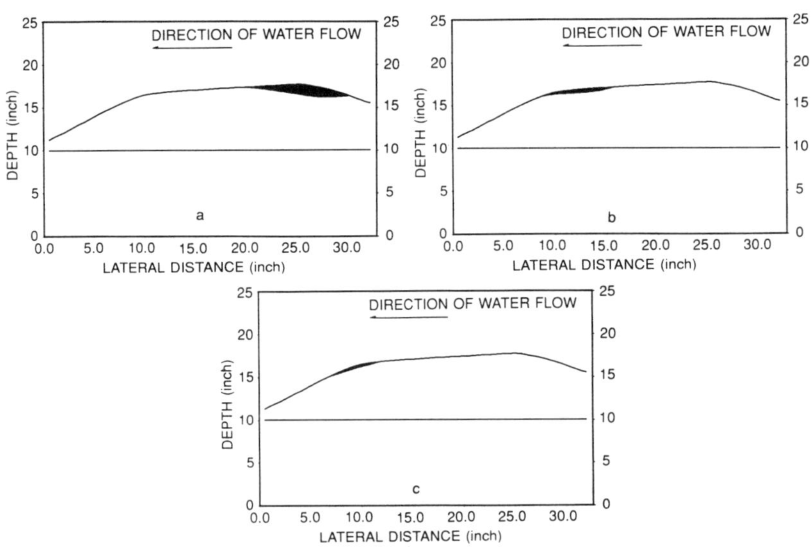

Figure 5.4. As for Fig. 5.3 except that the aquifer base has been moved to about 5 inches below the aquifer top.

Table for Figs. 5.5–5.10

Parameter varied	Value of parameter	Aquifer width: Oil 5″	Oil 15″	Gas 5″	Gas 15″
Permeability (mD)	10	5.9a	5.9c	5.10a	5.10c
	1000	5.9b	5.9d	5.10b	5.10c
Viscosity (cP)	0.6	5.7a	5.7c	5.8a	5.8c
	1.3	5.7b	5.7d	5.8b	5.8d
Surface tension (dyn/cm)	25	5.5a	5.5c	5.6a	5.6c
	60	5.5b	5.5d	5.6b	5.6d
Density (g/cm^3)	0.1			5.11a	
	0.3			5.11b	

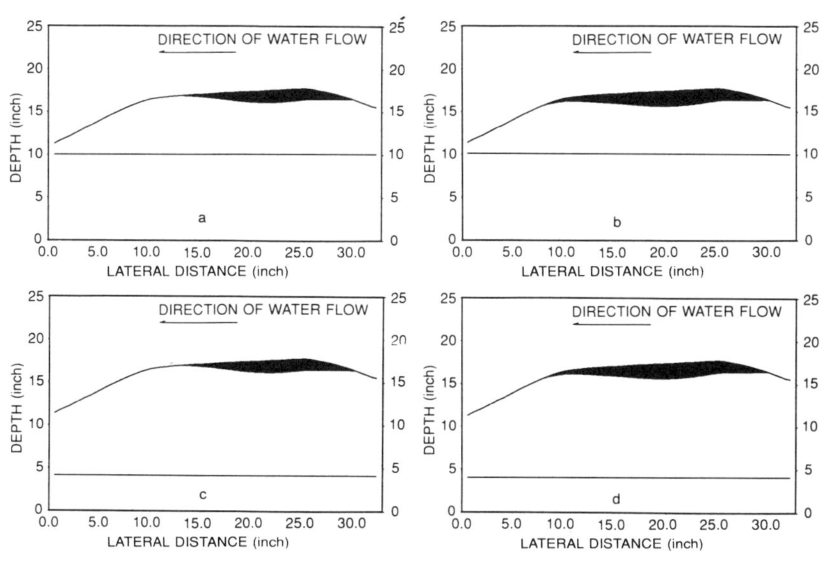

Figure 5.5.

Figures 5.5–5.10. General shapes of oil and gas stringers in a Hubbert style aquifer as the width decreases for a constant flow speed of 1 cm/yr. The legend for each figure can be obtained from the accompanying table, indicating the change of particular parameters from default values.

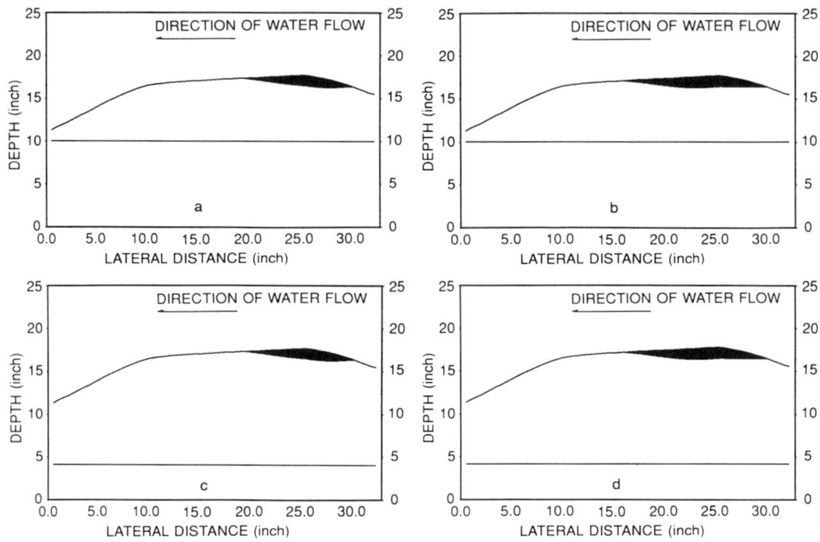

Figure 5.6.

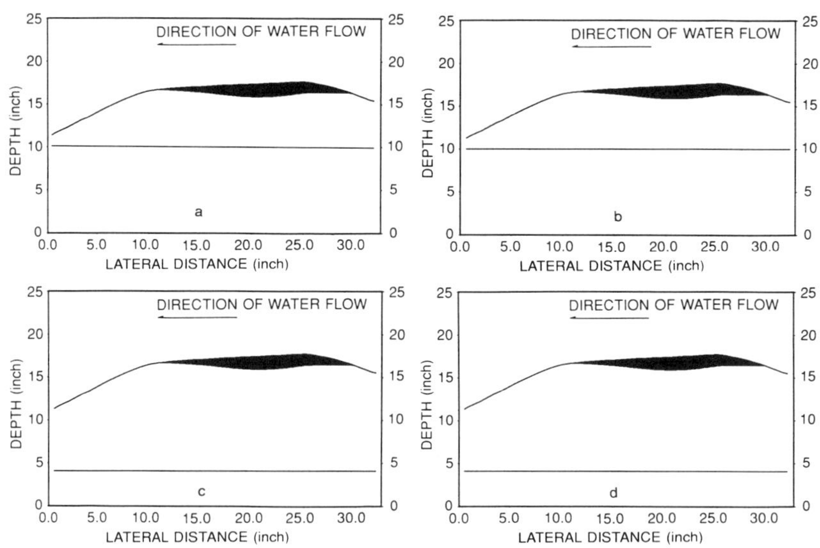

Figure 5.7.

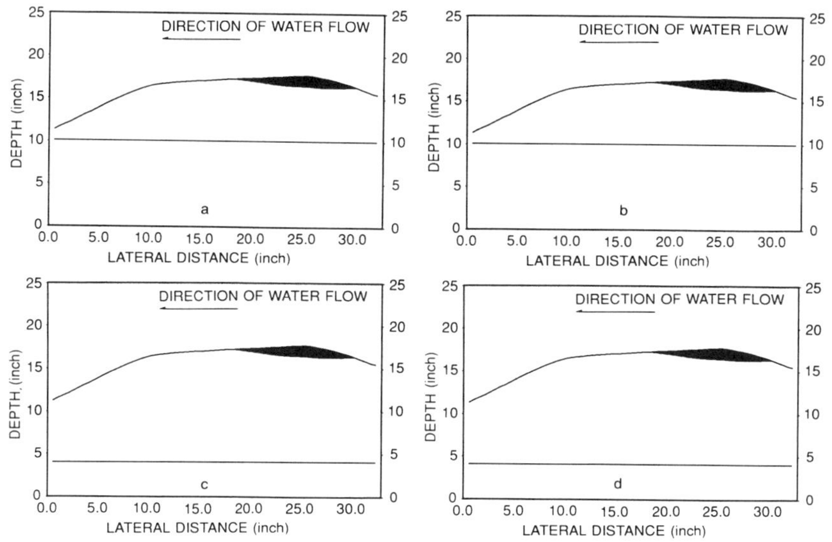

Figure 5.8.

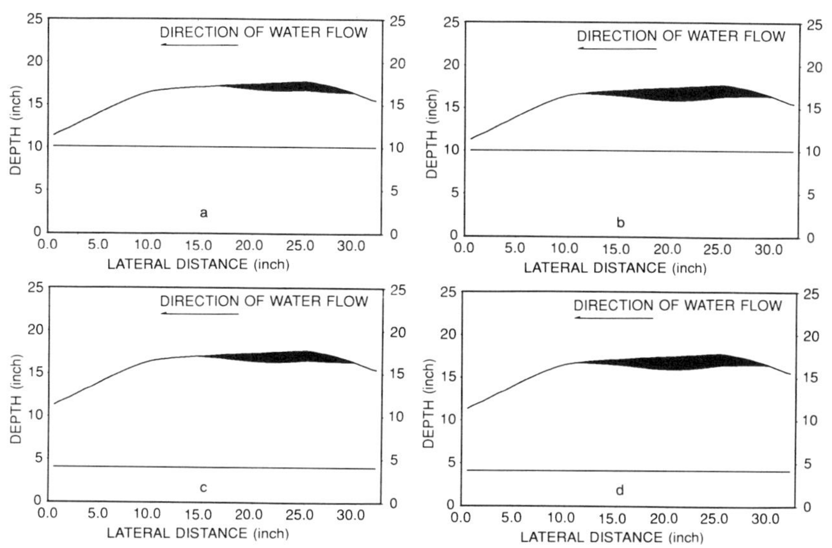

Figure 5.9.

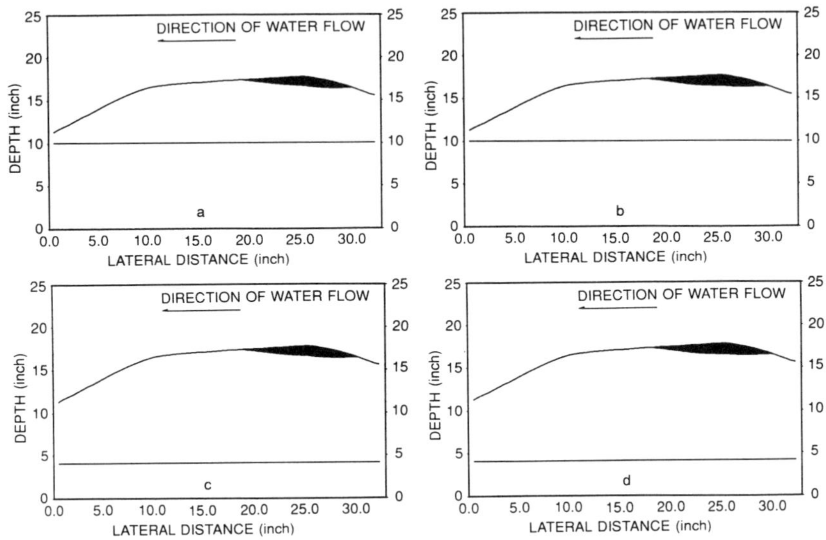

Figure 5.10.

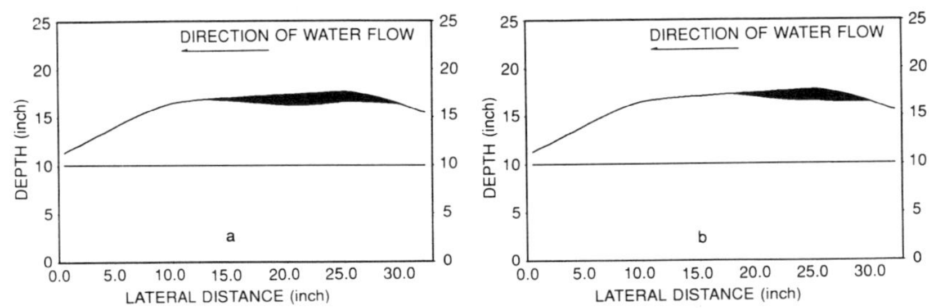

Figure 5.11. General shapes of gas stringer as density is varied from 0.1 g/cm³ to 0.3 g/cm³.

Table for Figs. 5.12–5.17

Location of "nose" of stringer	Flow rate (cm/yr)					
	Oil			Gas		
	1.0	100	200	1.0	100	200
10	5.12a	5.12b	5.12c	5.13a	5.13b	5.13c
20	5.14a	5.14b	5.14c	5.15a	5.15b	5.15c
30	5.16a	5.16b	5.16c	5.17a	5.17b	5.17c

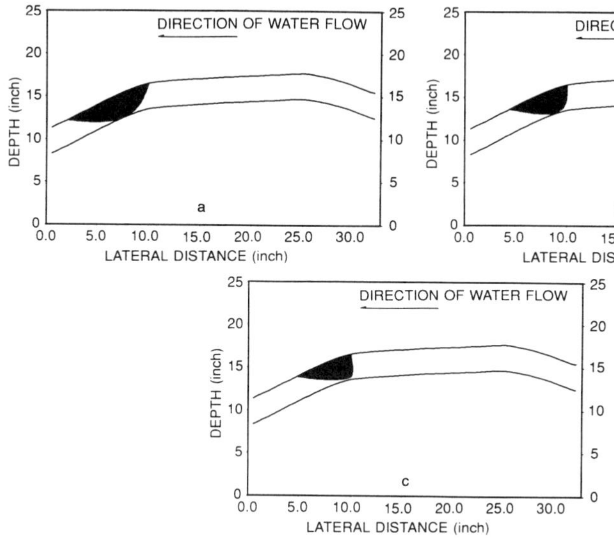

Figure 5.12.

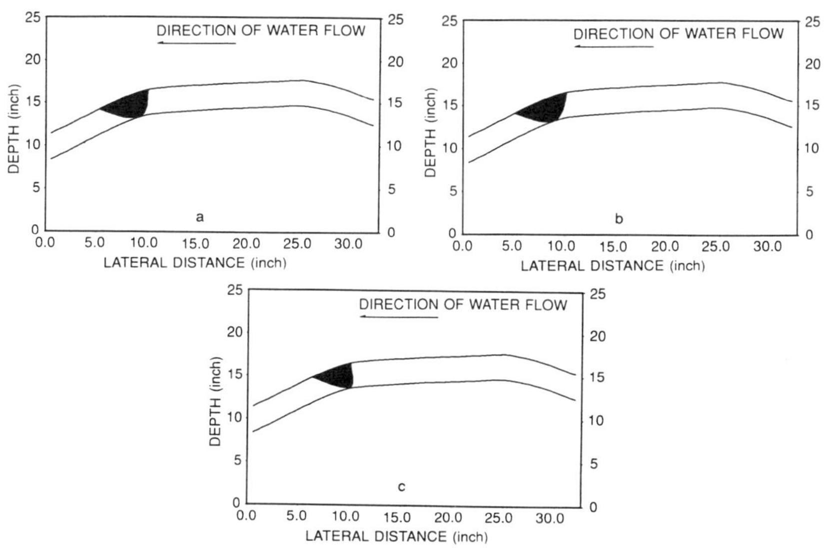

Figure 5.13.

Figures 5.12–5.17. General shapes of oil and gas stringers in a Hubble style aquifer. Flow speed: (a) $v = 1$ cm/yr; (b) $v = 100$ cm/yr; (c) $v = 200$ cm/yr. The location of the nose of the oil and gas stringers is progressively moved from location 10 (Figs. 5.12, 5.13) to location 20 (Figs. 5.14, 5.15) to location 30 (Figs. 5.16, 5.17) to illustrate positioning influence on shape of stringers.

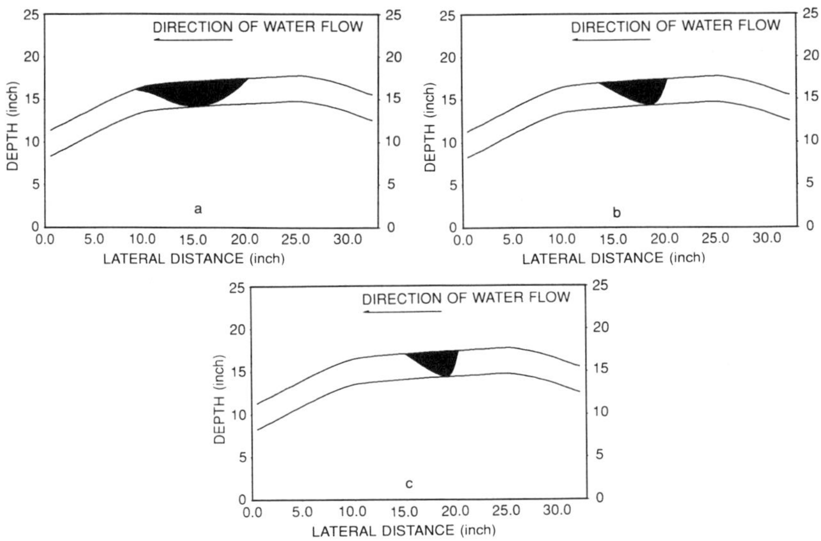

Figure 5.14.

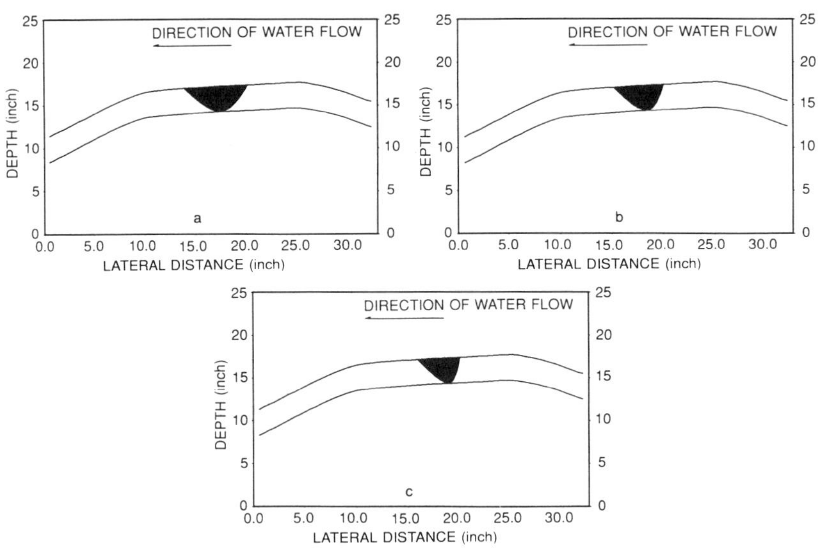

Figure 5.15.

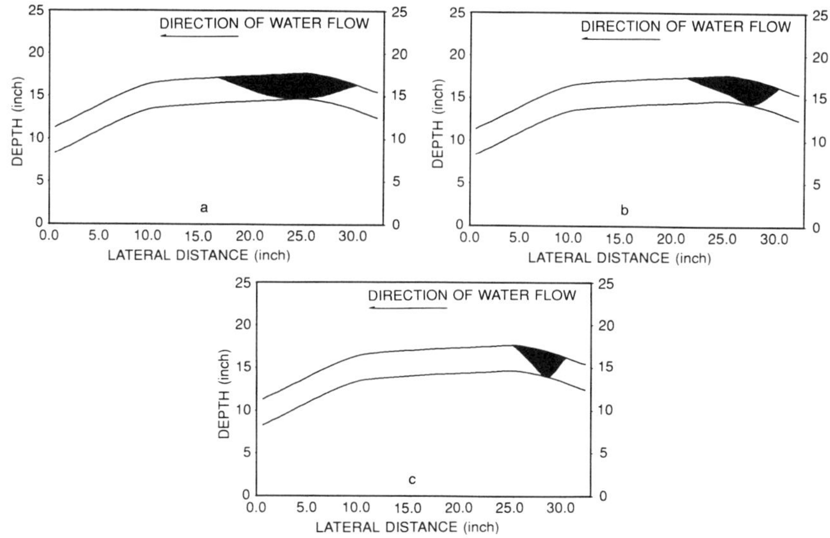

Figure 5.16.

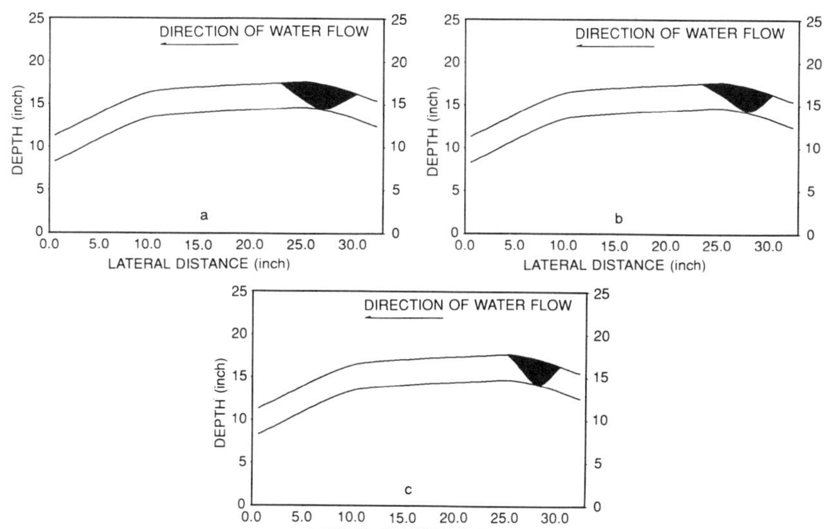

Figure 5.17.

6

Variable-Width, Constant-Angle Aquifers

Variable deposition of sediments in sedimentary basins and tectonic movement of structural units within a basin almost surely guarantee that an aquifer for hydrocarbon flow will be of nonconstant width and nonconstant angle along the length of the aquifer.

In this chapter we isolate the first of these two effects and consider an aquifer in which the position of either the top or the bottom boundary of the aquifer, or both, can vary relative to a fixed angle of the aquifer as a whole. The general effect of varying aquifer width will be to speed up the hydrodynamic flow in regions of constricted width and to slow down the flow in regions of expanded width. Thus, the hydrodynamic forces on their own will attempt to push buoyant hydrocarbons away from constrictions and toward the broader aquifer domains. However, the shape and positioning of a buoyant hydrocarbon mass will also be influenced by whether it is the basal horizon or the upper capping horizon of the aquifer that is varying relative to the fixed aquifer angle. Thus, the three situations depicted in Fig. 6.1 will all give rise to different hydrodynamic shaping and positioning of buoyant hydrocarbon masses, even though all three cases have effectively the same variation of aquifer width with distance along the aquifer. This difference in behavior is, of course, due to the buoyancy force by which the hydrocarbons tend to hug the upper surface of the aquifer.

For instance, Figs. 6.2a–c and 6.3a–c show the effect of "inverting" the geometry of the aquifer on the shape and positioning of an oil stringer, while Figs. 6.2d–f and 6.3d–f show the response of a gas stringer to the geometric inversion. Note the rather different shapes of stringers that arise as a consequence of the inversion.

6.1. Group I Behaviors

To provide some idea of the sorts of response patterns that are available, we consider numerically the following aquifer situations, which are classified as group I:

- A Top planar boundary converging onto a planar bottom boundary (Fig. 6.4a)
- B Bottom planar boundary converging onto a top planar boundary (Fig. 6.4b)
- C Top and bottom planar boundaries both converging (Fig. 6.4c)
- D Top planar boundary diverging from a planar bottom boundary (Fig. 6.4d)
- E Bottom planar boundary diverging from a top planar boundary (Fig. 6.4e)
- F Top and bottom planar boundaries both diverging (Fig. 6.4f)

In all cases considered, the hydrodynamic flow of water is maintained from upper right to lower left in Fig. 6.4 at an angle θ (dashed line) far ahead of the region where the aquifer distorts from a plane-parallel regime to a converging or diverging flow. In addition, the permeability of the aquifer is held fixed, as are the flux of water and the physical attributes (viscosity, surface tension, density, etc.) of the fluid components in the aquifer system. What we search for in this class of numerical illustrations is the variation in a hydrocarbon stringer with respect to initial position along the upper aquifer boundary.

Note that all of the group I situations have *no* hydrostatic analog; that is, if the down-dip drive of the hydrodynamic force is set to zero, then no steady-state stringer is permissible.

In the cases of a converging (diverging) aquifer, the water flow speed increases (decreases) toward (away from) the "throat" of the aquifer; there is then an ever increasing (decreasing) hydrodynamic force acting on a hydrocarbon stringer at any location.

As a general rule of thumb, the expectation is that the more strenuous the hydrodynamic pressure, the greater the buoyant force opposing the flow can be. Thus, there is the surmise that converging flow geometries should enable a greater areal mass of oil or gas to be accumulated as a stringer compared to that obtainable in a diverging flow geometry for the same hydrocarbon properties. Inspection of Figs. 6.5–6.18 shows the substantiation of this expectation in a broad sense.

There are, of course, differences in the individual behaviors even within converging or diverging flow geometries. These differences are due to the detailed "shape" of the aquifer, that is, whether it is the top or the bottom aquifer boundary, or both, that shows departures from the fixed angle θ, representing the far distant flow direction.

The fundamental reason for these individual variations around a main theme is due to the buoyancy. The upward drive of buoyancy, forcing the stable hydrocarbon states to hug the top aquifer boundary, means that changing the angle of the top boundary while retaining the total geometric shape of the aquifer (converging or diverging) produces a different pattern of response than does keeping the angle of the top aquifer boundary fixed and varying the angle of the bottom aquifer boundary. These differences in effect are clearly portrayed for both converging and diverging aquifers in Figs. 6.5–6.18 for oil and gas stringers.

6.2. Group II Behaviors

Here the individual variations of behavior are not as easily grouped into generic classes of responses as were the group I patterns. Perforce, therefore, we consider each behavioral pattern individually.

6.2.1. "Wiggly" Aquifers of Constant Angle

We illustrate the behavior of allowed hydrocarbon stringers, of maximum areal mass, using the "wiggly" shapes sketched in Fig. 6.1 for the cases of both oil and gas. To be noted from the resulting numerical sequences (as shown in Figs. 6.19a–d and 6.19e–h) are the rather larger variations in oil and/or gas accumulations with respect to position along the aquifer. Also to be readily observed on Figs. 6.19a–d and 6.19e–h is the marked contrast in patterns of behavior for the stringers depending on whether oil or gas is being considered. Just as for the planar converging and diverging aquifers previously examined in this chapter, the fundamental reason for the marked variations is due to the buoyancy force. For regions of the upper aquifer boundary which are closer to horizontal, not as much of the buoyancy drive has to be exerted to impact the downdip flushing tendency of the hydrodynamic force. Thus, a larger region is available for hydrocarbon accumulation. The converse is, of course, the case in regions of more verticality of the upper aquifer boundary.

The case of a planar top boundary and a "wiggly" lower boundary is not shown because the lower boundary has little effect. Thus, the "wiggly" top and "wiggly" bottom boundary case looks exactly like the case exhibited in Fig. 6.19.

6.2.2. Increasing Water Flow

With all other conditions held fixed, an increased water flux provides a larger hydrodynamic force and so a smaller maximum areal mass for a hydrocarbon stringer than would be the case at lower flow conditions. Exceptions to this rule are certainly those situations in which no steady-state hydrocarbon stringer can exist in the absence of a flow. The fixed-angle, but variable cross section, aquifers examined in this chapter qualify as exceptions under this rule. It is these cases which we now examine.

Three geometric cases will suffice to illustrate the rich diversity of allowable behaviors. We consider the cases of (a) a converging aquifer, (b) a diverging aquifer, and (c) a "wiggly" aquifer with the top surface doing the wiggling. In each case, we examine the maximum areal mass allowed for a stringer as a function of position along the aquifer when the hydrocarbons are oil or gas, and we examine the behavior of the accumulations when the hydrodynamic water flux is increased.

Figures 6.5–6.19 illustrate the three geometrical conditions for oil and gas stringers as the flow is increased.

Some general consequences are fairly obvious. Because of the larger buoyancy of gas relative to that of oil, a larger maximum areal extent of gas can be held in place by balancing the higher buoyancy pressure with a higher water flow. Alternatively, we can note that a low flow has a smaller hydrodynamic pressure so not as much gas (relative to oil) can be held in place as the flow speed is lowered.

The combined effect is to show that the oil accumulations allowed generally maximize at a lower flow rate for a given geometry than do gas accumulations and, equally, are more rapidly depleted at higher flow speeds than a corresponding areal extent of gas.

6.2.3. Different, Constant, Aquifer Angles

If the aquifer's geometric shape, the physical attributes, and water flux are maintained at constant values, there is then the question of which orientation of an aquifer is best for maximizing the accumulation of hydrocarbons. Clearly, a vertical, or nearly vertical, aquifer has great difficulty in maintaining a significant hydrocarbon stringer because the slightest increase or decrease in the flux of water will cause a given hydrocarbon mass to be swept either down dip or up dip. On the other hand, a horizontal aquifer suffers from the opposite problem. The buoyancy force acts vertically so that the slightest flow along the horizontal aquifer will cause an immediate flushing of any hydrocarbon accumulation so that no stable stringer is possible. Thus, there must be an angle at which, all other conditions being equal, a maximum areal extent of a stringer can be maintained.

To illustrate the range of possible behaviors, several case studies are presented.

First, for a fixed-width planar aquifer, the orientation angle is varied. The resulting shapes of the maximum areal masses of oil and gas stringers are exhibited in Figs. 6.11 and 6.12.

Second, behaviors of oil and gas stringers in a converging and a diverging aquifer are examined with respect to angle of orientation of the aquifer, with all other parameters held fixed as in the planar aquifer cases so that a direct comparison of results is possible. Figs. 6.13–6.18 show how the stringers now depend not only on the orientation of the aquifer but also on the position of the stringer along the aquifer.

Third, for a "wiggly" aquifer, Fig. 6.19 shows the variation of hydrocarbon stringer behaviors as the angle of the aquifer varies. Again we see a dependence of the stringer area (for both oil and gas) on both the aquifer angle of orientation and the wiggly nature of the aquifer, representing a combination of the effects illustrated in the previous two limiting geometric situations.

6.2.4. Variations in Capillarity and Surface Tension

Lowering the surface tension (capillarity) has the effect of requiring that hydrodynamic forces are compensated for more by buoyancy forces and capillary (surface tension) forces. As the balance of forces shifts, so, too, do the shape and position of hydrocarbon stringers in a given aquifer under otherwise constant conditions.

To illustrate briefly some of these considerations, we examine in the next chapter the behavior of oil and gas stringers in a plane-parallel, fixed-angle aquifer under several conditions for each case.

However, to illuminate some of the peculiar effects that can be obtained, Fig. 6.20 presents some representative cases of "bubbles" of oil with variable wetting angles and variable surface tension attached to the flow direction side of an anticline where, without surface tension and wetting angle considerations, no steady-state stringers can exist. Note from Fig. 6.20a–d, calculated for a fixed wetting angle of 20°, that as the surface tension decreases from 40 dyn/cm (Fig. 6.20d) to 10 dyn/cm (Fig. 6.20a), there is a decrease in the size of the "bubble" that can be held by surface tension and capillarity forces against the hydrodynamic and buoyancy drives (both of which would prefer to have the oil at the top of the anticline).

As the wetting angle is increased to 40°, the corresponding size of the bubble is increased at a given surface tension (Fig. 6.20e–h) relative to its size for the same surface tension value at the lower wetting angle of 20° (Fig. 6.20a–d). Note also that as the surface tension decreases from 40 dyn/cm (Fig. 6.20h) to 10 dyn/cm (Fig. 6.20e) at the fixed wetting angle of 40°, there is a corresponding diminution in the size of the oil stringer. For surface tension values much below 10 dyn/cm, the capability to hold a bubble of oil in place on the down-dip side, shown in Fig. 6.20a–h, decreases to the point where no numerical solution is found—indicating the fact that no balance of forces can be struck at that location.

Clearly, as the capillary wetting angle, θ, is increased, or as the capillary pressure is increased (with the appropriate sign of capillary pressure), the capillary term $p_c\ dh/dl$ ($\approx p_c \sin\theta$) increases so that a larger bubble of oil can be held in place at a given location.

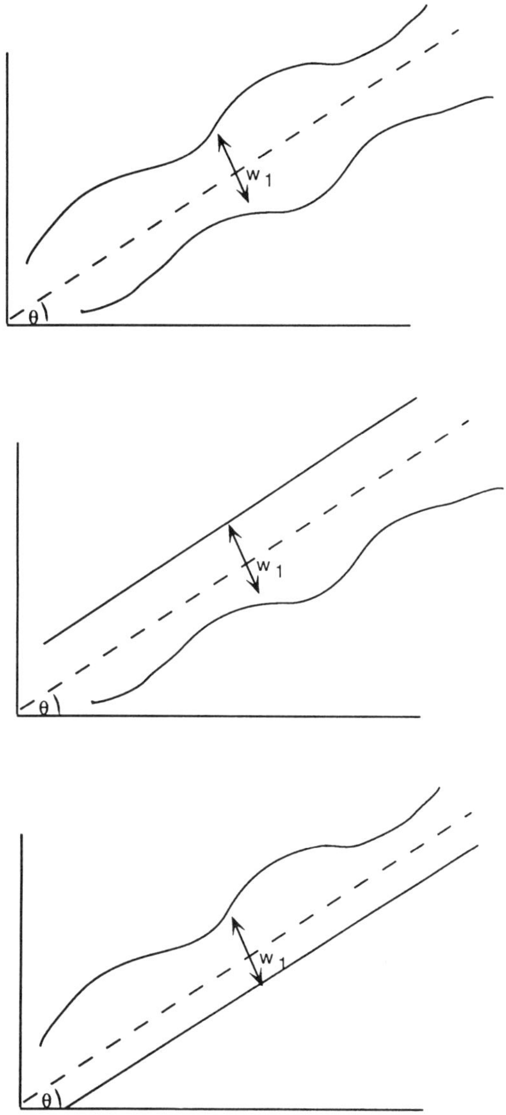

Figure 6.1. Sketch of three aquifers having the same variation of aquifer width with position along the aquifer but which give rise to different shaping and positioning of hydrocarbon stringers.

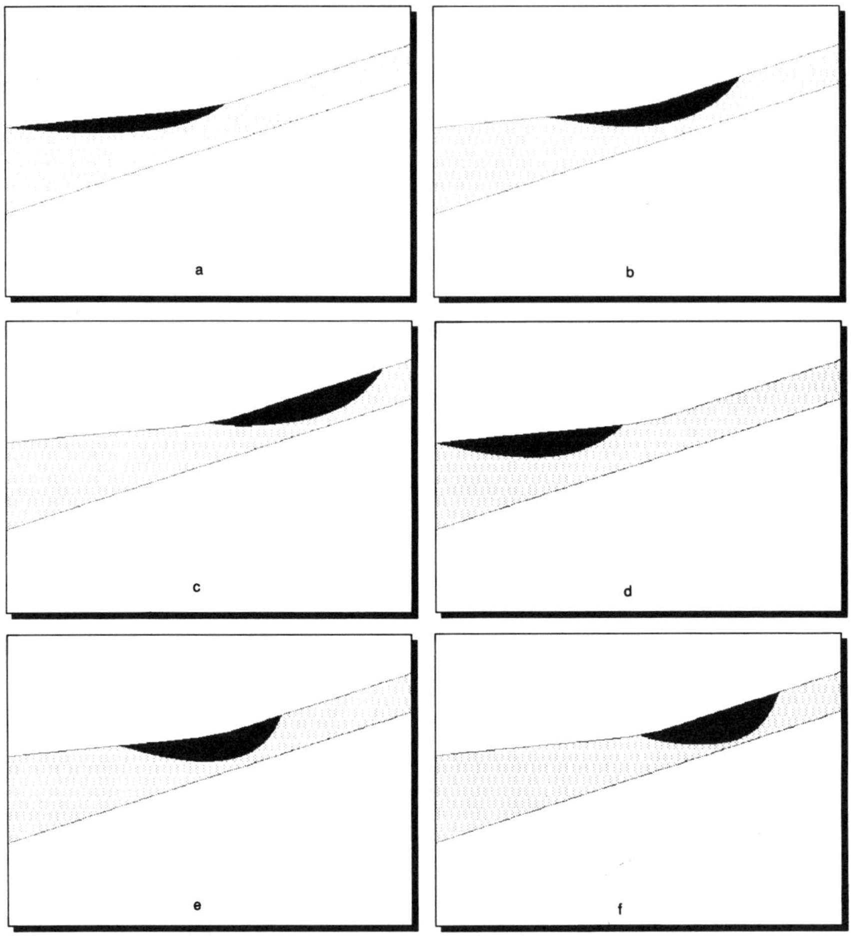

Figure 6.2. Representations of oil (a–c) and gas (d–f) accumulations under the geometrical conditions sketched for different required positions of the leading edge of the stringer. The flow speed is 2 cm/yr.

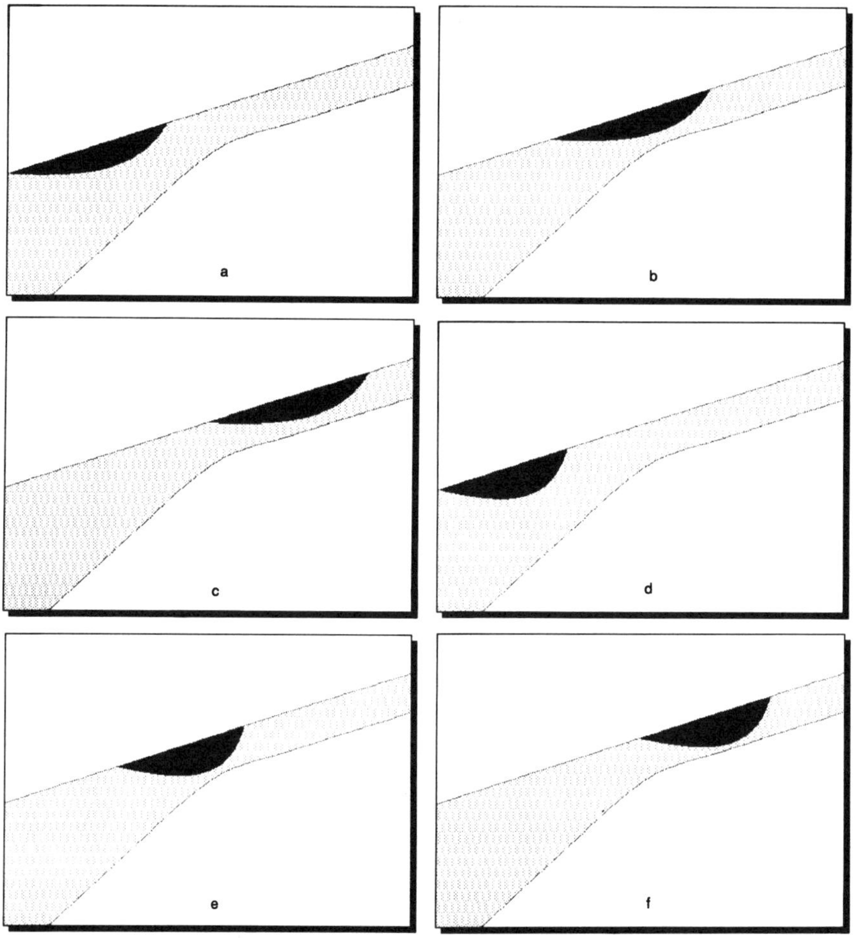

Figure 6.3. As for Fig. 6.2 but with the geometry "inverted."

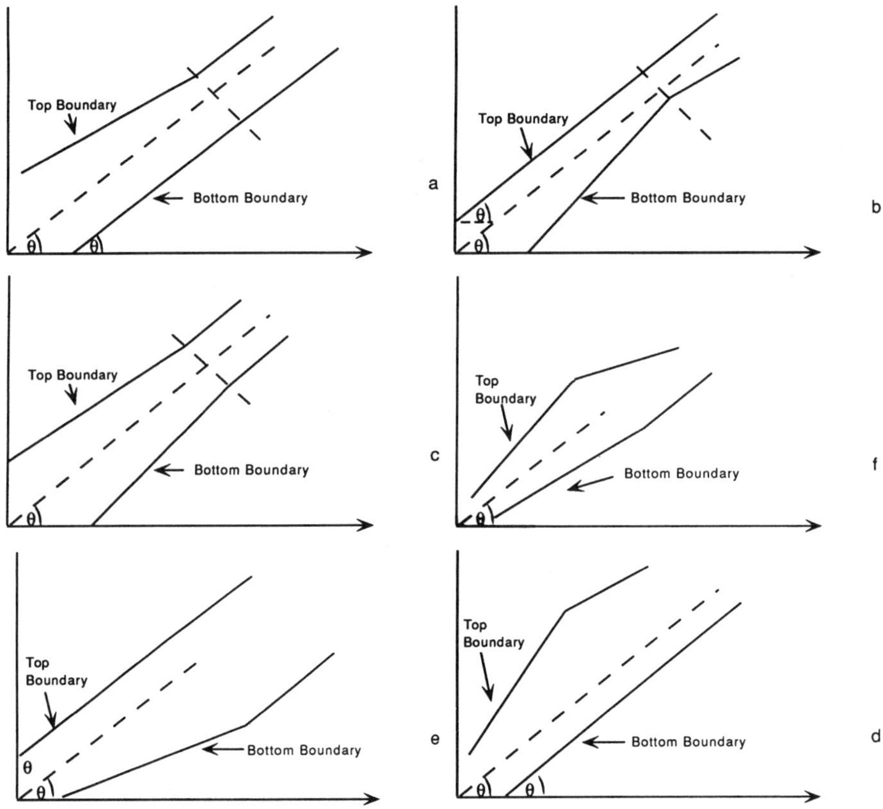

Figure 6.4. Various aquifer geometries that are to be modeled in order to examine their effects on steady-state hydrocarbon accumulations at different locations within the geometrical configurations illustrated.

Table for Figs. 6.5–6.10

	Position along aquifer of "nose" of stringer					
	Oil			Gas		
Aquifer geometry	0	25	45	0	25	45
Converging top, planar base	6.5a	6.5b	6.5c	6.5d	6.5e	6.5f
Planar top, converging base	6.6a	6.6b	6.6c	6.6d	6.6e	—
Converging top and base	6.7a	6.7b	6.7c	6.7d	6.7e	6.7f
Diverging top, planar base	6.8a	6.8b	6.8c	6.8d	6.8e	6.8f
Planar top, diverging base	6.9a	6.9b	6.9c	6.9d	6.9e	6.9f
Diverging top and base	6.10a	6.10b	6.10c	6.10d	6.10e	6.10f

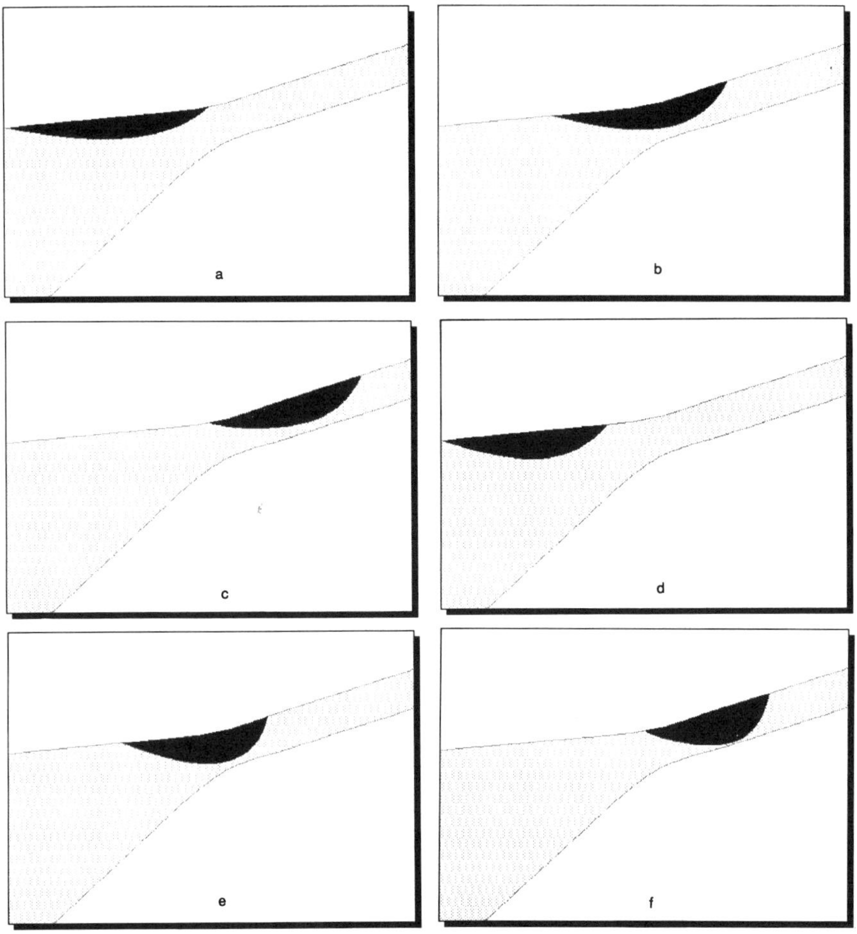

Figures 6.5–6.19. These groups of figures show variations in oil and gas stringer shapes and locations as the flow speed, geometry, and overall vertical nature of the aquifer change. Flow speeds range from 0.2 to 20 cm/yr; the permeability of the aquifer is maintained at 100 mD throughout, and the oil and gas densities are 0.8 g/cm^3 and 0.2 g/cm^3, respectively. The legend for each group of figures can be obtained from the accompanying tables.

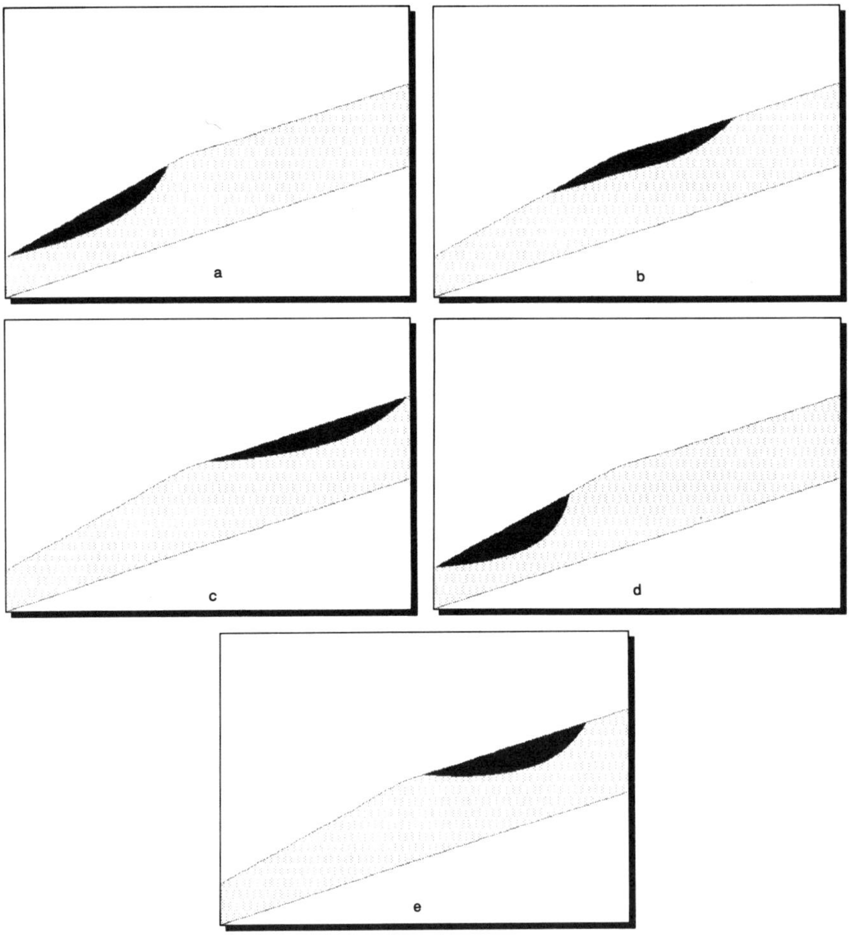

Figure 6.6.

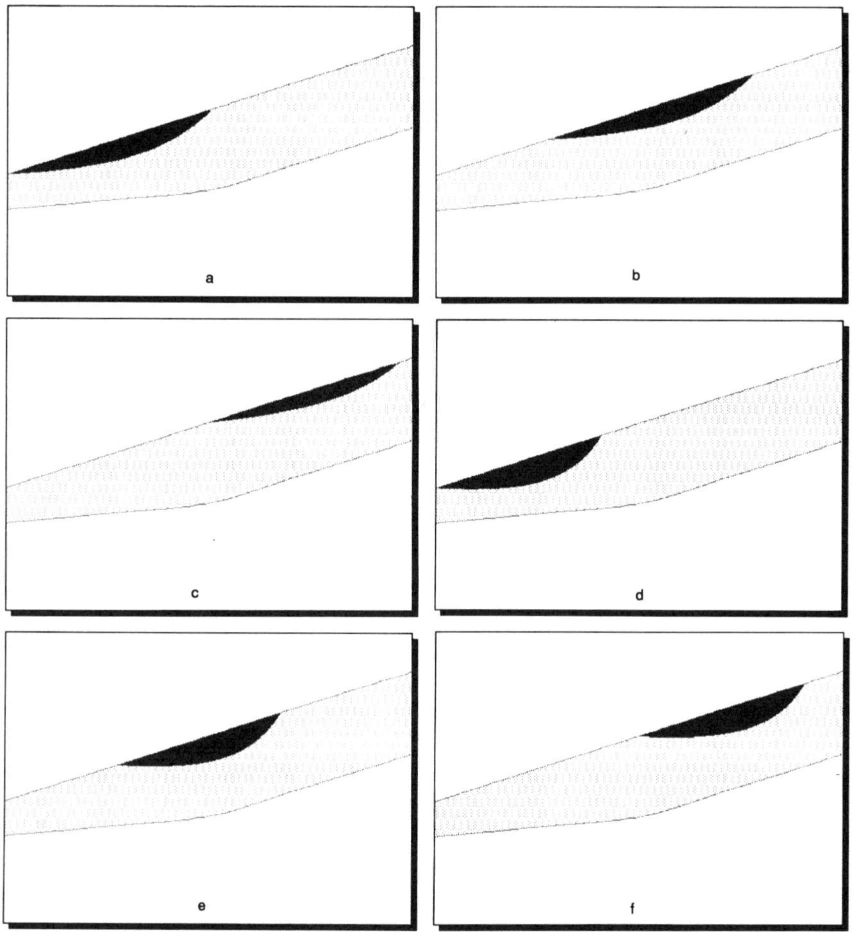

Figure 6.7.

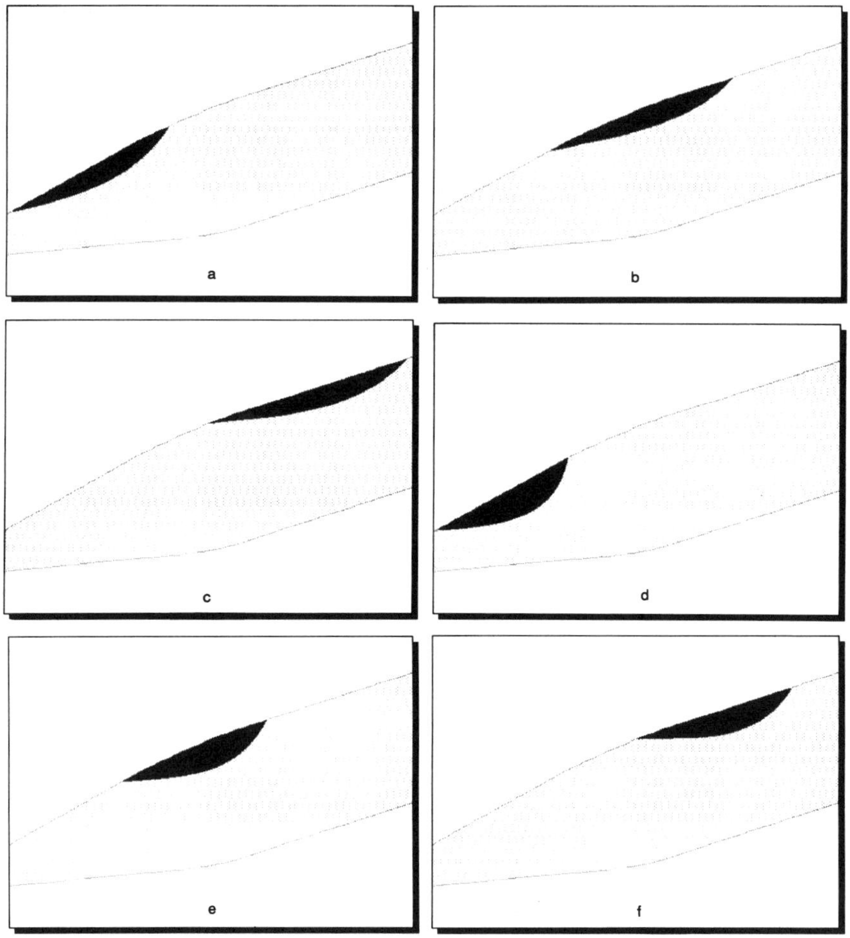

Figure 6.8.

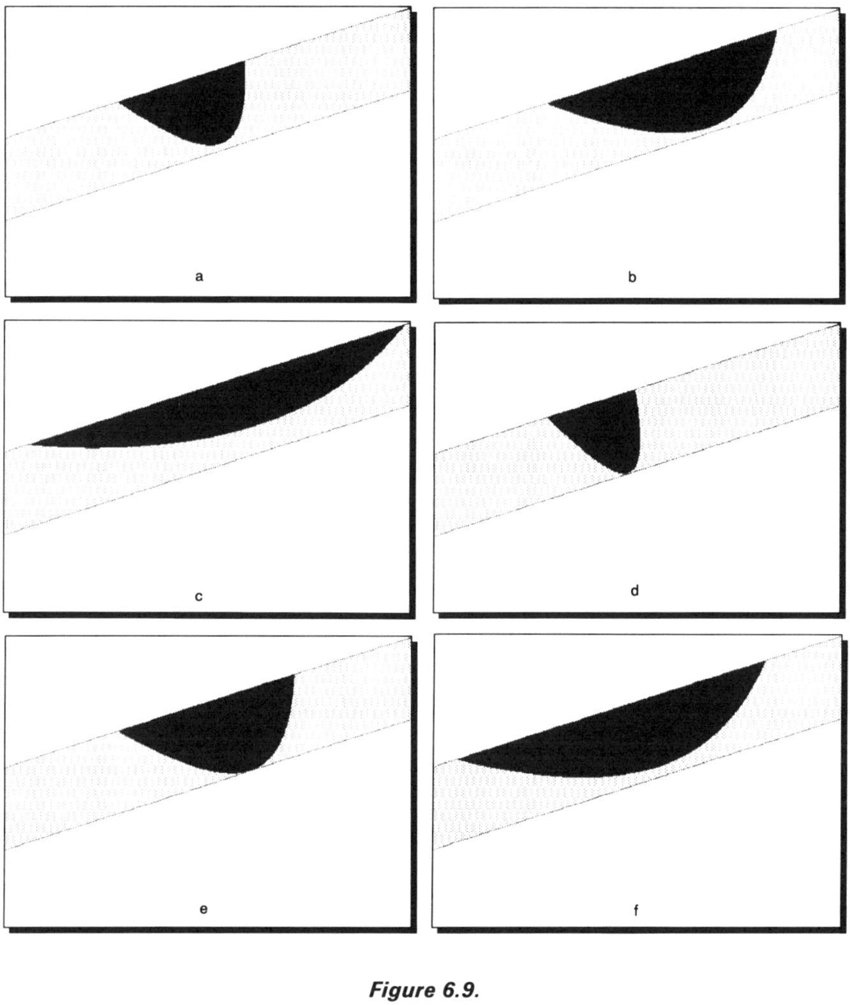

Figure 6.9.

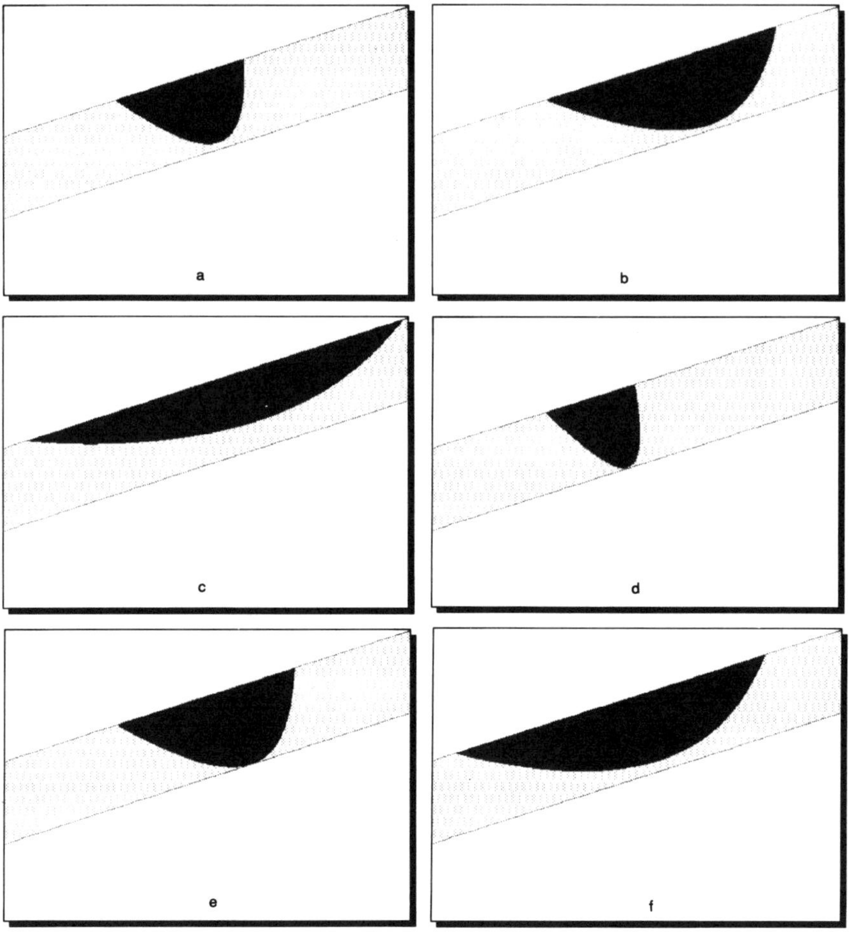

Figure 6.10.

Table for Figs. 6.11 and 6.12. Aquifer Geometry: Planar Top and Base

Flow speed from right (cm/yr)	Oil		Gas	
	Low angle	High angle	Low angle	High angle
20	6.11a	6.12a	6.11d	6.12d
2	6.11b	6.12b	6.11e	6.12e
0.2	6.11c	6.12c	6.11f	6.12f

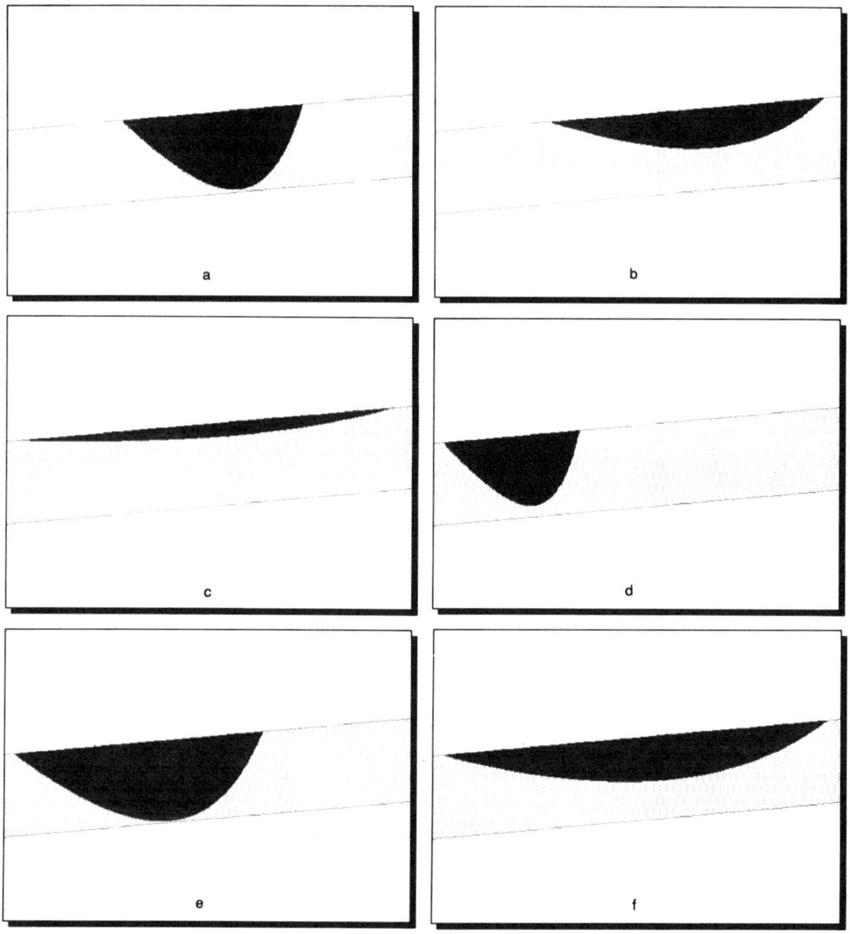

Figure 6.11.

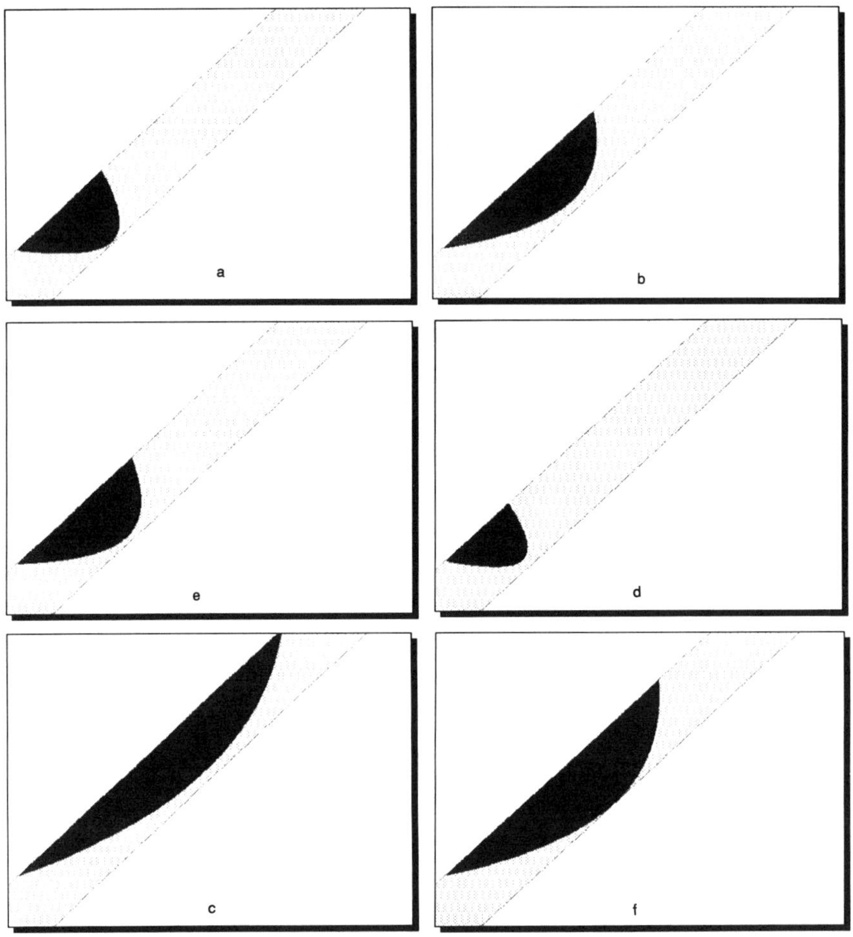

Figure 6.12.

Table for Figs. 6.13–6.18

Flow speed from right (cm/yr)	Position	Oil			Gas		
		Medium angle	Low angle	High angle	Medium angle	Low angle	High angle
20	0, 25, 45	6.13a, b, c	6.14a, b, c	6.15a, b, c	6.13d, e, f	6.14d, e, f	6.15d, e, f
2	0, 25, 45	6.13g, h, i	6.14g, h, i	6.15g, h, i	6.13j, k, l	6.14j, k, l	6.15j, k, l
0.2	0, 25, 45	6.13m, n	6.14m, n	6.15m, n	6.13o, p	6.14o, p	6.15o, p
20	0, 25, 45	6.16a, b, c	6.17a, b, c	6.18a, b, c	6.16d, e, f	6.17d, e, f	6.18d, e, f
2	0, 25, 45	6.16g, h, i	6.17g, h, i	6.18g, h, i	6.16j, k, l	6.17j, k, l	6.18j, k, l
0.2	0, 25, 45	6.16m, n, o	6.17m, n	6.18m, n, o	6.16p, q, r	6.17o, p	6.18p, q, r

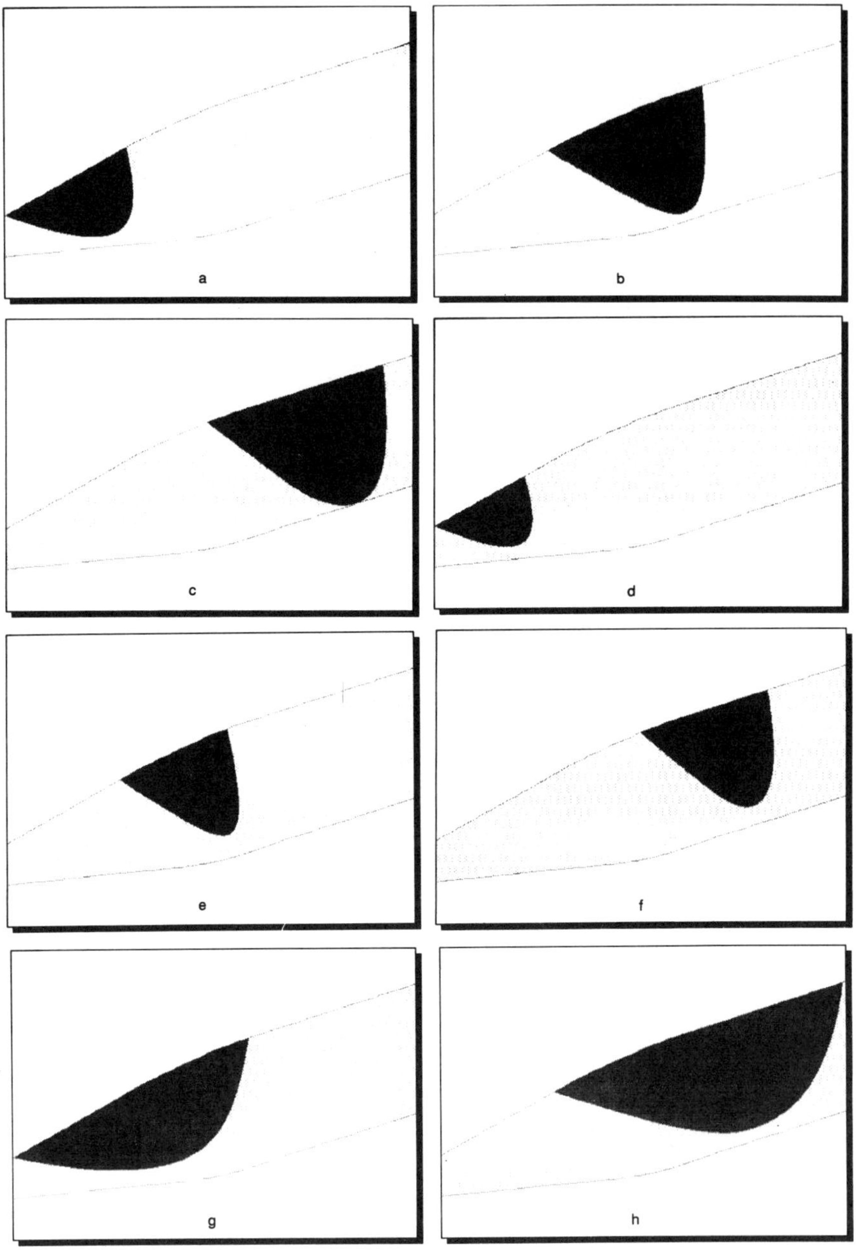

Figure 6.13.

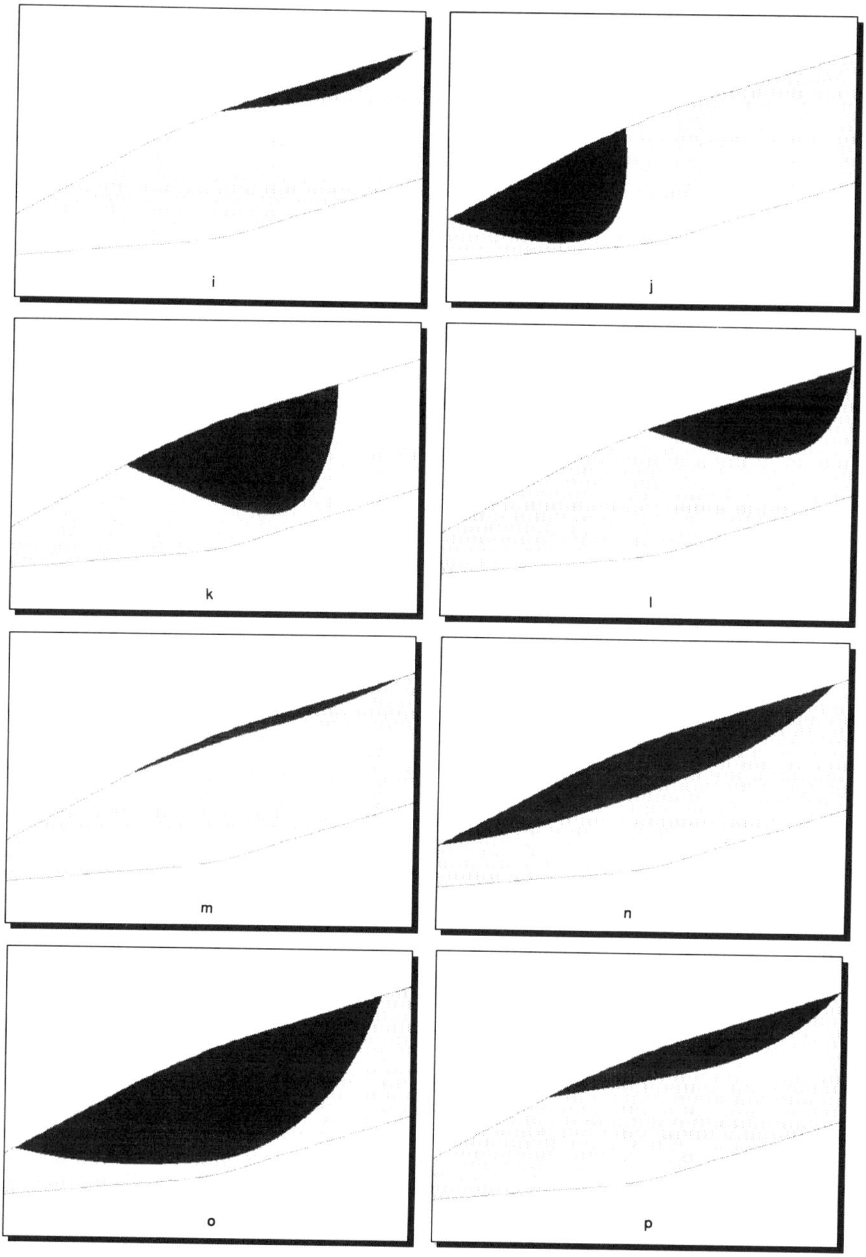

Figure 6.13. (*Continued*)

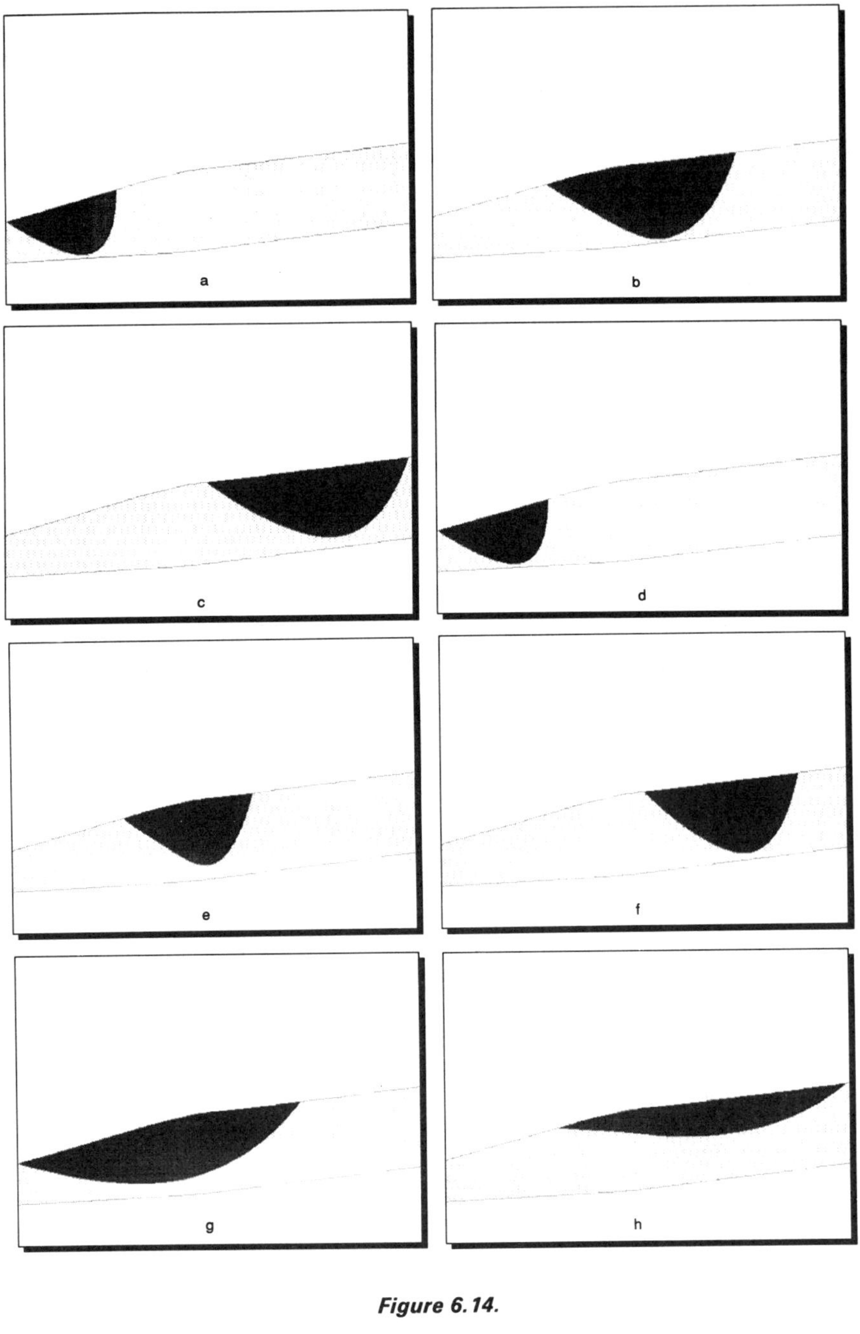

Figure 6.14.

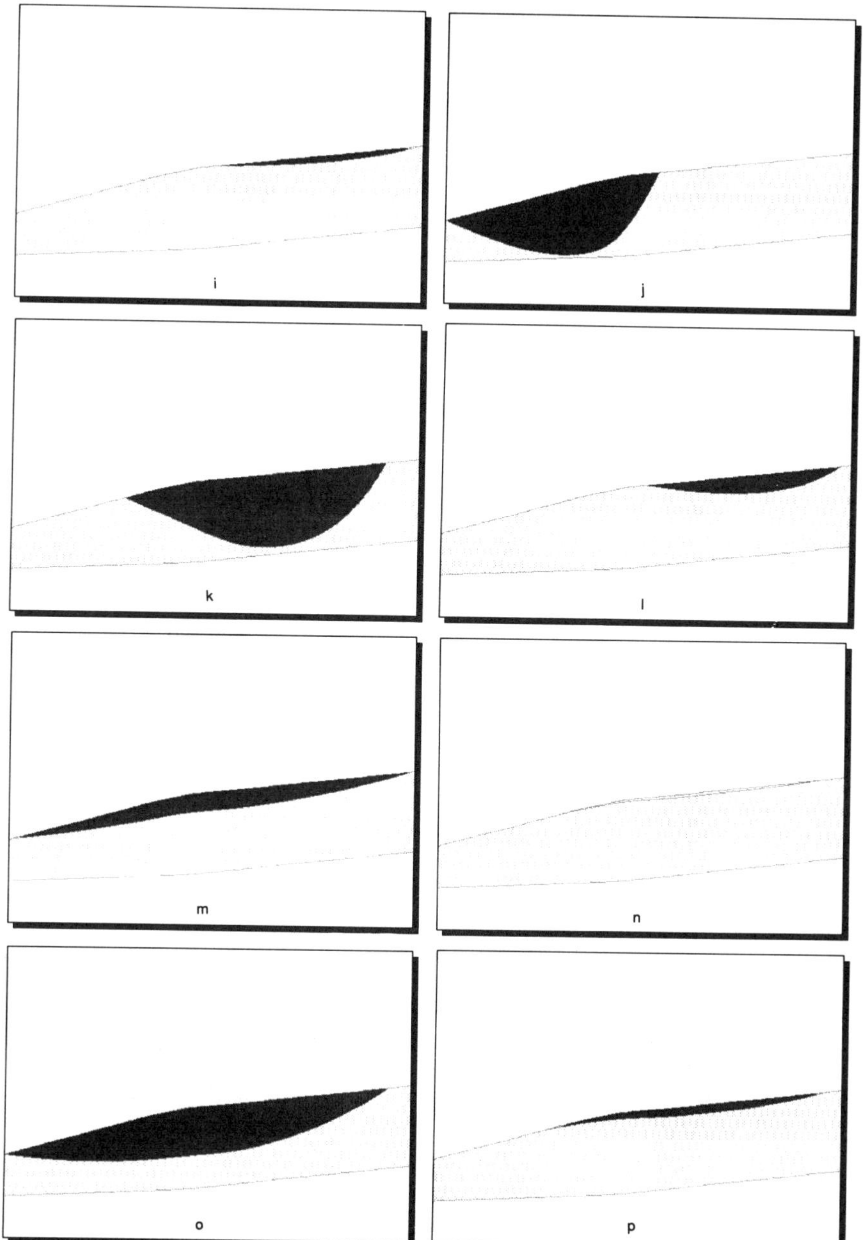

Figure 6.14. (*Continued*)

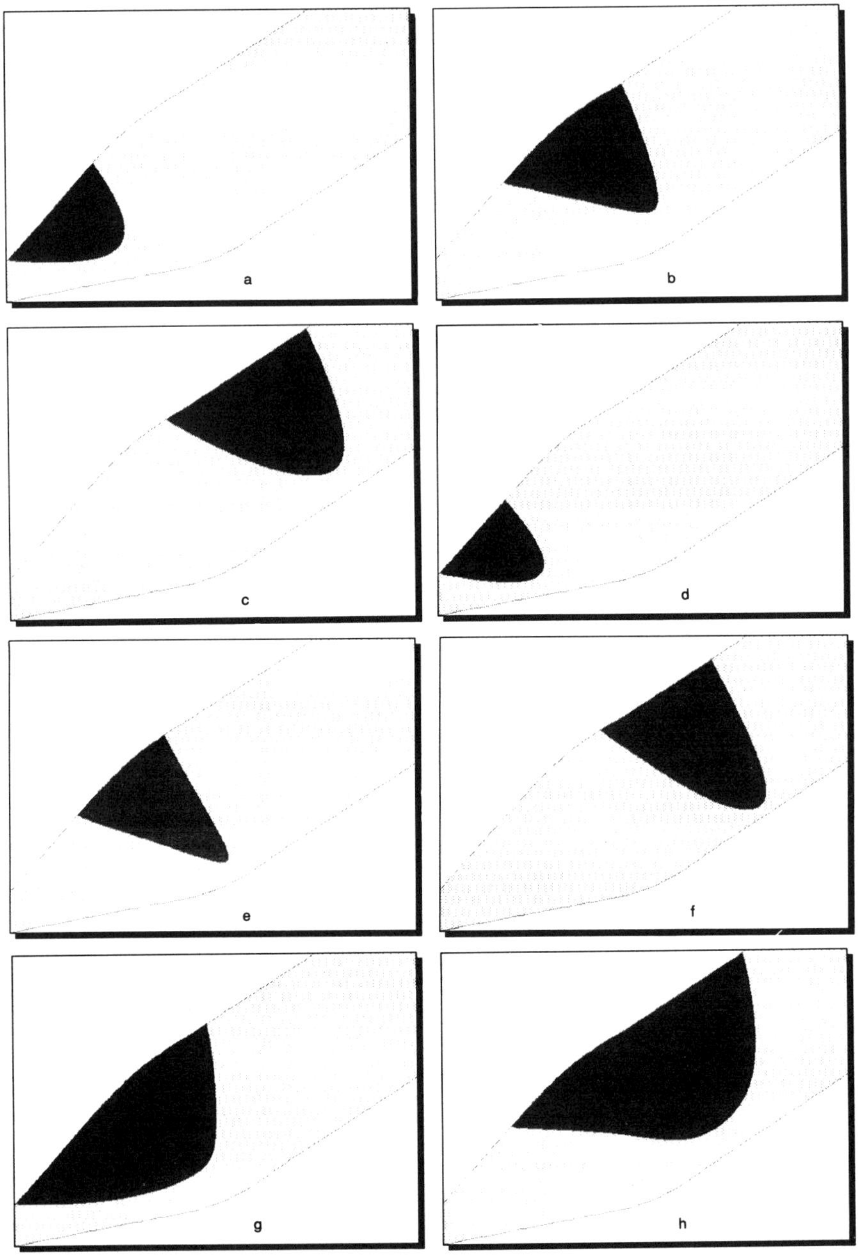

Figure 6.15.

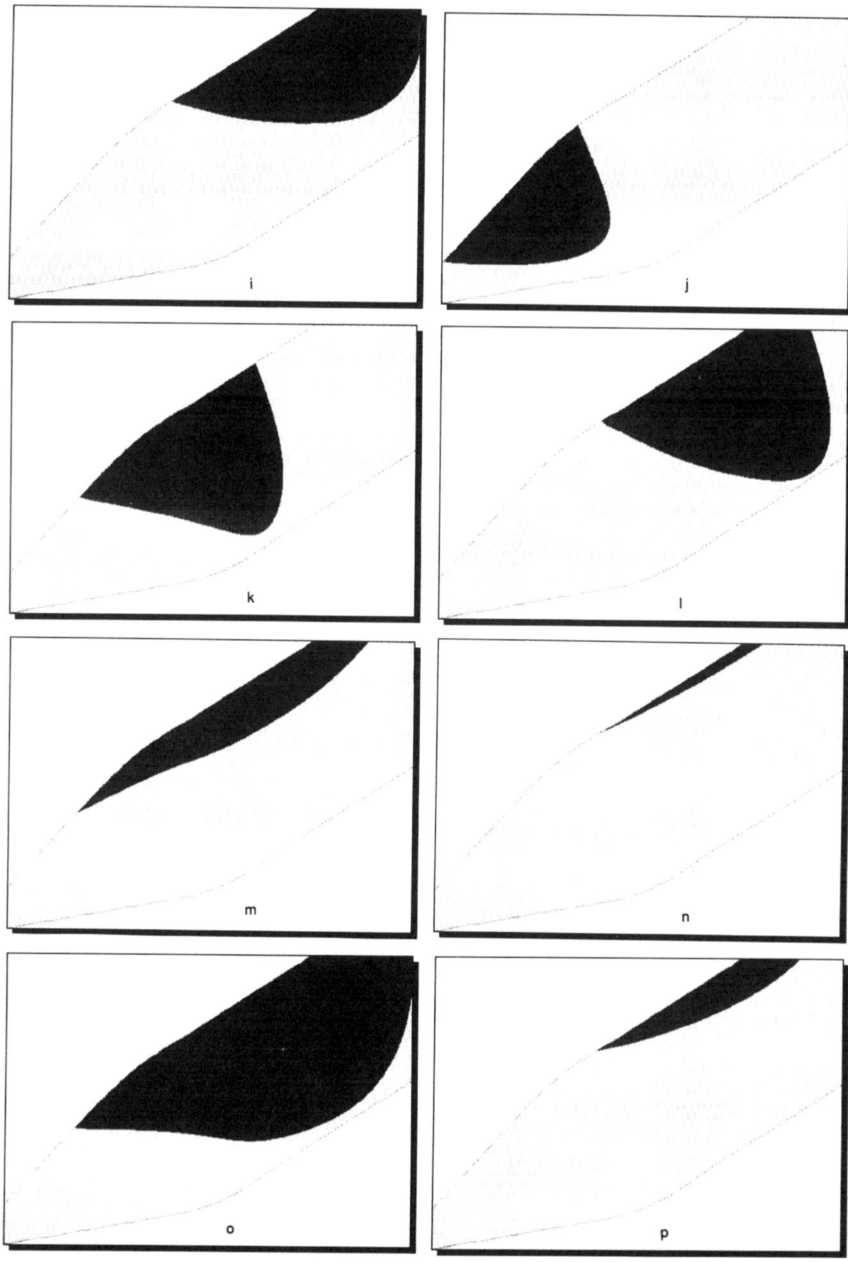

Figure 6.15. (*Continued*)

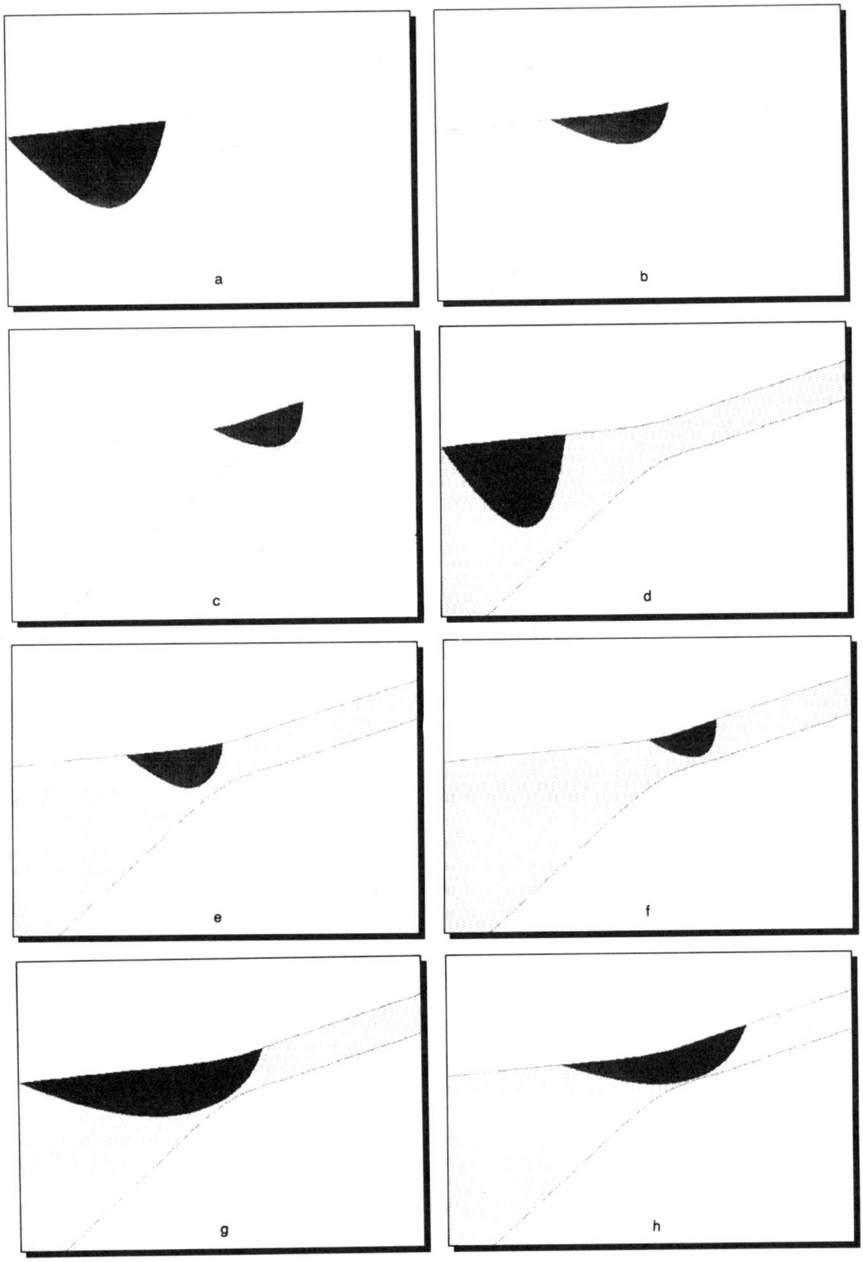

Figure 6.16.

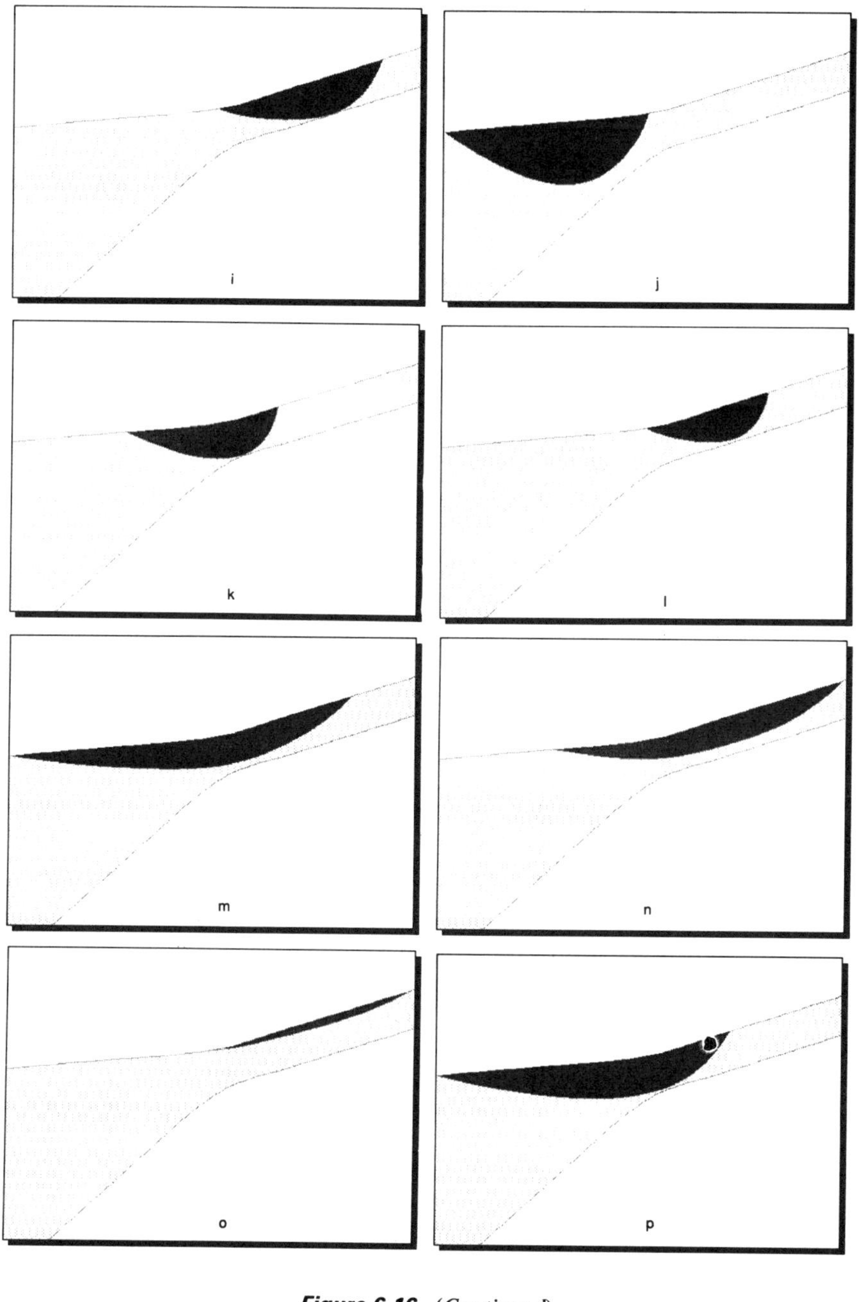

Figure 6.16. (*Continued*)

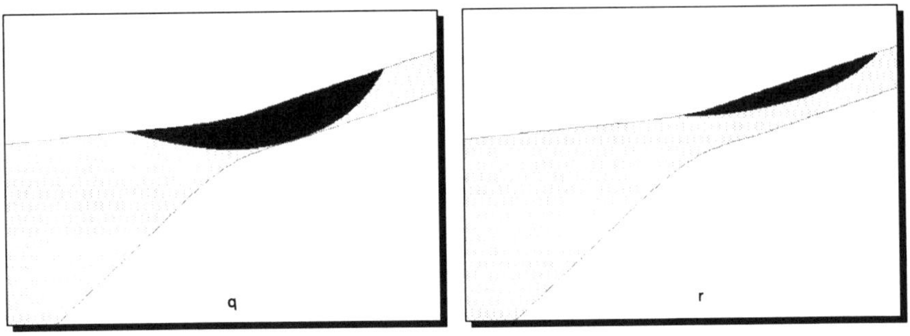

Figure 6.16. (*Continued*)

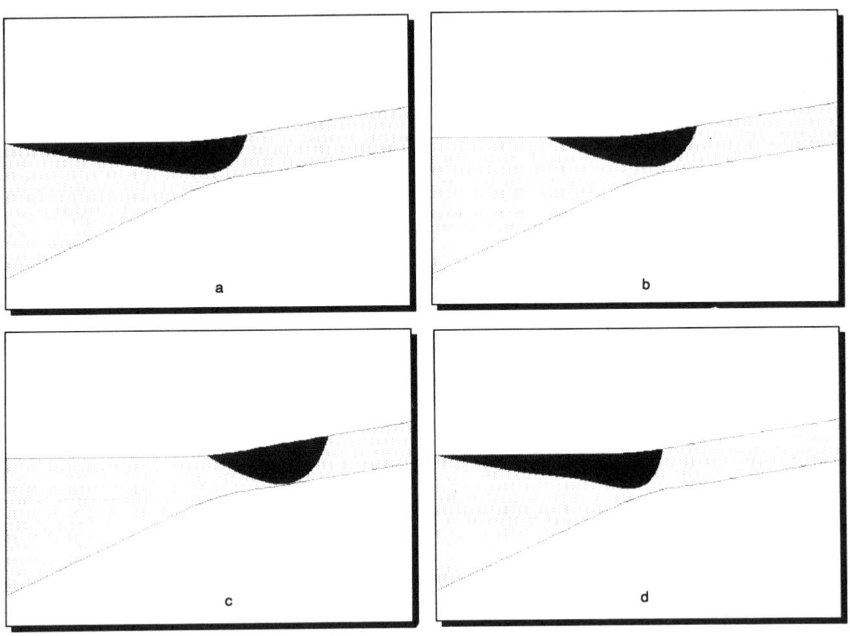

Figure 6.17.

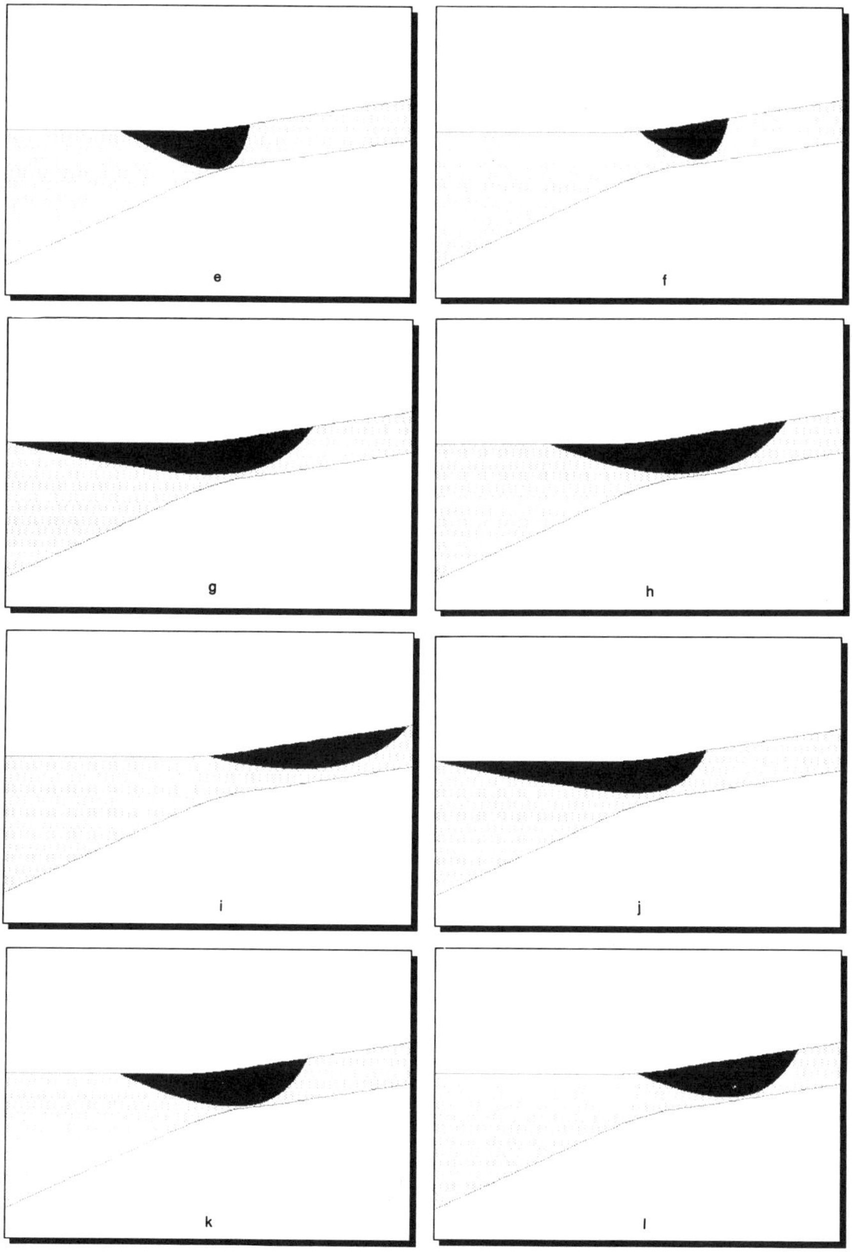

Figure 6.17. *(Continued)*

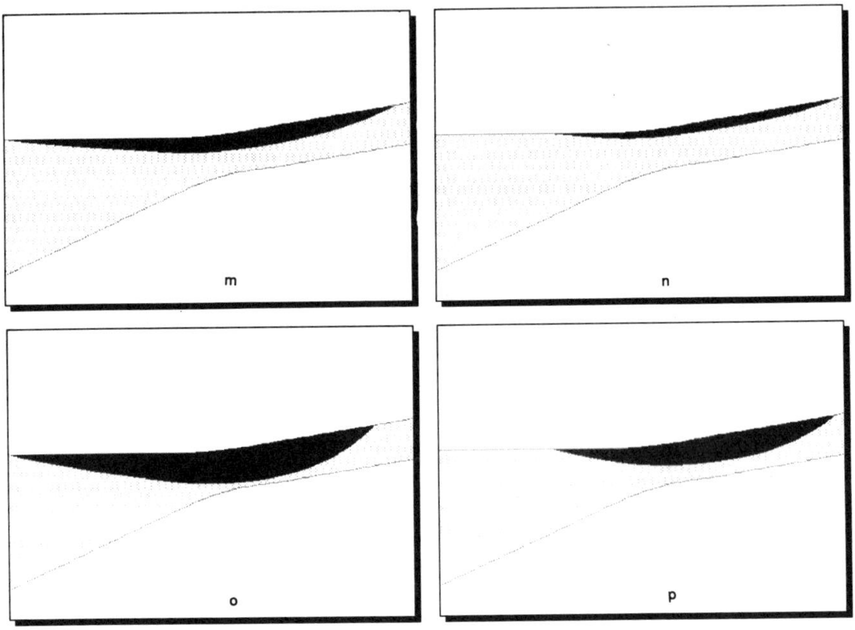

Figure 6.17. (*Continued*)

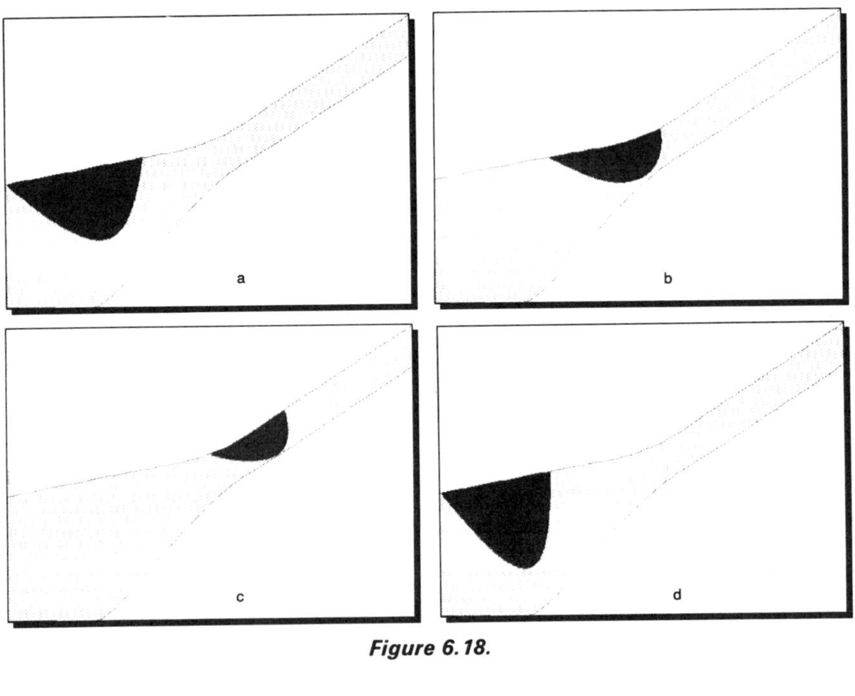

Figure 6.18.

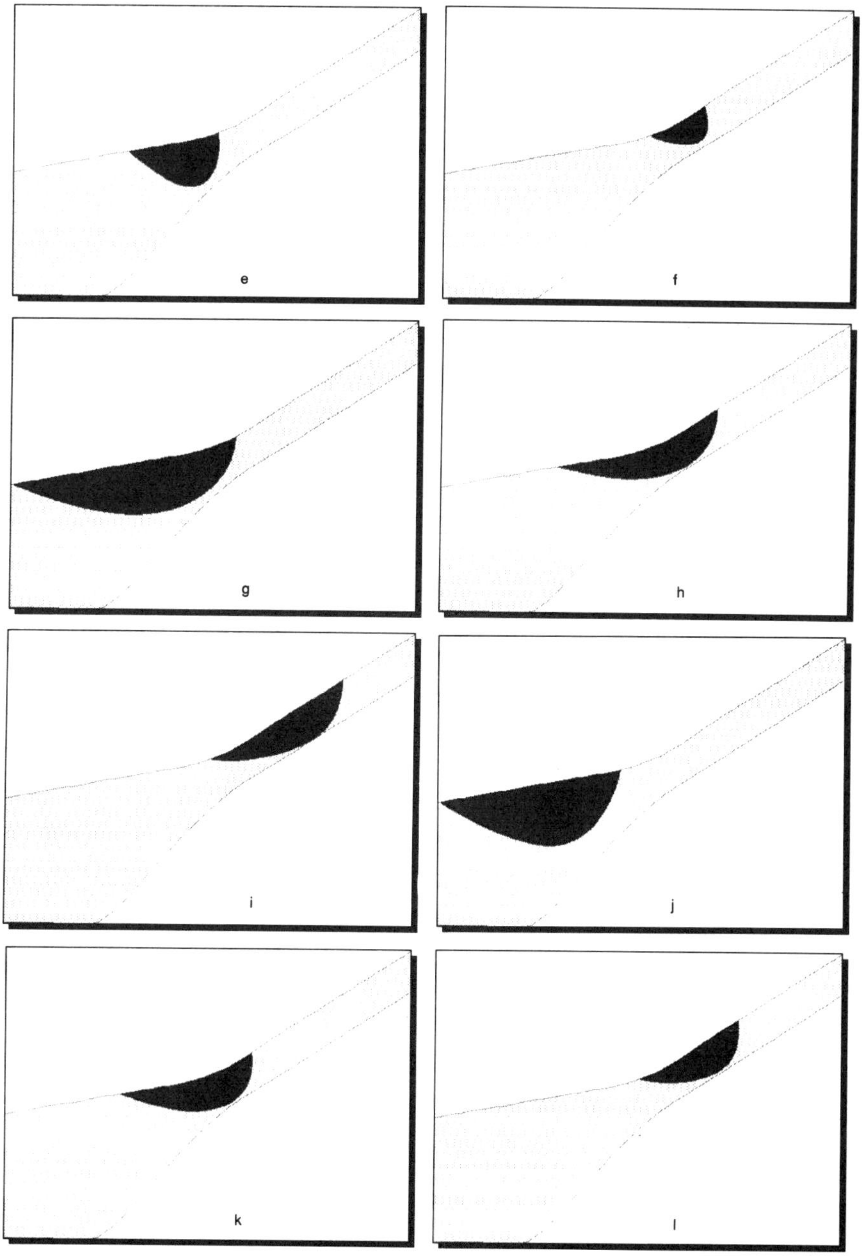

Figure 6.18. *(Continued)*

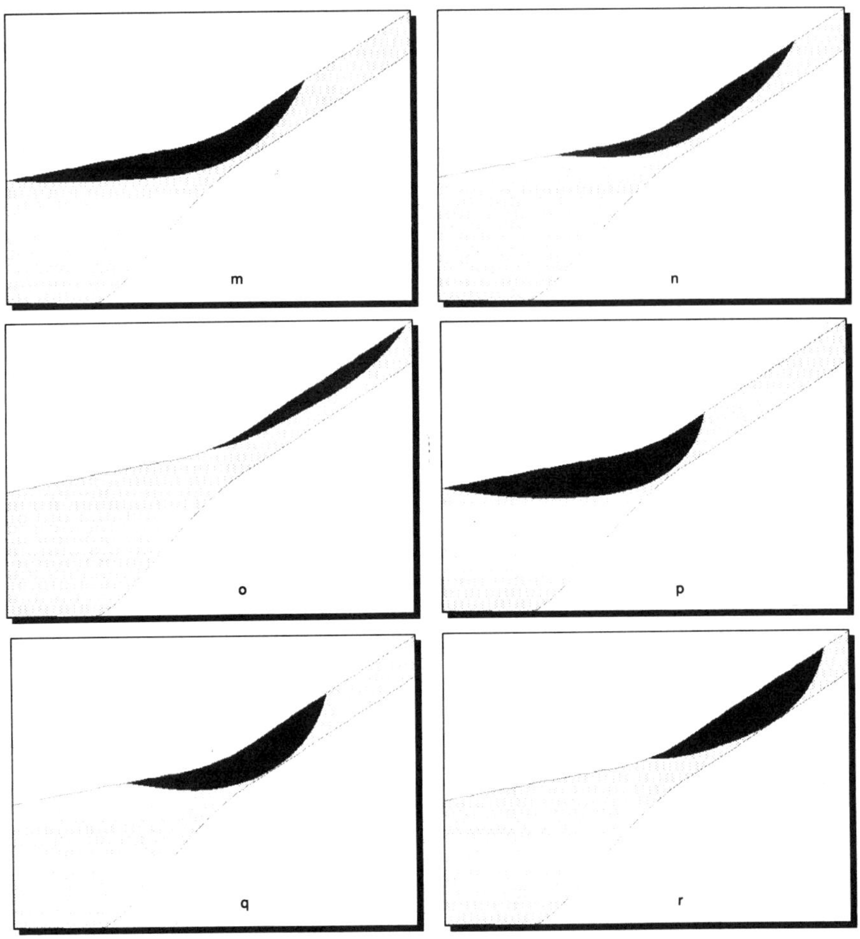

Figure 6.18. (*Continued*)

Table for Fig. 6.19. Aquifer Geometry: "Wavy" Top, Planar Base

Position along aquifer of "nose" of stringer	Oil	Gas
0	6.19a	6.19e
5	6.19b	6.19f
10	6.19c	6.19g
15	6.19d	6.19h

[a]Flow speed, 20 cm/yr from right.

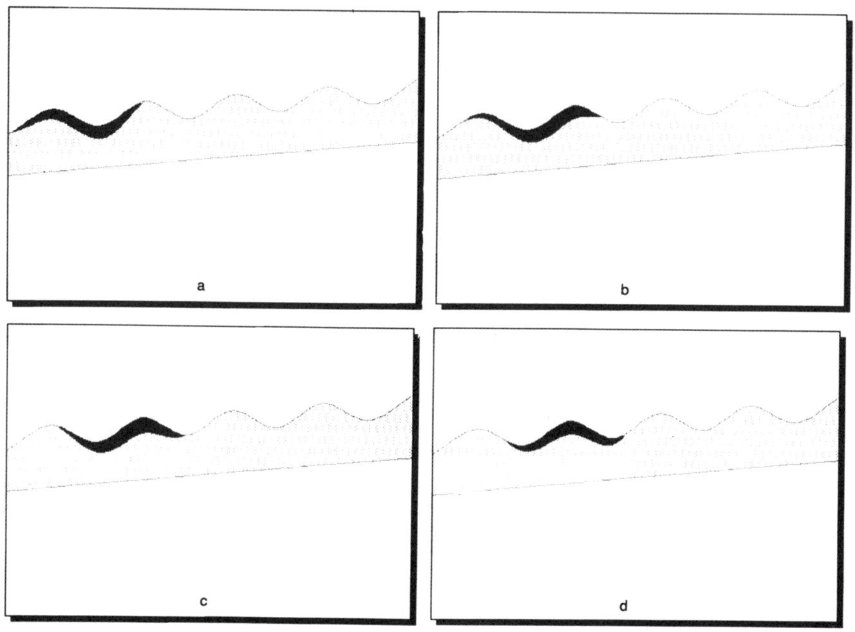

Figure 6.19.

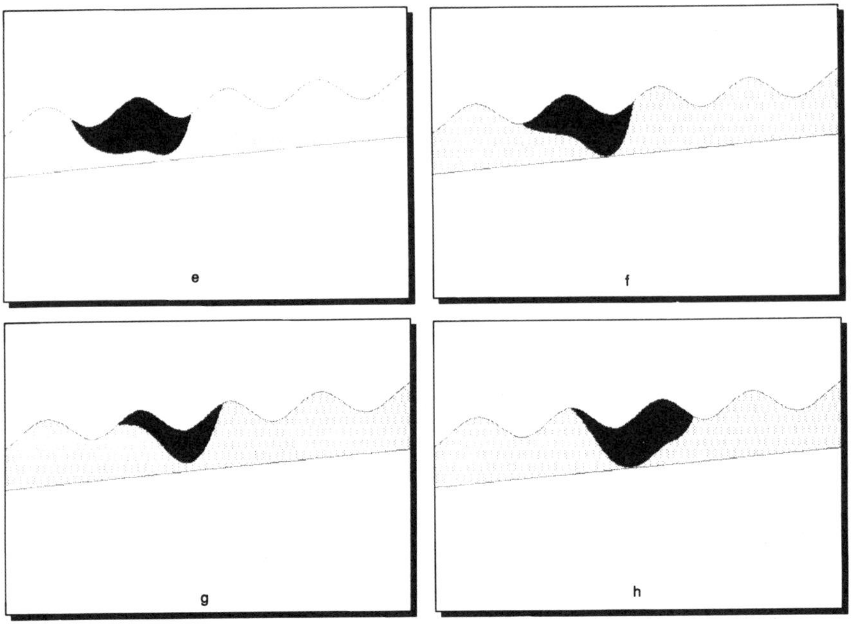

Figure 6.19. *(Continued)*

Table for Fig. 6.20. Aquifer Geometry: Single Anticline, with Structurally Deeper Connecting Throats

	Surface tension (dyn/cm)			
Wetting angle	10	20	30	40
20°	6.20a	6.20b	6.20c	6.20d
40°	6.20e	6.20f	6.20g	6.20h

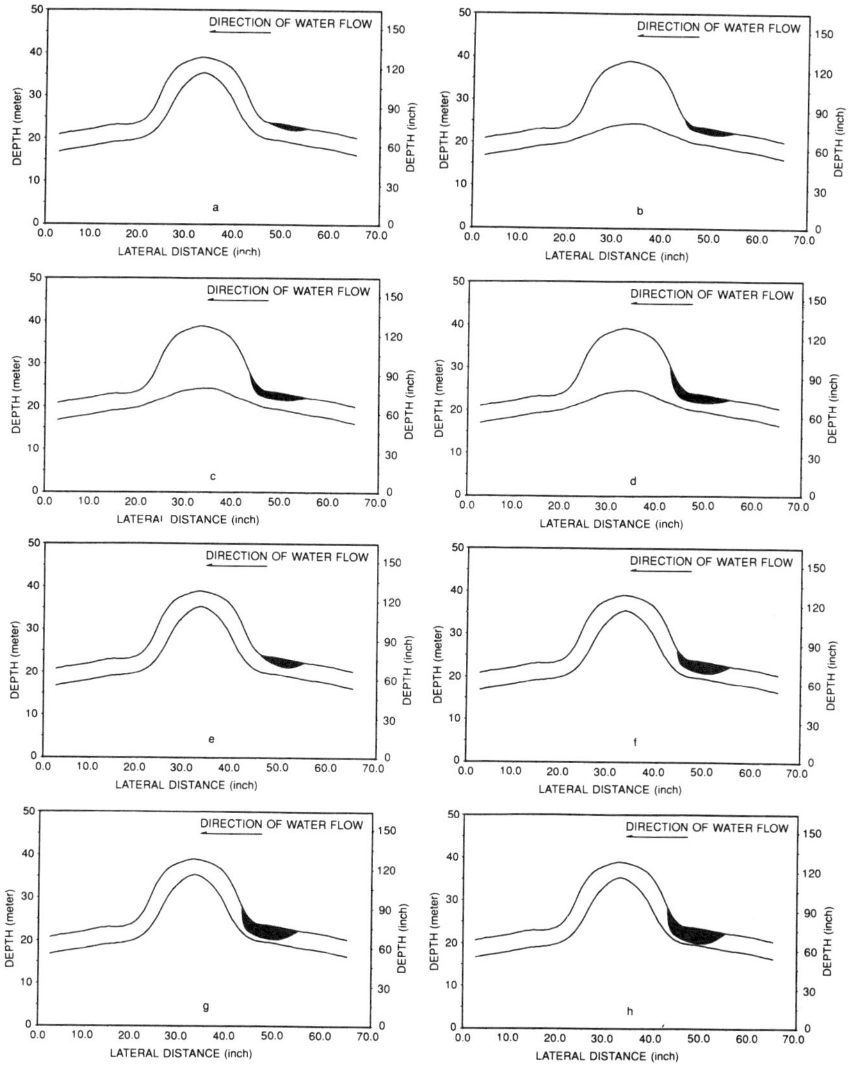

Figure 6.20. Illustrations of the effect of changing the surface tension on the hydrodynamic shaping of an oil stringer, with the wetting angle of the "nose" (right-hand side) of the stringer set at different values. Water flow is from the right at 100 cm/yr, oil density is 0.8 g/cm^3, and the aquifer permeability is 1000 mD. The values of the surface tension and wetting angle for each panel in the figure are given in the accompanying table. Panels a–d refer to a fixed wetting angle of 20° and variable surface tension, while panels e–h refer to a 40° wetting angle and variable surface tension.

7

Fixed-Width, Variable-Angle Aquifers

The manifest presence of oil-bearing and gas-bearing anticlines, and of synclines, in sedimentary basins strongly suggests that an investigation be undertaken of the effect of hydrodynamic flow on the amount, shape, and positioning of buoyant hydrocarbon accumulations as the global geometric shape of the aquifer is distorted. In this chapter we isolate the geometric bending of an aquifer, keeping aquifer width and permeability fixed.

A variety of cases are examined to illustrate the diversity of steady-state response patterns to changing aquifer angle for both oil and gas stringers. The simplest set of cases involves an inclined aquifer of fixed dip, and the hydrocarbon behavior as both the hydrodynamic flux and the dip angle are changed, is investigated for aquifers of fixed, but variable, width. The second set of cases is analogous to the first set, but with the aquifer changed from one of fixed dip to one of a single synclinal shape of varying radius of curvature. By way of contrast, the third set of cases considers a single anticlinal geometry but again covers the range of parameter variations used in the first set of cases. The fourth set of cases involves synclinal and anticlinal aquifers that are contiguously related. As we shall see, the multiple trapping nature of this sort of geometrical configuration is quite different under hydrodynamic flow conditions than under purely hydrostatic conditions. In particular, it is possible to have oil and gas trapped in *synclinal* geometries with the upward buoyancy drive being held in check by the down-thrusting hydrodynamic forces.

7.1. Fixed-Dip Aquifers

While individual facets of fixed-dip aquifers have been examined in previous chapters, a systematic set of operations is considered in this section, some of which reproduce results from prior chapters. A deliberate sequence is followed because the next sections will also consider the same sequence of operations but under different and gradually more complex geometrical aquifer shapes. In this way, the simple cases of this section provide a set of templates within a familiar framework so that the development of the later more complex situations is more readily traced back to one or more of the basic cases shown here.

In each section three situations are sequentially considered for both oil and gas stringers. They are (i) increasing flow, (ii) increasing dip angle, and (iii) increasing aquifer width.

7.1.1. Increasing Flow

As we have seen in the previous chapters, there are no steady-state oil or gas stringers in the absence of a confining hydrodynamic pressure for fixed-dip aquifers.

Figures 7.1a–d and 7.2a–c illustrate the areal mass that can be confined for hydrocarbon stringers comprised of oil and gas, respectively, as the flow speed increases. The differences in shape behavior of the oil and gas stringers under otherwise identical conditions have been shown in previous chapters to be due to the difference in buoyancy pressure. There is no need to repeat these arguments here. Note that in the cases of both oil and gas there is an absolute maximum area containable as the flow speed increases.

7.1.2. Increasing Dip Angle

For fixed flow conditions, Fig. 7.3a–r shows the variation in oil and gas stringer shapes as the angle of the aquifer increases, as also recorded in the previous chapter.

7.1.3. Increasing Width

As the aquifer width is systematically increased, with angle and flow speed kept constant, there is less and less increase in aquifer flow speed as the water transits past the hydrocarbon stringer. In particular, when the aquifer width much exceeds the maximum width of the hydrocarbon stringer, the shape of the stringer then depends only on the flow speed and not on the mass flux of water, as shown theoretically in Chapter 3. Thus, there is a limiting behavior of the hydrocarbon stringer as the aquifer width increases. This effect is clearly seen in both the oil and gas stringer cases as exhibited in Fig. 7.4a–f.

7.2. A Single Syncline

To some extent, a synclinal aquifer under a hydrodynamic drive has features in common with fixed-angle aquifers. First, in the absence of any hydrodynamic drive, there is no steady-state stringer possible. Second, under a hydrodynamic drive on the down-flow side of the aquifer, there is the possibility of balancing the down-driving hydrodynamic force against the upward tendency of the buoyancy force, and thus of constructing a steady-state hydrocarbon stringer. However, on the up-flow side of the syncline this possibility no longer exists because both buoyancy and hydrodynamic drive are then upward-propelling forces. The similarities and differences are illustrated in the following subsections.

7.2.1. Increasing Flow

From Fig. 7.5a–t we see that as the flow speed increases, there is a critical speed reached at which no stringer (either oil or gas) can be maintained. Basically, the hydrocarbon buoyancy for any shape is insufficient to overcome the hydrodynamic drive, and the hydrocarbons are flushed beyond the syncline's nadir when both buoyancy and hydrodynamic forces act together, driving the upward removal. The critical flow speed is different for oil and gas stringers, reflecting the difference in buoyancy.

7.2.2. Increasing Synclinal Radius

As the syncline flattens and gives the appearance of being closer to a horizontal layer, the component perpendicular to the aquifer top of buoyancy capable of resisting a given hydrodynamic driving pressure decreases. Accordingly, lesser areal masses of oil and/or gas can be retained in flatter synclines. Again, we see from Fig. 7.6a–x that the effect is most noticeable for an oil stringer because of the lesser buoyancy of oil.

7.2.3. Increasing Aquifer Width

As with the planar inclined aquifers, the effect of increasing the width of a synclinal aquifer is to lessen the dependence of the hydrodynamic forces on aquifer width so that, all other factors being equal, the oil and gas stringers tend to shapes that depend solely on the water flow speed as opposed to the water flux. This behavior is noticed for both oil and gas stringers, as exhibited in Fig. 7.7a–p.

7.3. A Single Anticline

If we invert exactly the syncline geometry considered in Section 7.2, then a single-apex anticline is created. A fundamental difference in behavior is then obtained. First, note that, unlike the synclinal situation, the anticline *does* have a hydrostatic solution for a hydrocarbon accumulation. Second, note that the flux of water from the right of the anticline is now up dip, rather than down dip as in the syncline case. The difference in flow direction then implies that steady-state hydrocarbon stringers can be constructed with the leading edge of the stringer to either the left or the right of the apex of the anticline, depending upon physical attributes and flow speed.

7.3.1. Increasing Flow

As the flow speed is increased, there comes a point where the buoyancy pressure attempting to retain the hydrocarbon stringer at or near the anticlinal apex cannot overcome the hydrodynamic drive. The stringers are then flushed leftward, down dip from the apex. This situation is similar to that occurring in the cases of both finite-width and infinitely wide Hubbert aquifers as sketched in earlier chapters. Also of interest, as shown in Fig. 7.8a–p, is the ability of the hydrocarbon stringer to exist partially to the right of the apex where buoyancy upthrust is communicated from the fraction of the hydrocarbon stringer to the left of the apex to overcome the upward, left-seeking buoyancy and the hydrodynamic drive from the right side of the apex. This phenomenon is seen for both oil and gas stringers as exhibited in Fig. 7.8a–p, but there is less of an observed departure from the hydrostatic behavior for gas than for oil under the same flow conditions because of the higher buoyancy of gas.

7.3.2. Increasing Anticlinal Radius

As the radius of the anticline increases, the anticline flattens, becoming more nearly horizontal. In the limit of an infinite radius, there is no stable stringer—the slightest hydrodynamic force will cause immediate lateral motion of a hydrocarbon stringer. Thus, under given flow conditions, the maximum areal mass that can be sustained decreases as the radius of the anticline increases.

Figure 7.9a–x illustrates this effect for both oil and gas stringers. Note again that the higher buoyancy of gas relative to that of oil implies that a larger areal mass of gas can be sustained at a given radius.

7.3.3. Increasing Aquifer Width

As the anticlinal width of an aquifer increases, the shape of the hydrocarbon stringers once again becomes more controlled by the water flow speed and not by the water flux. Thus, an oil or gas stringer shape eventually reaches a pattern that is independent of the aquifer width.

This behavior is cataloged in Fig. 7.10a–k and 7.10l–w for oil and gas stringers, respectively. Again, we see a difference in the behaviors of oil and gas, dominantly due to the difference in buoyancy.

7.4. Complex Syncline/Anticline Geometries

A variety of unexpected response patterns can be coaxed out of the basic hydrodynamic hydrocarbon stringer equations as the geometry is changed to more realistic patterns. In this section we provide a sampling of behaviors for a few geometries.

7.4.1. Syncline/Anticline Pair

As sketched in Fig. 7.11a–e, we consider a constant-width single syncline/anticline pair with a flow of fluid from right to left. For both gas and oil stringers we plot the areal masses permitted at each location along the aquifer. Notice that whereas the two situations are geometric mirror images, they are not dynamical mirror images because, in the case of the syncline part of the aquifer, buoyancy opposes hydrodynamic drive, while, in the case of the anticline part of the aquifer, buoyancy and hydrodynamic drive act in concert.

As the surface tension of the stringer in the anticline/syncline pair is increased, the hydrodynamic trapping of hydrocarbon stringers changes its character, as shown in Fig. 7.11a–e, with a greater areal mass being trapped near the apex of the anticline and a lesser amount being trapped near the side of the syncline. This effect is more marked for gas than for oil due to the greater buoyancy of gas.

For a given amplitude of the anticline/syncline pair, note that as the flow of fluid is increased, there is an increasing tendency to flush the hydrocarbon stringer away from the apex of the anticline as shown in Fig. 7.11a–e.

7.4.2. An Oscillating, Rising Aquifer

In the case of an oscillating, rising aquifer, as shown in Fig. 7.12a–h and 6.19a–h, there is a periodic behavior to the trapping of hydrocarbons (oil and gas) with respect to position along the aquifer. Indeed, with respect to position we see a dynamic range of over a factor of two in total hydrocarbons trapped (for the numerical parameters given), which is a significant variation.

As the flow speed is increased, the areal mass distribution along the aquifer changes, and the areal mass with respect to position also changes as sketched in Fig. 7.12a–h.

7.4.3. An Isolated Anticline

Here we consider some simple cases in which the anticline width is comparable to, much narrower than, and much broader than the aquifer bed leading to the anticline, as shown in Fig. 7.13 under fixed water flow conditions. Note the differences in behavior of the areal mass with respect to position along the aquifer for both oil and gas stringers.

As the angle of the juxtaposed aquifer bed is raised or lowered with respect to the horizontal, and all other conditions are held fixed, note, from Fig. 7.13a–ee, the change in the shapes and positions of the hydrocarbon stringers with respect to the previous case.

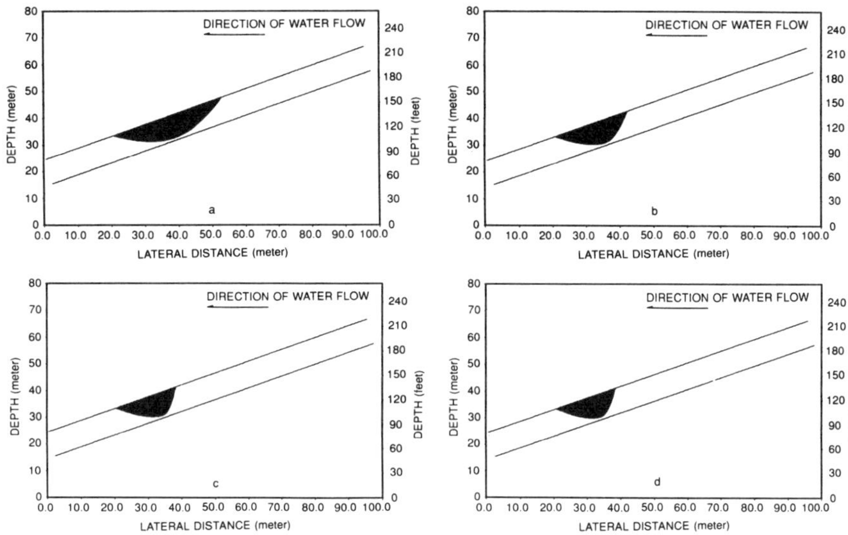

Figure 7.1. Hydrodynamic shapes of oil stringers in a fixed-angle, fixed-width aquifer with a permeability of 1000 mD as the speed of water flow changes. The density of the oil stringers is 0.8 g/cm³, and the water flow speeds are 1 (a), 100 (b), 150 (c), and 200 cm/yr (d).

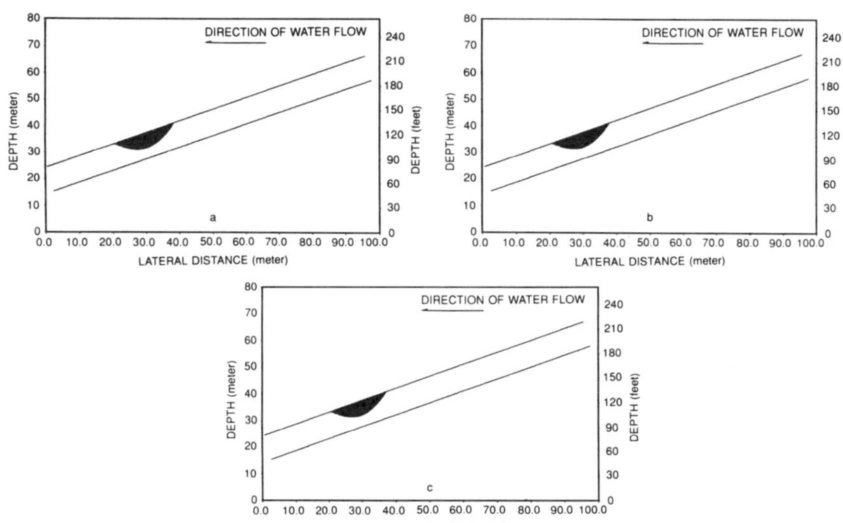

Figure 7.2. As for Fig. 7.1 but for gas stringers of density 0.1 g/cm³. The water flow speeds are 1 (a), 100 (b), and 200 cm/yr (c).

Table for Fig. 7.3. Aquifer Geometry: Fixed Angle, Fixed Width

	Flow rate (cm/yr)					
	Oil			Gas		
Aquifer angle	1	100	200	1	200	500
20°	7.3a	7.3b	7.3c	7.3j	7.3k	7.3l
40°	7.3d	7.3e	7.3f	7.3m	7.3n	7.3o
60°	7.3g	7.3h	7.3i	7.3p	7.3q	7.3r

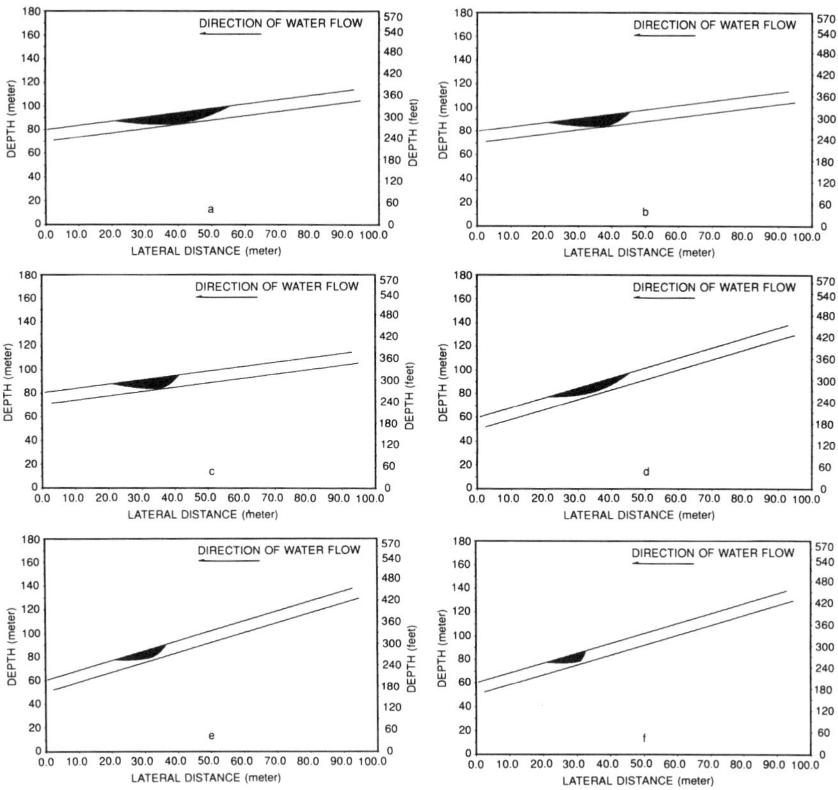

Figure 7.3. General shapes of oil and gas stringers in a fixed-width, fixed-angle aquifer as both the angle of the aquifer and the water flow rate vary. The values of these parameters for each panel in the figure are given in the accompanying table.

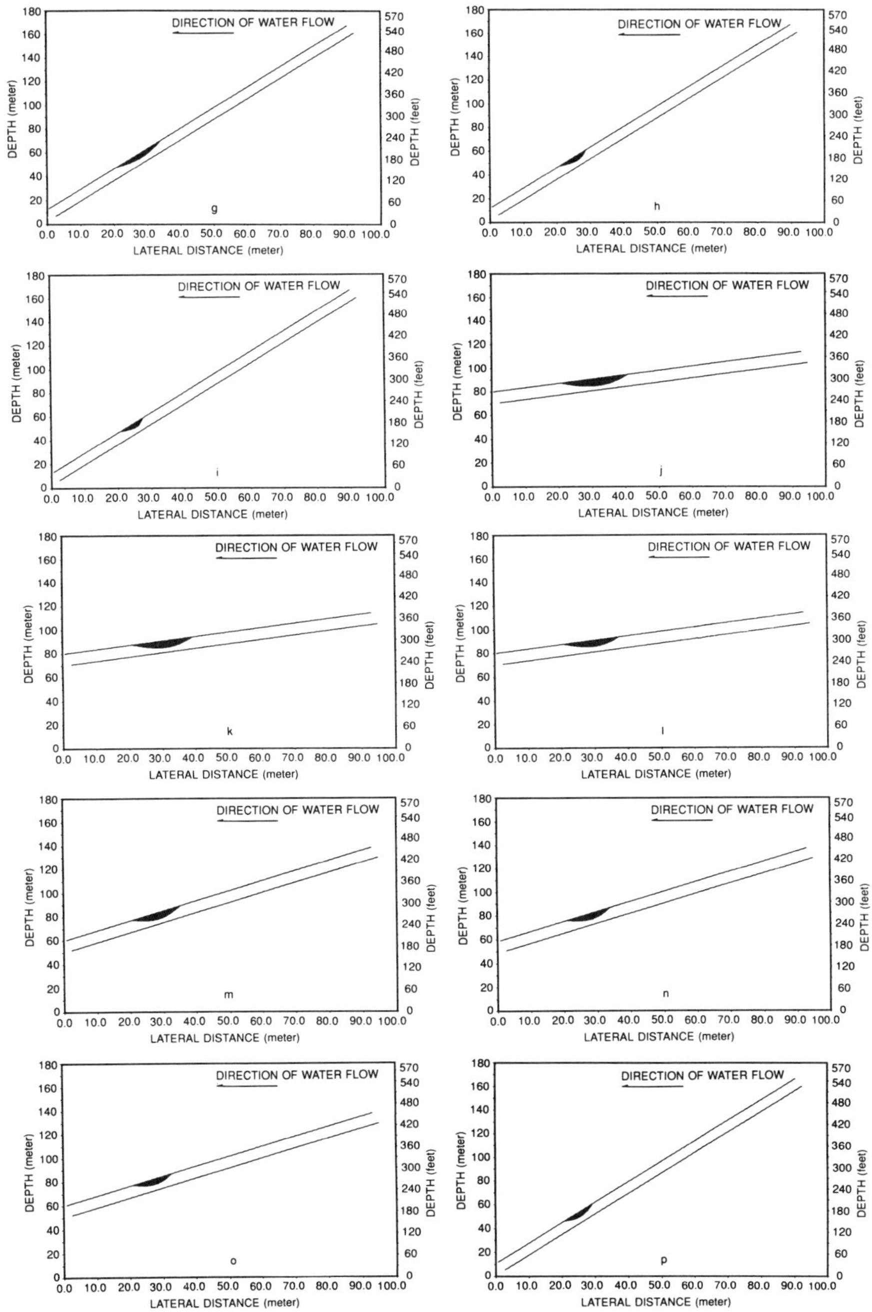

Figure 7.3. *(Continued)*

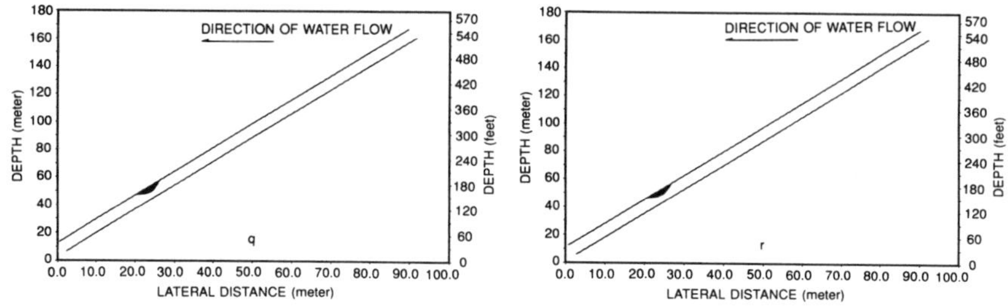

Figure 7.3. *(Continued)*

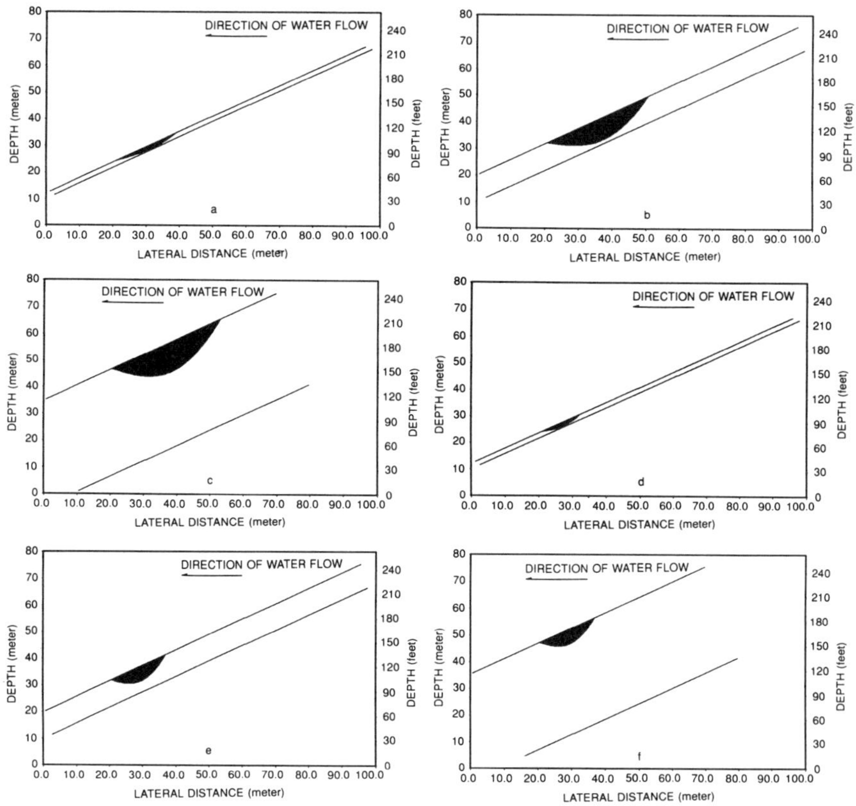

Figure 7.4. Hydrodynamic shapes of oil and gas stringers in a fixed-angle, variable-width aquifer with a constant permeability of 100 mD under constant water flow rate (100 cm/yr) as the aquifer width increases. Panels a–c refer to an oil stringer of density 0.8 g/cm^3 in aquifers having widths of 2 (a), 10 (b), and 40 m (c); panels d–f refer to a gas stringer of density 0.1 g/cm^3 in aquifers having widths of 2 (d), 10 (e), and 40 m (f).

Table for Fig. 7.5. Aquifer Geometry: Single Syncline of Fixed Width and Fixed Curvature

	Water flow rate (cm/yr)				
	Oil			Gas	
Position along aquifer	100	300	500	100	500
24	7.5a	7.5e	7.5i	7.5m	7.5q
26	7.5b	7.5f	7.5j	7.5n	7.5r
30	7.5c	7.5g	7.5k	7.5o	7.5s
32	7.5d	7.5h	7.5l	7.5p	7.5t

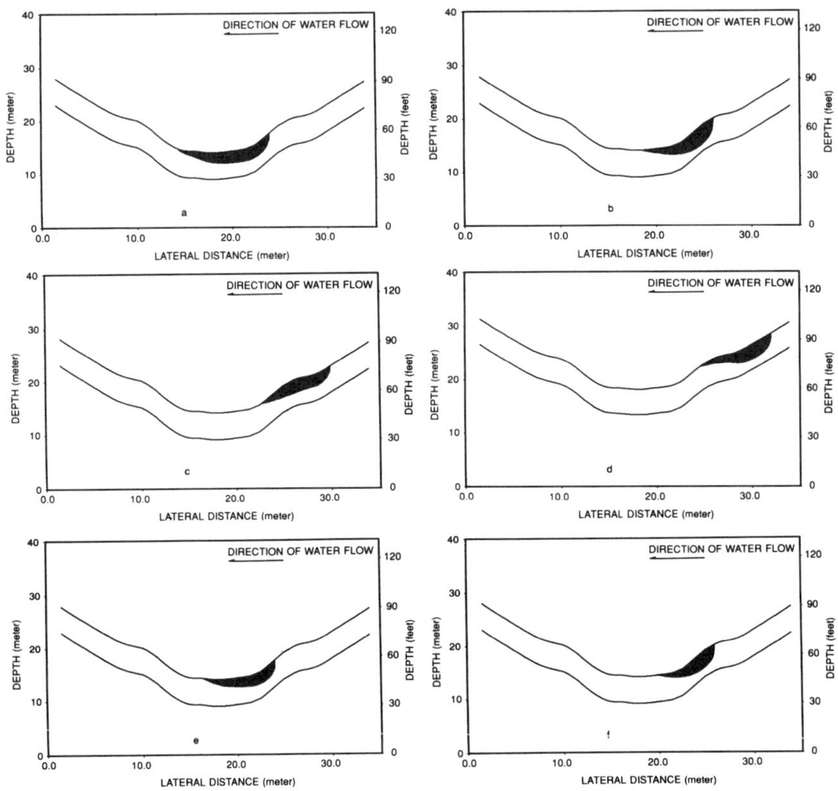

Figure 7.5. Hydrodynamic shapes of steady-state oil and gas stringers in a single syncline of fixed curvature and fixed width and with a permeability 1000 mD as the water flow speed increases and the "nose" of the stringer is chosen at different locations. The values of these parameters for each panel in the figure are provided in the accompanying table. Panels a–l refer to oil stringers (density 0.8 g/cm^3), and panels m–t refer to gas stringers (density 0.1 g/cm^3).

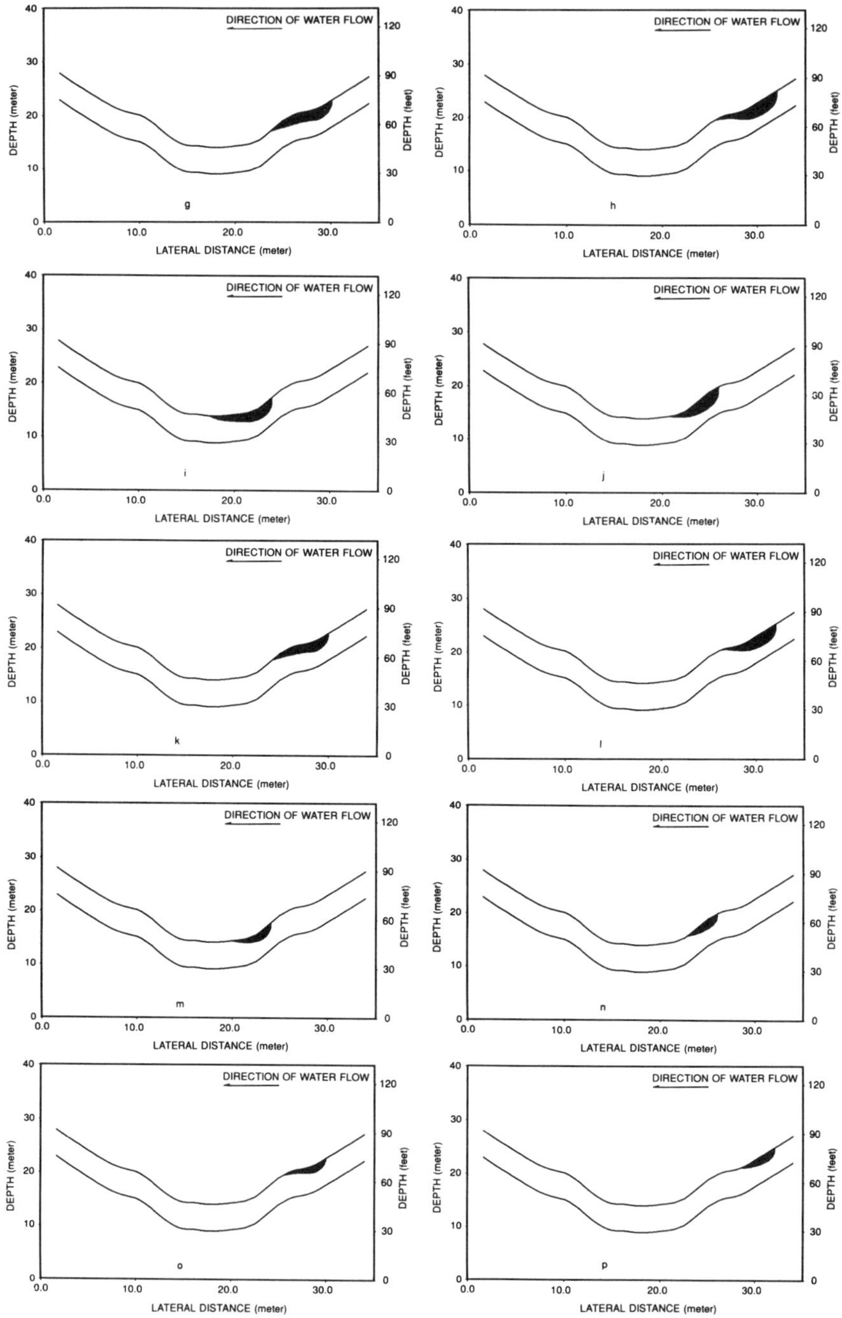

Figure 7.5. (Continued)

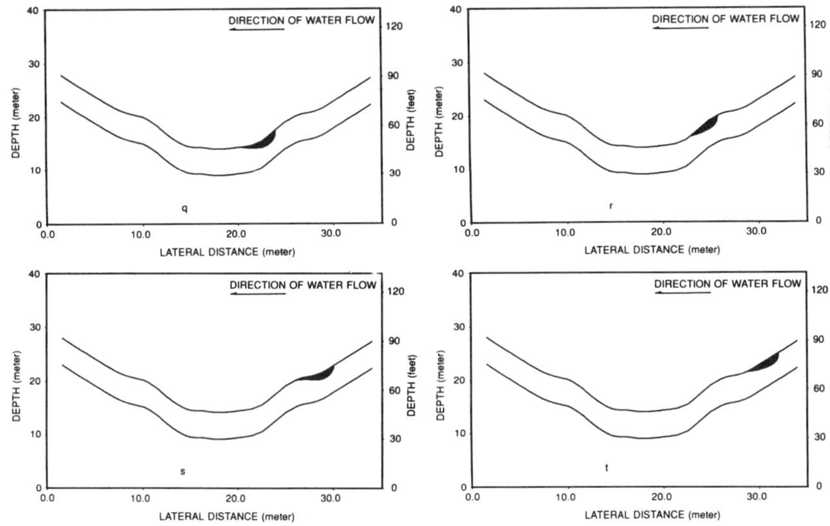

Figure 7.5. *(Continued)*

Table for Fig. 7.6. Aquifer Geometry: Single Syncline of High, Medium, and Low Curvature

Position along aquifer	Curvature					
	Oil			Gas		
	High	Medium	Low	High	Medium	Low
25	7.6a	7.6e	7.6i	7.6m	7.6q	7.6u
28	7.6b	7.6f	7.6j	7.6n	7.6r	7.6v
30	7.6c	7.6g	7.6k	7.6o	7.6s	7.6w
32	7.6d	7.6h	7.6l	7.6p	7.6t	7.6x

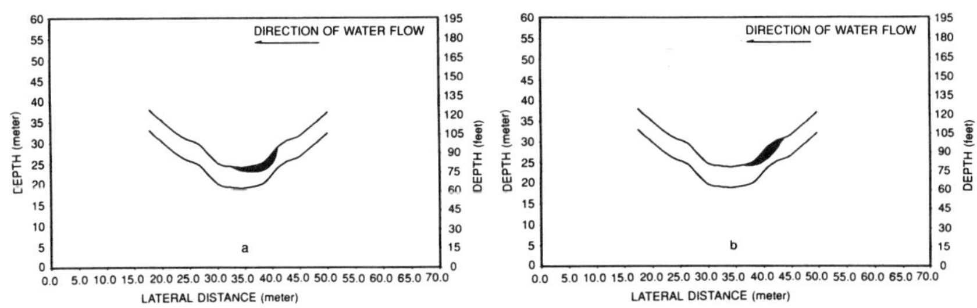

Figure 7.6. Hydrodynamic shapes of steady-state oil and gas stringers in a single syncline of permeability 1000 mD, under constant water flow conditions of 100 cm/yr, as the curvature of the syncline is decreased. Different locations for the "nose" of the stringer yield different behaviors for the shapes of the stringers. The value of this parameter and the curvature of the syncline for each panel in the figure are provided in the accompanying table. Panels a–l refer to oil stringers (density 0.8 g/cm^3), and panels m–x refer to gas stringers (density 0.1 g/cm^3).

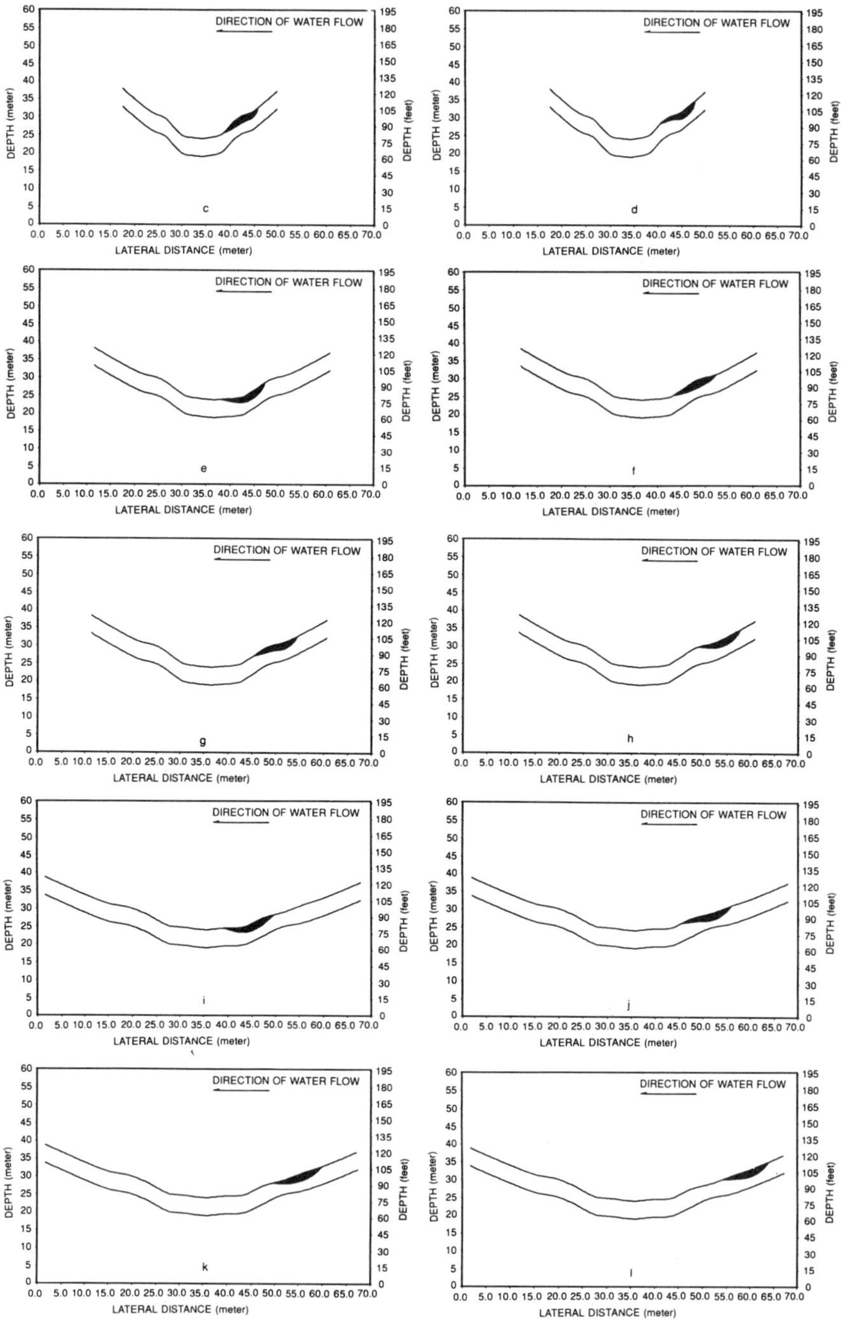

Figure 7.6. (*Continued*)

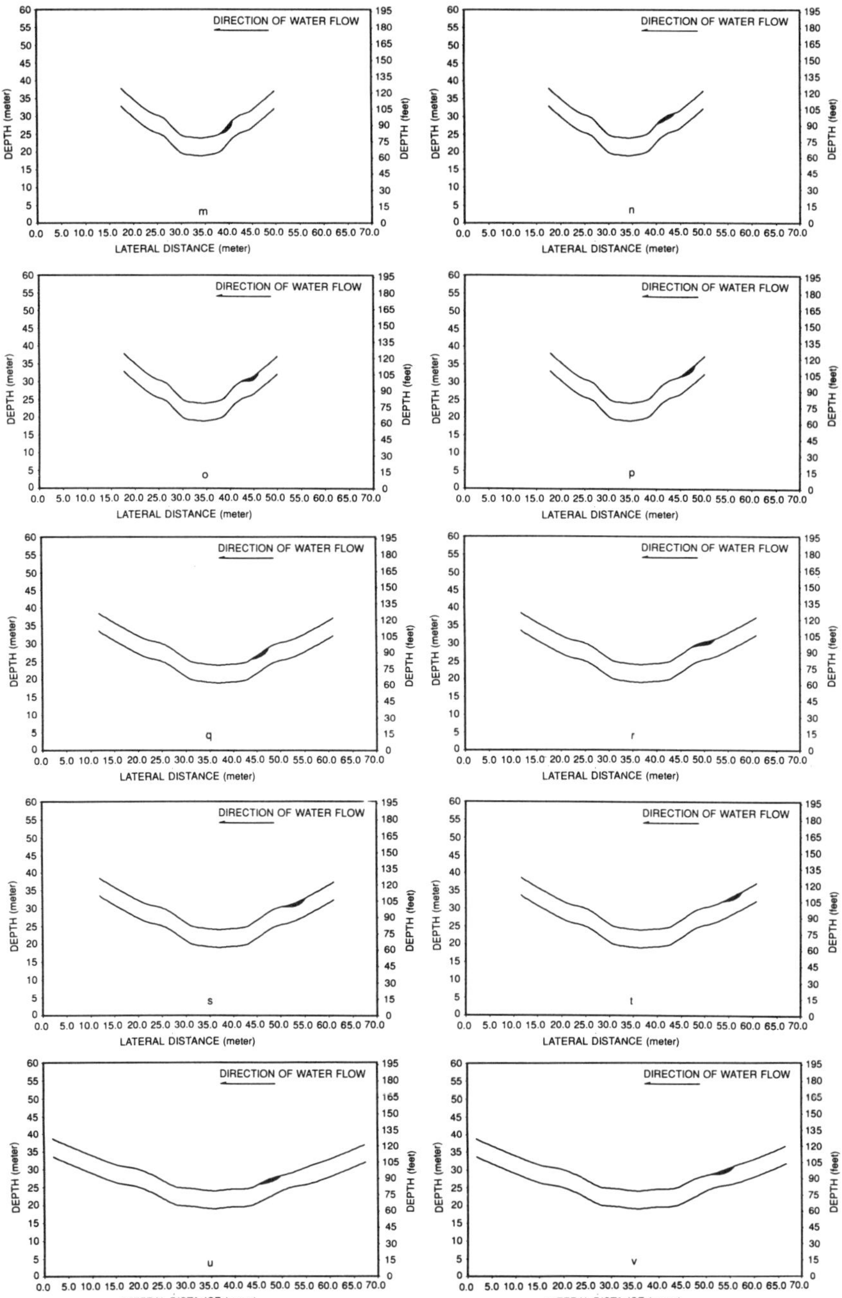

Figure 7.6. *(Continued)*

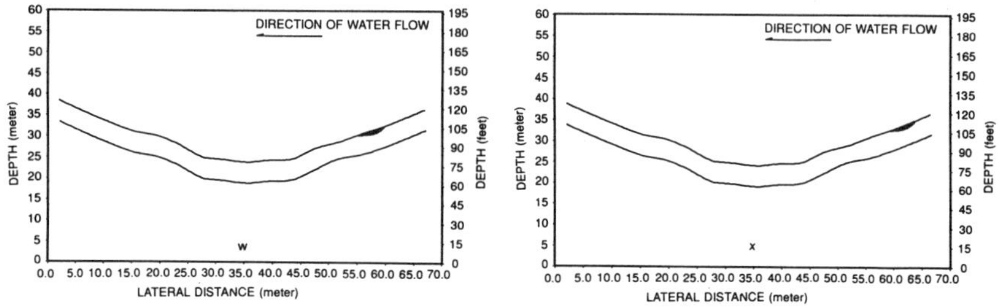

Figure 7.6. *(Continued)*

Table for Fig. 7.7. Aquifer Geometry: Single Syncline of Fixed Curvature but Large and Small Width

Position along aquifer	Aquifer width (m)			
	Oil		Gas	
	20	2	20	2
24	7.7a	7.7e	7.7i	7.7m
26	7.7b	7.7f	7.7j	7.7n
30	7.7c	7.7g	7.7k	7.7o
32	7.7d	7.7h	7.7l	7.7p

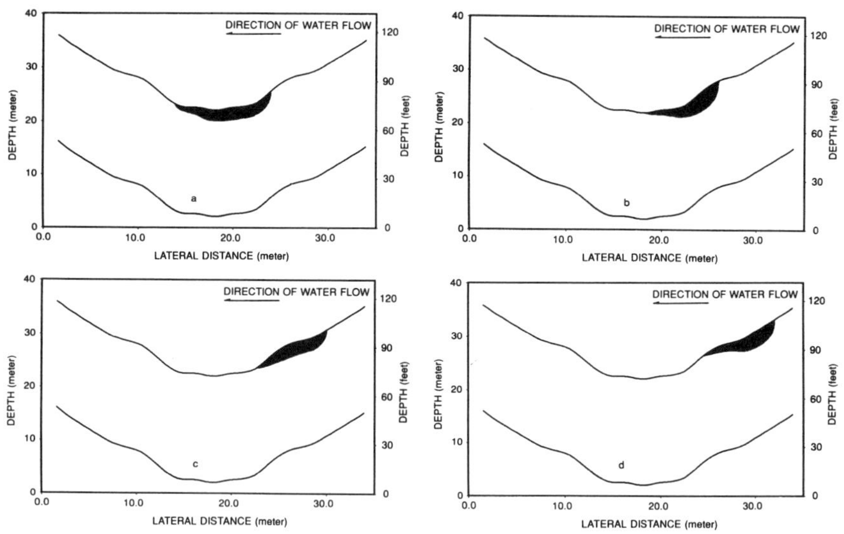

Figure 7.7. Hydrodynamic shapes of steady-state oil and gas stringers in a single syncline of fixed curvature but variable width and with a permeability of 1000 mD, under constant water flow speed of 100 cm/yr, as the width varies and the "nose" of the stringer is chosen at different locations. The values of the parameters for each panel in the figure are provided in the accompanying table. Panels a–h refer to oil stringers (density 0.8 g/cm^3), and panels i–p refer to gas stringers (density 0.1 g/cm^3).

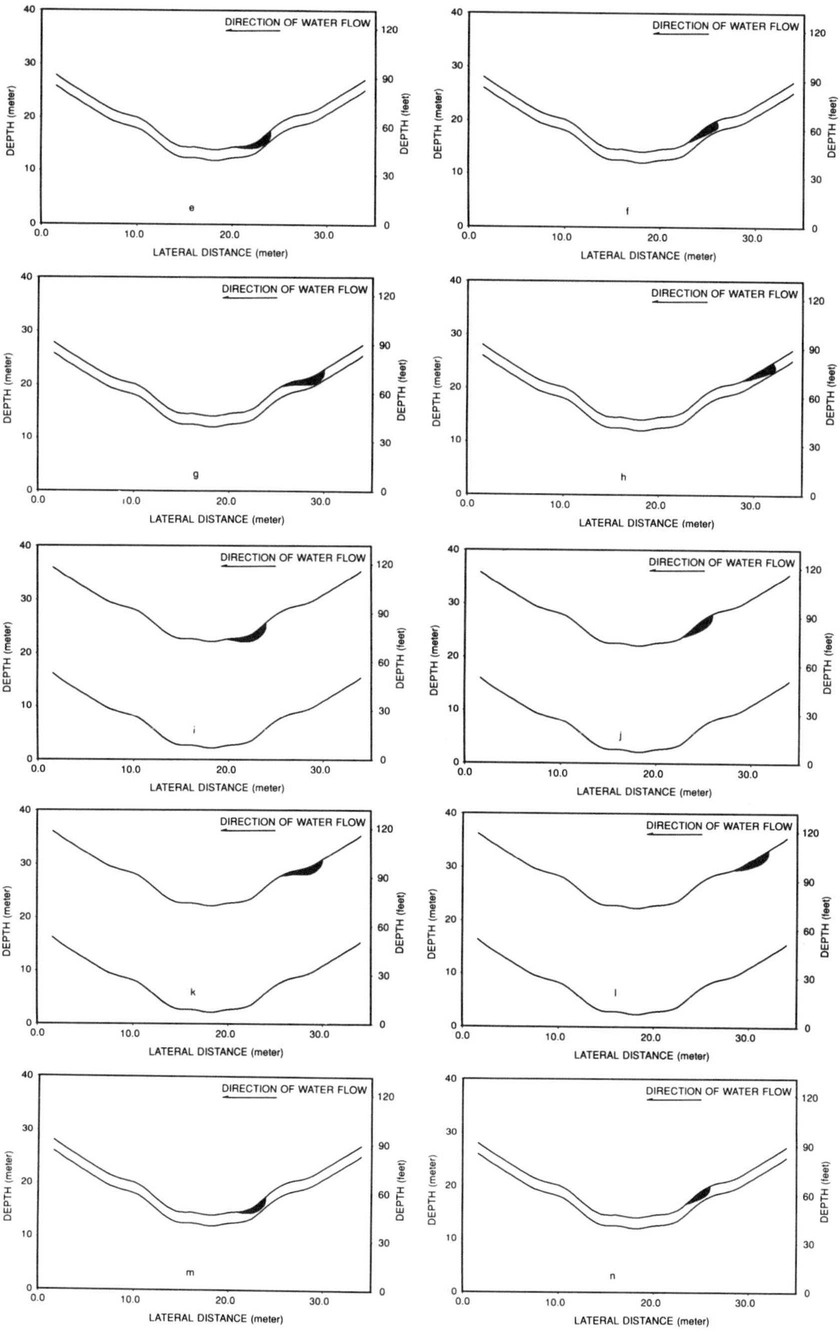

Figure 7.7. (*Continued*)

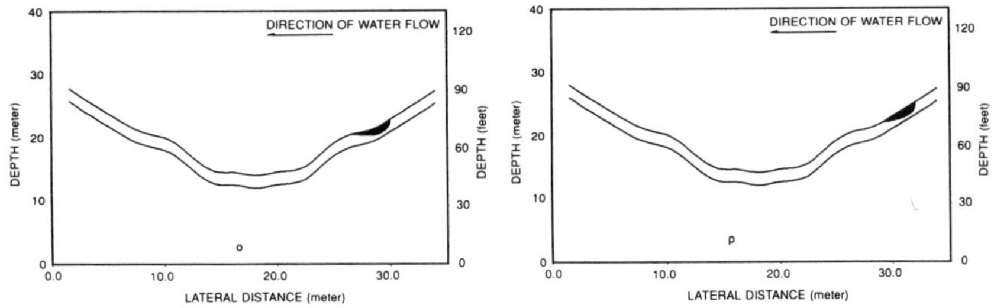

Figure 7.7. (*Continued*)

Table for Fig. 7.8. Aquifer Geometry: Single Anticline of Fixed Width and Fixed Curvature

Position along aquifer	Water flow rate (cm/yr) Oil 100	Oil 300	Gas 100	Gas 300
47	7.8a	7.8e	7.8i	7.8m
37	7.8b	7.8f	7.8j	7.8n
27	7.8c	7.8g	7.8k	7.8o
17	7.8d	7.8h	7.8l	7.8p

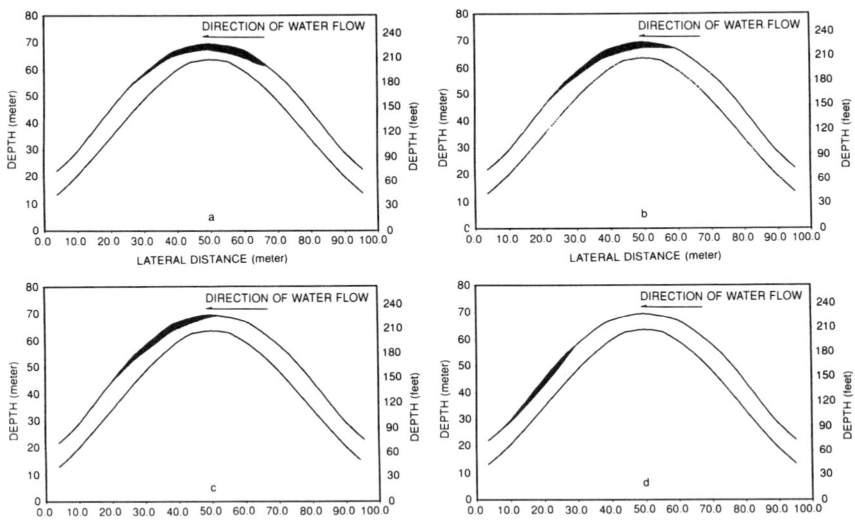

Figure 7.8. Hydrodynamic shapes of steady-state oil and gas stringers in a single anticline of fixed width and fixed curvature and with a permeability of 1000 mD as the water flow speed increases and the "nose" of the stringer is chosen at different locations. The values of these parameters for each panel in the figure are provided in the accompanying table. Panels a–h refer to oil stringers (density 0.8 g/cm^3) and panels i–p refer to gas stringers (density 0.1 g/cm^3).

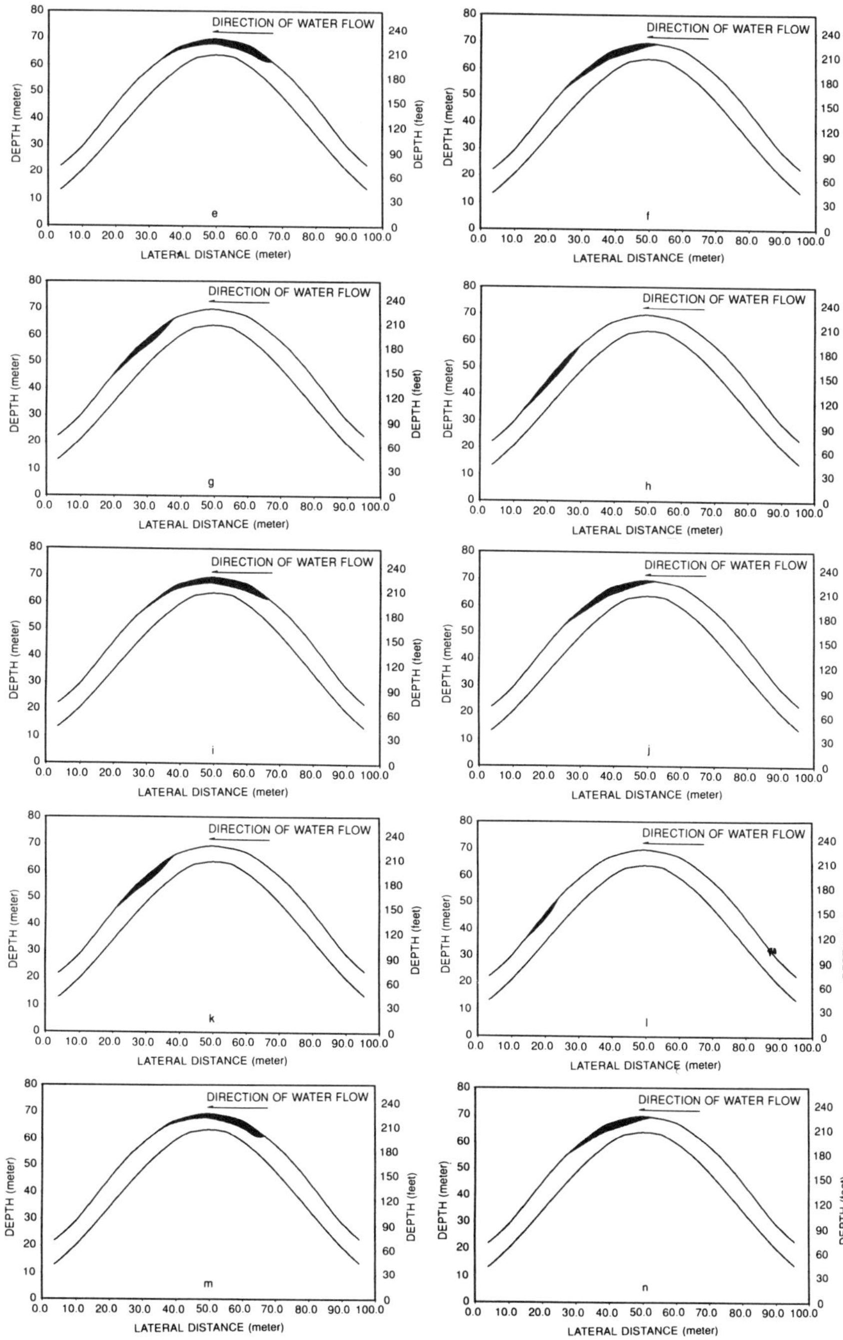

Figure 7.8. *(Continued)*

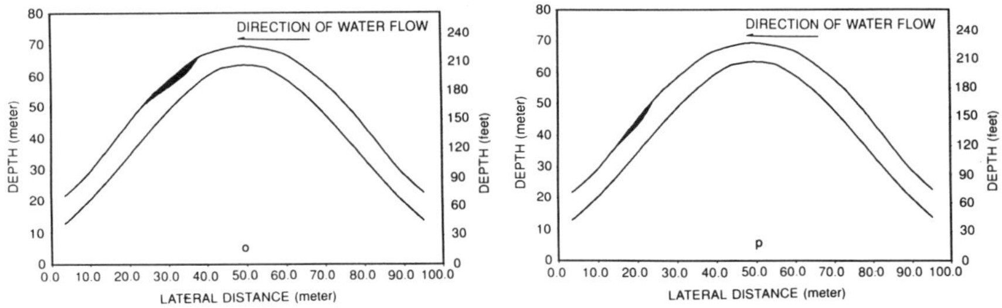

Figure 7.8. *(Continued)*

Table for Fig. 7.9. Aquifer Geometry: Single Anticline of Low, Medium, and High Curvature

Position along aquifer	Curvature					
	Oil			Gas		
	Low	Medium	High	Low	Medium	High
15	7.9a	7.9e	7.9i	7.9m	7.9q	7.9u
25	7.9b	7.9f	7.9j	7.9n	7.9r	7.9v
32	7.9c	7.9g	7.9k	7.9o	7.9s	7.9w
40	7.9d	7.9h	7.9l	7.9p	7.9t	7.9x

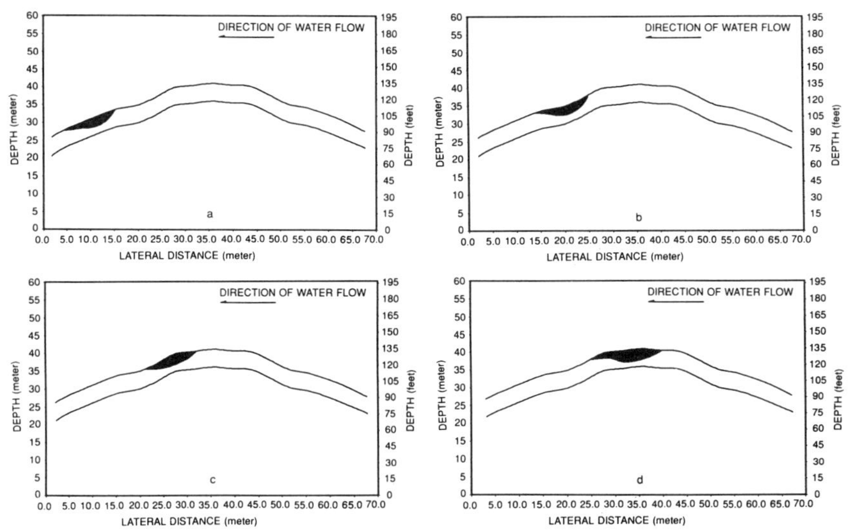

Figure 7.9. Hydrodynamic shapes of steady-state oil and gas stringers in a single anticline of permeability 1000 mD, under constant water flow conditions of 100 cm/yr, as the curvature of the anticline is increased. Different locations for the "nose" of the stringer yield different behaviors for the shapes of the stringers. The value of this parameter and the curvature of the anticline for each panel in the figure are provided in the accompanying table. Panels a–l refer to oil stringers (density 0.8 g/cm^3), and panels m–x refer to gas stringers (density 0.1 g/cm^3).

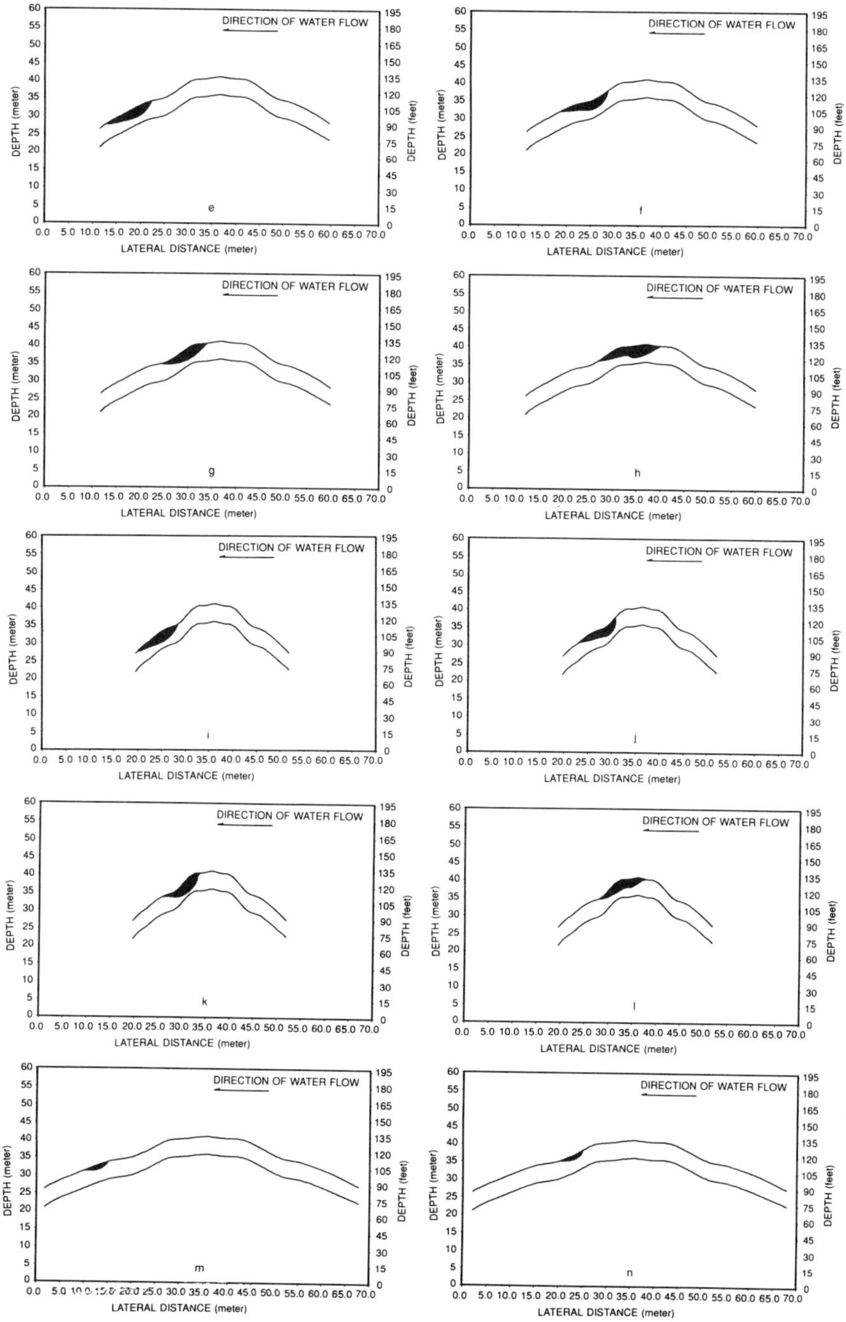

Figure 7.9. (*Continued*)

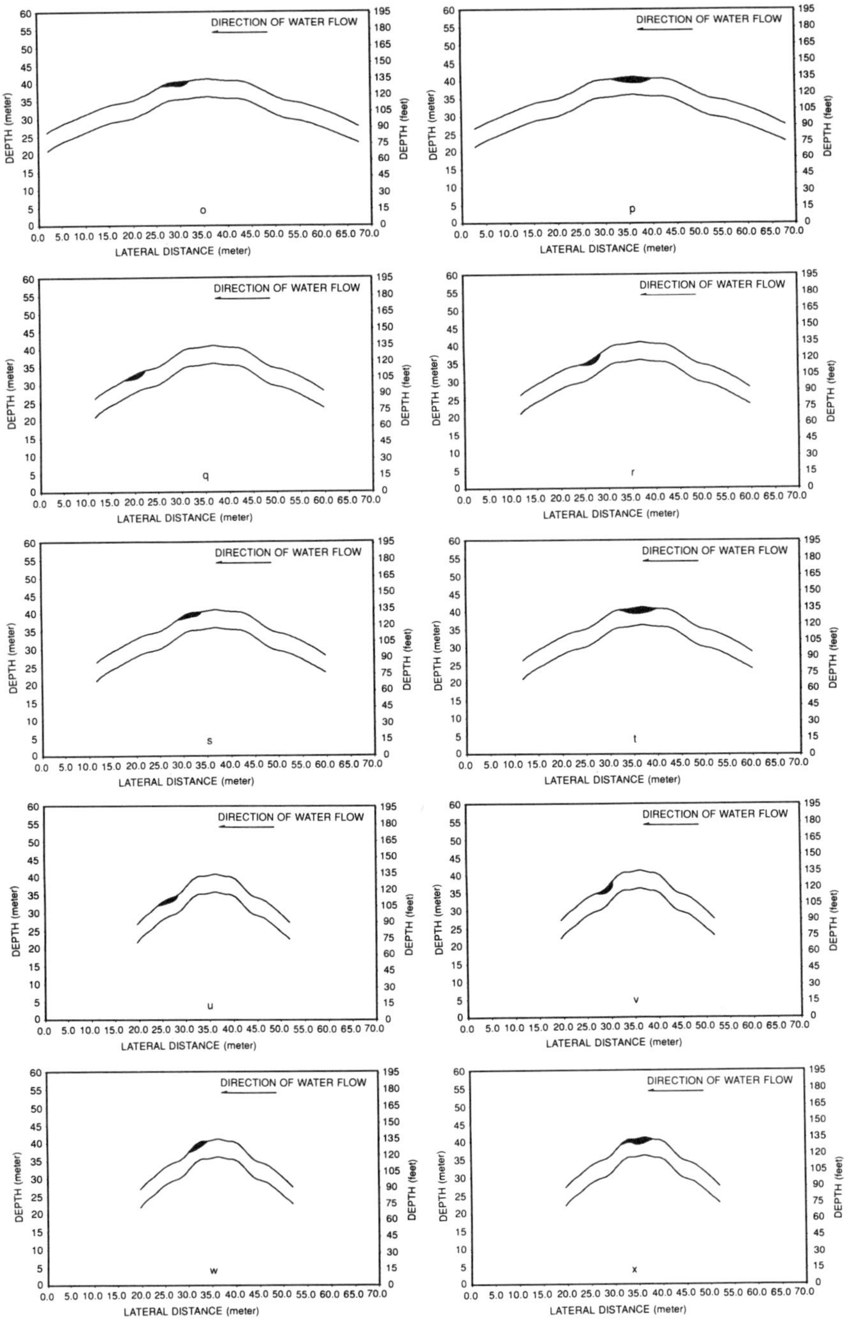

Figure 7.9. *(Continued)*

Table for Fig. 7.10. Aquifer Geometry: Single Anticline of Fixed Curvature but Narrow, Medium, and Broad Widths

Position along aquifer	Aquifer width: Oil			Aquifer width: Gas		
	Narrow	Medium	Broad	Narrow	Medium	Broad
47	—	7.10e	7.10i	—	—	—
42	7.10a	7.10f	—	7.10l	7.10p	7.10t
37	7.10b	7.10g	—	7.10m	7.10q	7.10u
32	7.10c	7.10h	7.10j	7.10n	7.10r	7.10v
27	7.10d	—	—	7.10o	7.10s	7.10w
17	—	—	7.10k	—	—	—

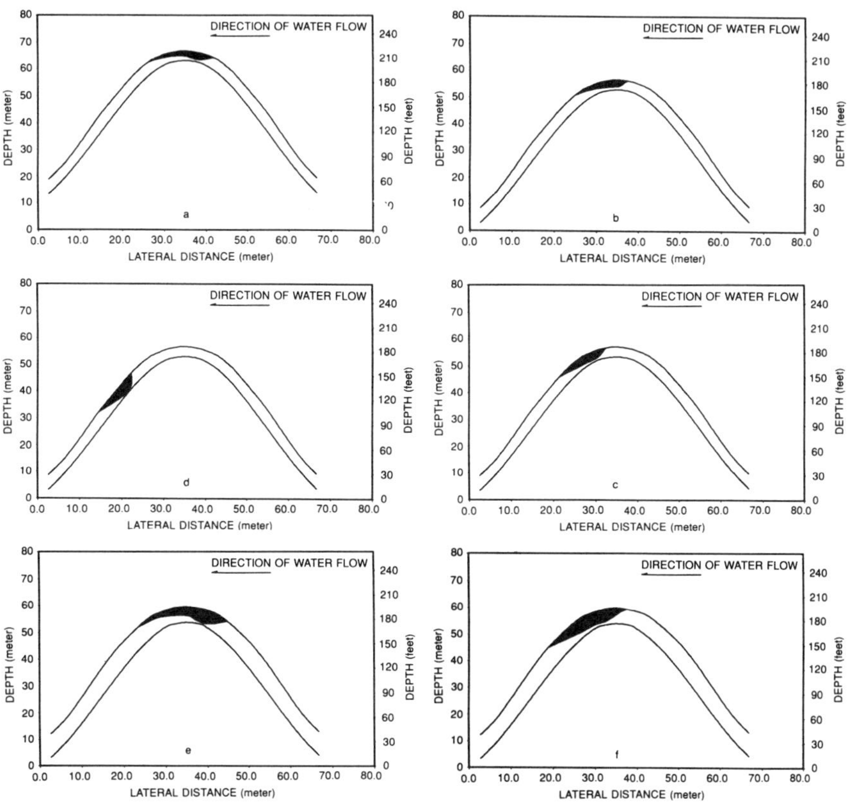

Figure 7.10. Hydrodynamic shapes of steady-state oil and gas stringers in a single anticline of fixed curvature but variable width and with a permeability of 1000 mD under constant water flow speed of 100 cm/yr as the "nose" of the stringer is chosen at different locations. The value of the parameter and the aquifer width for each panel in the figure are provided in the accompanying table. Panels a–k refer to oil stringers (density 0.8 g/cm^3), and panels l–w refer to gas stringers (density 0.1 g/cm^3).

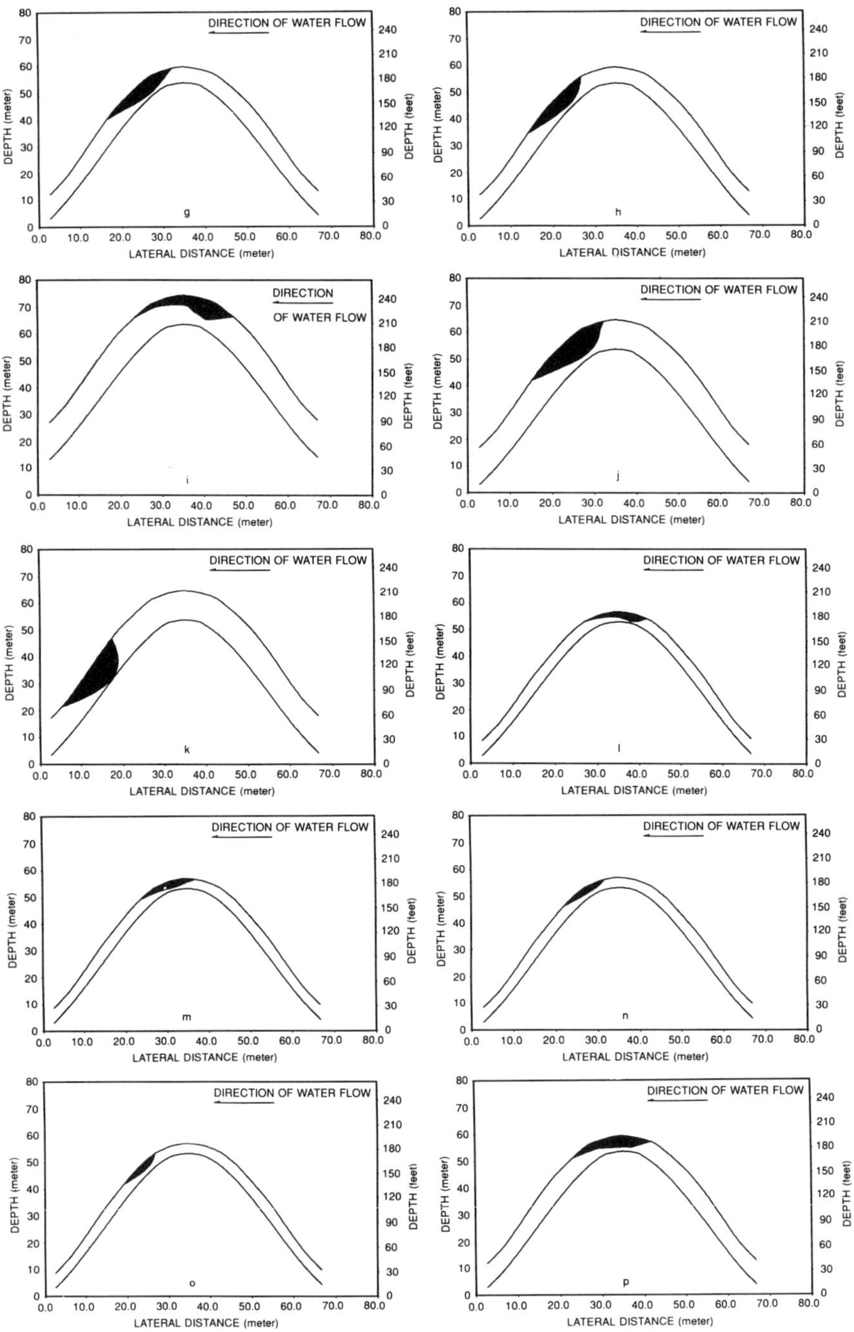

Figure 7.10. *(Continued)*

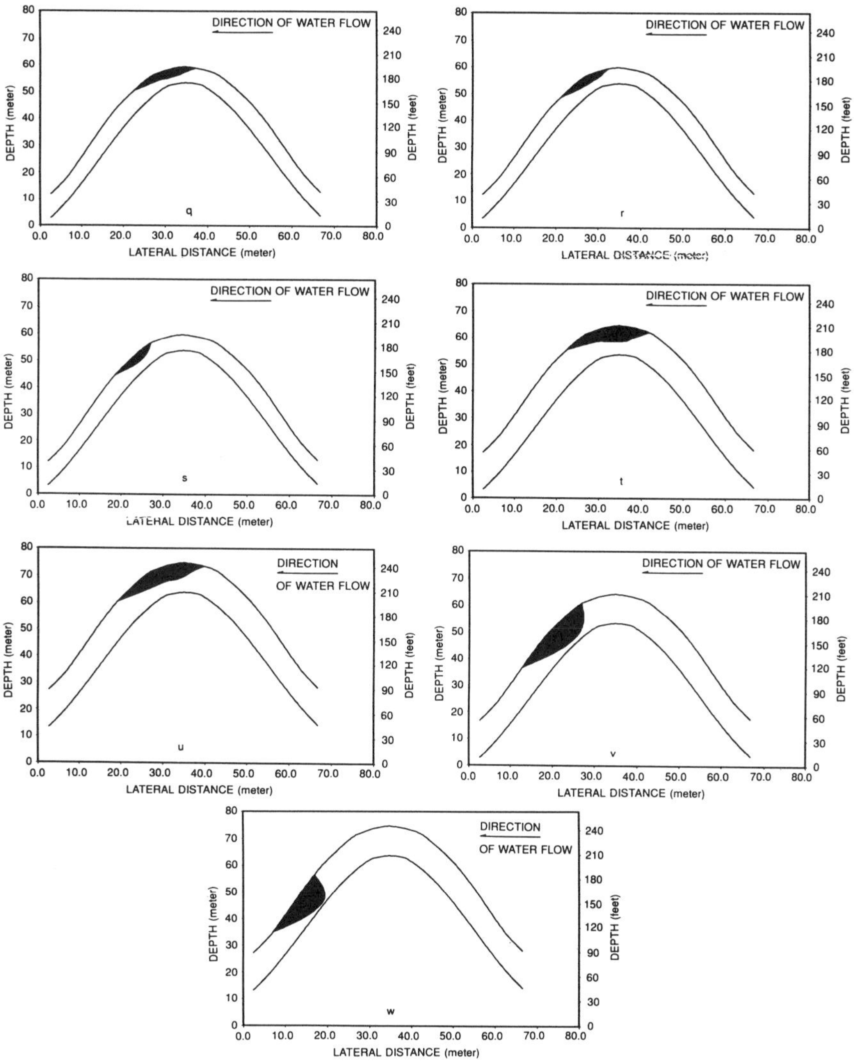

Figure 7.10. *(Continued)*

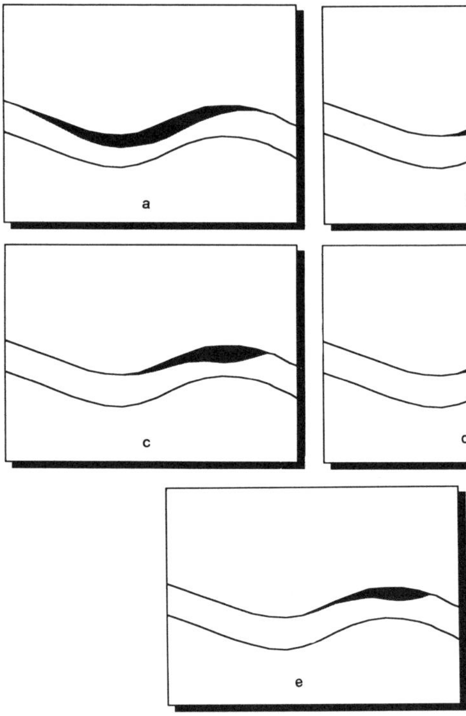

Figure 7.11. Hydrodynamic shapes of oil and gas stringers in a syncline/anticline pair for a fixed permeability of 100 mD. The vertical scale of each panel in the figure is 120 m, and the horizontal scale is 175 m. (a) Flow speed, 2 cm/yr; oil density 0.8 g/cm^3; surface tension, 10 dyn/cm; (b) as for (a) but with the surface tension increased to 35 dyn/cm; (c) as for (a) but with the density decreased to 0.6 g/cm^3 and the surface tension set at 35 dyn/cm; (d) as for (b) but with the flow speed decreased to 0.2 cm/yr; (e) as for (d) but with the density set at 0.6 g/cm^3 and the surface tension at 35 dyn/cm.

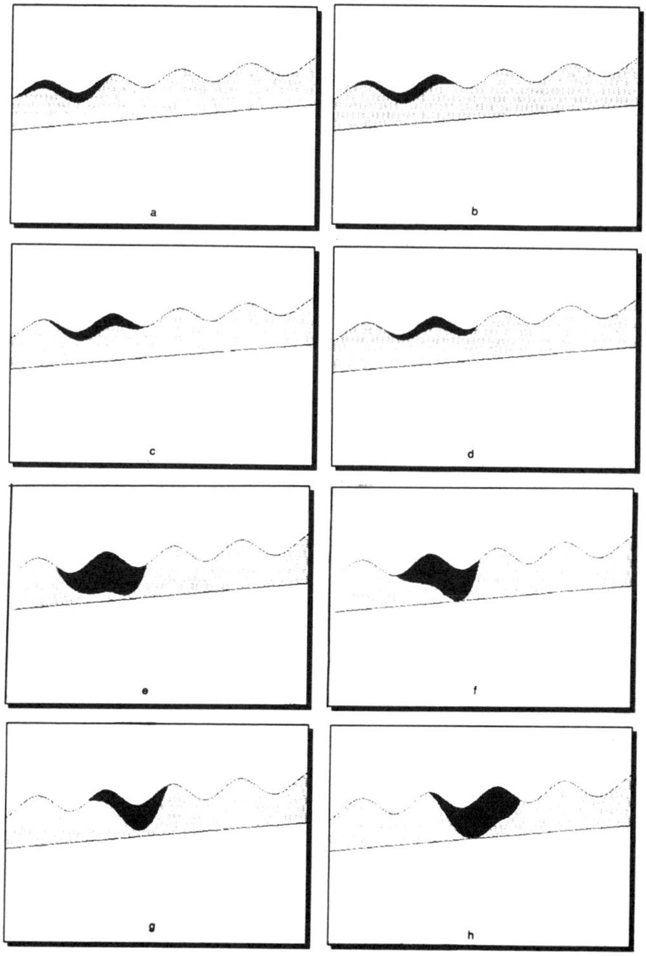

Figure 7.12. Hydrocarbon accumulations at different locations in wiggly, uprising aquifers.

Table for Fig. 7.13. Aquifer Geometry: Single Anticline with Down-Going or Up-Going Edges (D or U) and with Narrow or Wide (N or W) Central Section

Position along aquifer	Type of aquifer: Oil DN	Oil DW	Oil UN	Oil UW	Gas DN	Gas DW	Gas UN	Gas UW
17	7.12d	7.12h	—	—	7.12s	7.12w	—	—
23	7.12c	7.12g	7.12k	7.12o	7.12r	7.12v	7.12aa	7.12ee
30	7.12b	7.12f	7.12j	7.12n	7.12g	7.12u	7.12z	7.12dd
40	7.12a	7.12e	7.12i	7.12m	7.12p	7.12t	7.12y	7.12cc
55	—	—	—	7.12l	—	—	7.12x	7.12bb

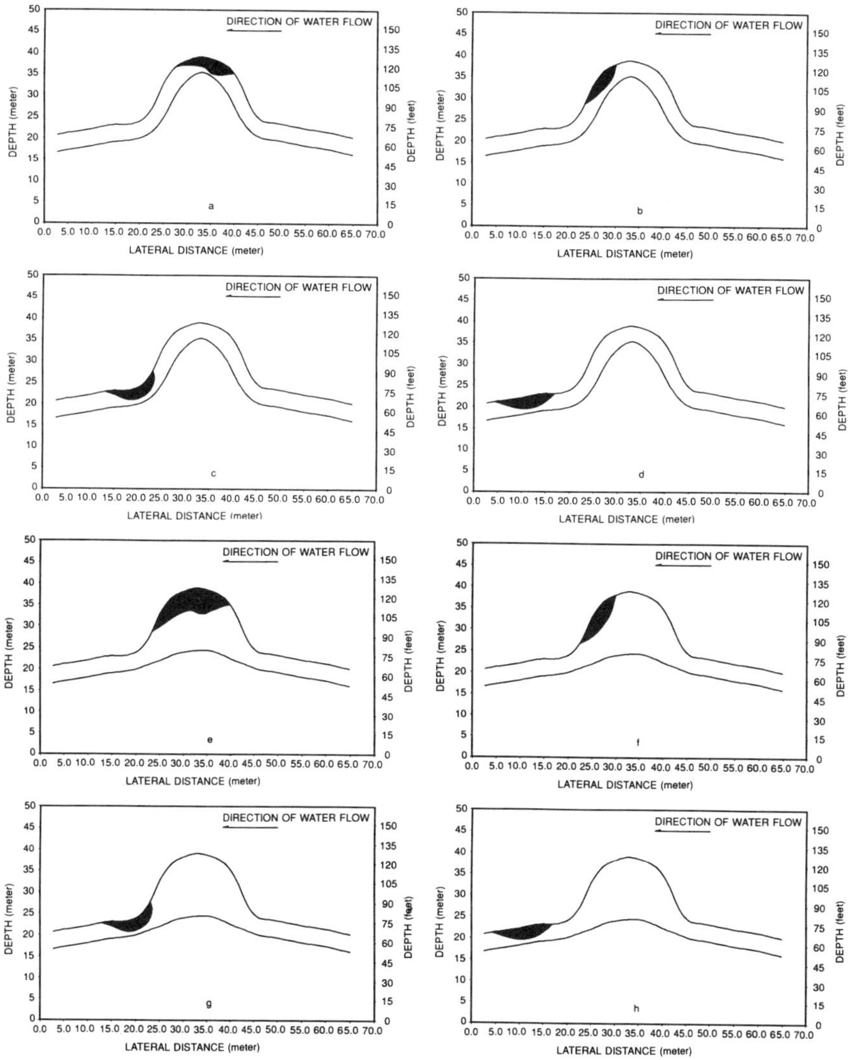

Figure 7.13. Hydrodynamic shapes of steady-state oil and gas stringers in a single anticline under fixed water flow (100 cm/yr) and with fixed permeability (1000 mD) as both the shape of the edges of the anticline and the choice of "nose" position of the stringer are varied. The four geometries examined keep the central apex of the top of the anticline fixed but vary the base of the central section of the aquifer in relation to the top and also vary the lateral edges away from the central apex from down-going to up-going as shown in the figure. The aquifer geometry and the "nose" position of the stringer for each panel in the figure are provided in the accompanying table. Panels a–o refer to oil stringers (density 0.8 g/cm^3), and panels p–ee refer to gas stringers (density 0.1 g/cm^3).

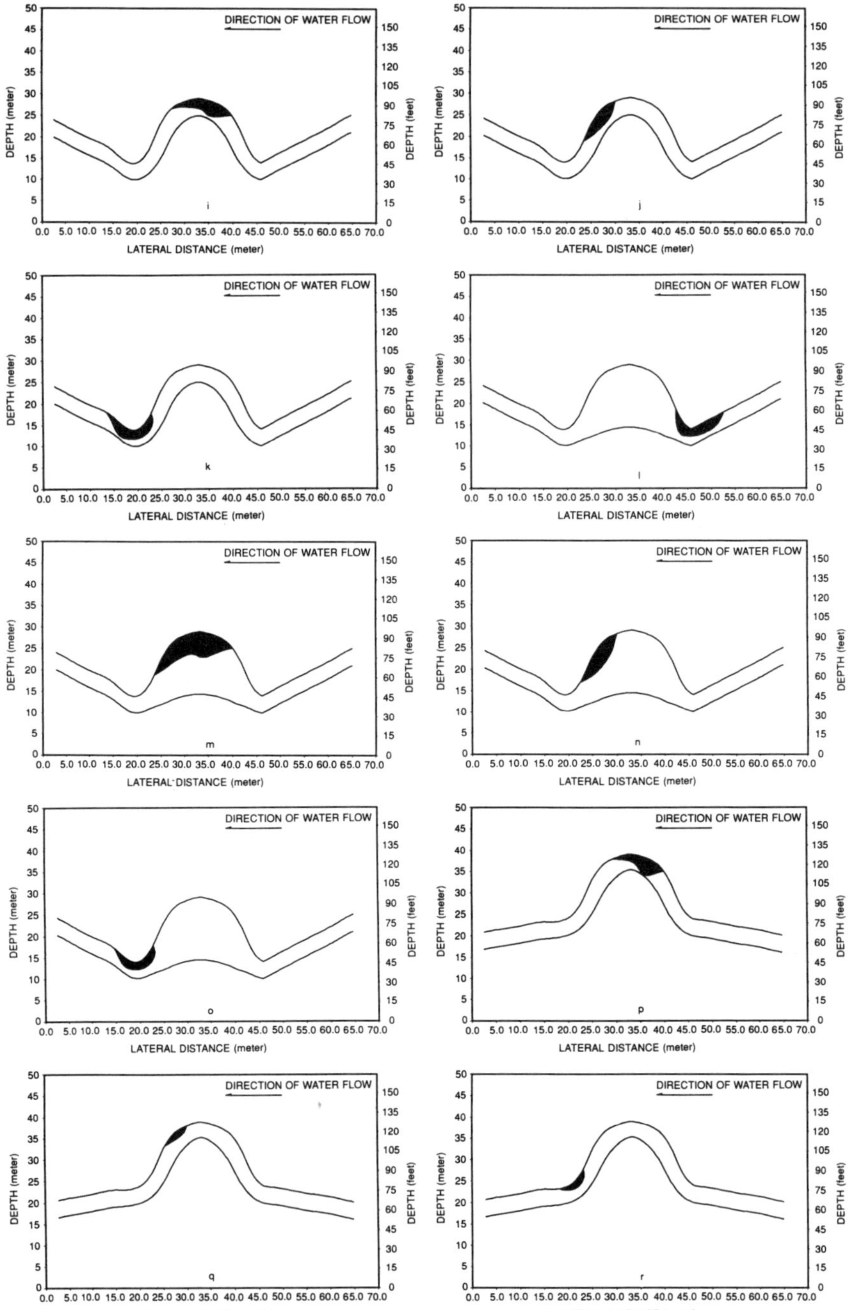

Figure 7.13. *(Continued)*

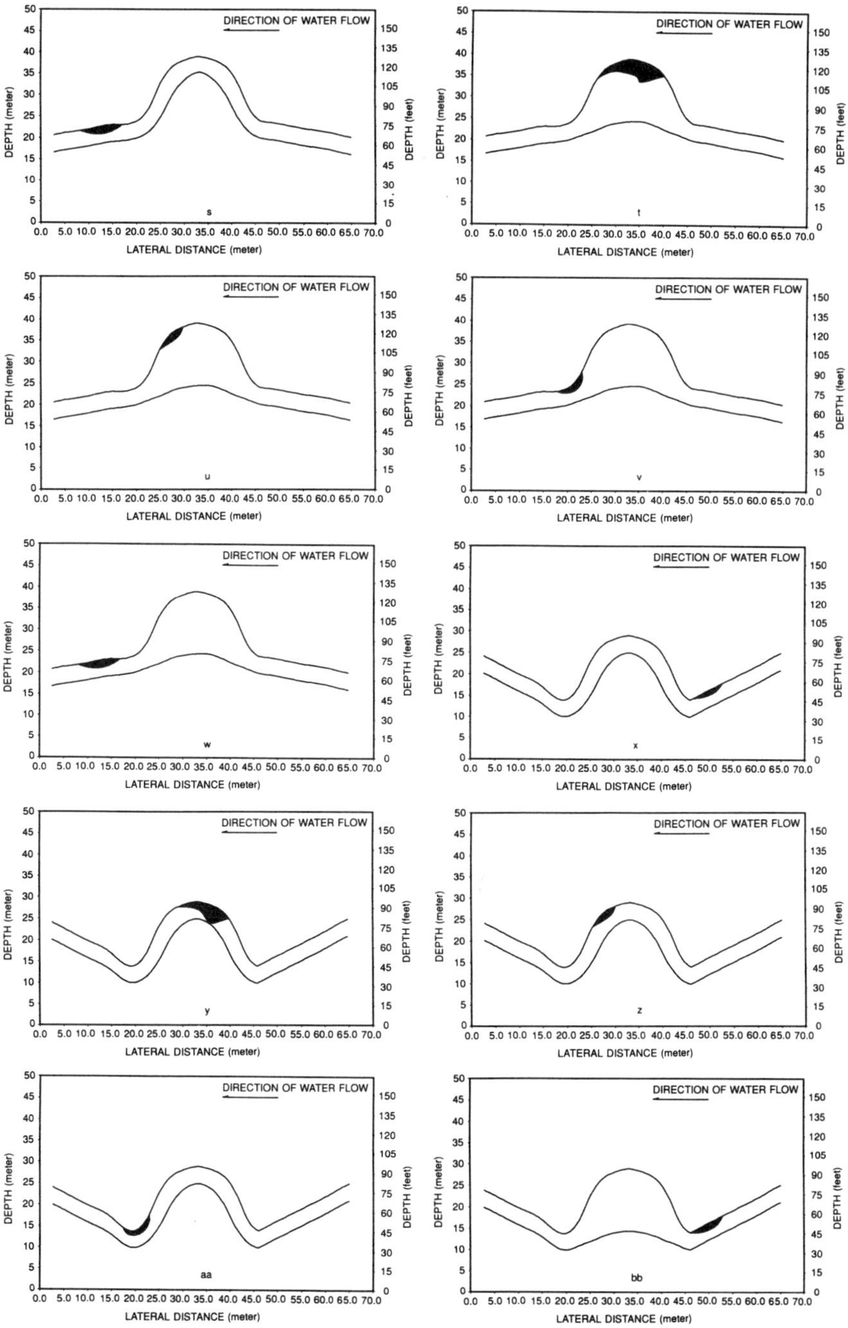

Figure 7.13. (*Continued*)

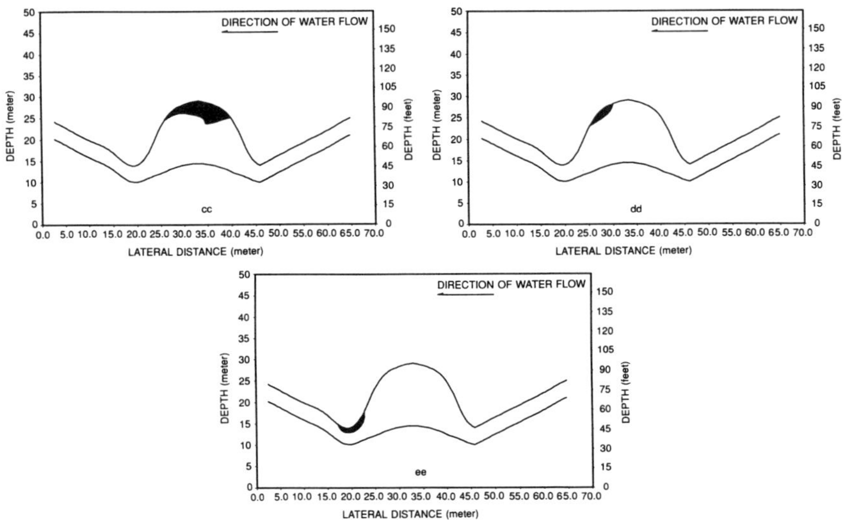

Figure 7.13. (*Continued*)

8

Effects Due to Variable Permeability

The permeability along an aquifer can change for a variety of reasons: the grain size distribution may change from fine to coarse, the lithology may change from sand to shale, diagenetic cementation may increase or decrease, fracturing may become less or more prevalent, and so on. In an effort to understand the impact that permeability variations have on the positioning and shaping of hydrocarbon accumulations in relation to variations in aquifer width, angle, flow, and variable hydrocarbon buoyancy (oil or gas), in this chapter we provide a compendium of case studies illustrating some of the patterns of behavior.

Because of the wide array of possibilities, we will restrict our attention to the two simple cases where the permeability along the aquifer increases (decreases) monotonically away from the upper end of the hydrocarbon accumulation. The scale distance along the aquifer over which the permeability changes is variable.

We split the behaviors into two groups depending on whether oil or gas is the trapped hydrocarbon:

A. Oil accumulations
   1. Permeability increasing up dip (Fig. 8.1)
      a. Increasing aquifer angle
      b. Increasing flow speed
      c. Increasing aquifer width
   2. Permeability decreasing up dip (Fig. 8.2)
      a. Increasing aquifer angle
      b. Increasing flow speed
      c. Increasing aquifer width

B. Gas accumulations
   1. Permeability increasing up dip (Fig. 8.1)
      a. Increasing aquifer angle

b. Increasing flow speed
c. Increasing aquifer width
2. Permeability decreasing up dip (Fig. 8.2)
a. Increasing aquifer angle
b. Increasing flow speed
c. Increasing aquifer width

In addition, a set of particular cases is examined to illustrate some of the more complex behaviors that can be encountered:

C. Oil or gas accumulations
1. Permeability oscillations along an aquifer (High–Low–High–Low . . .)
2. Anticline with permeability variations
3. Syncline with permeability variations

Let us consider each set of behaviors in turn. For the variation in permeability along the aquifer, we take the high value to be 1000 mD, representative of a clean sand, while the low value is set at 1 mD, more representative of a fine-grained or cemented sand. This thousandfold variation is sufficient to enhance the portrayal of effects at, or near, the changeover from high to low values.

In addition, we treat the behaviors listed under items A and B above with a planar, constant-width aquifer, making an angle θ with respect to the horizontal. It is also important to remember, as Hubbert (1953) noted, that the capillary pressure gradient will change as a consequence of permeability variations, be such variations due to changes in grain size, cementation, lithology, or other causes along the aquifer. Clearly, a model dependence of capillary pressure on permeability needs to be specified in order that the effect can be examined.

For illustrative purposes only, we model the capillary pressure, p_c, along the aquifer by the addition of two factors: one due to the contrast between aquifer boundary and aquifer volumetric properties, and the other due to changes in the bulk permeability variations along the aquifer. Thus, at position ℓ along the aquifer we write the effective capillary pressure, $p_c(\ell)$, as $p_c = p_{c0}\,[\delta\kappa(\ell)/\kappa(\ell) + \tan\psi\ \mathrm{wall}(\ell)]$, where the change in permeability $\delta\kappa(\ell)$ at position ℓ is given by

$$\frac{\delta\kappa(\ell)}{\kappa(\ell)} \cong \frac{3[\kappa(\ell + \Delta\ell) - 2\kappa(\ell) + \kappa(\ell - \Delta\ell)]}{\kappa(\ell + \Delta\ell) + 2\kappa(\ell) + \kappa(\ell - \Delta\ell)}$$

where $\Delta\ell$ is the step size along the aquifer. The wall term, which accounts for the contact angle of the hydrocarbon stringer at the leading and trailing edges of the stringer, we write as (Shorthose, 1936)

$$\text{wall}(\ell) = \exp\left[-|h(\ell)|\,|\Delta\rho g \tan\theta_0|\,|p_{c0}|^{-1}\left(\frac{\kappa(\ell)}{\kappa_w} - 1\right)\right]$$

where $\tan\theta_0$ is the initial input slope angle for the oil stringer, and κ_w is the wall permeability [much less than the permeability $\kappa(\ell)$ along the aquifer]. Note that the surficial wall capillary pressure needs to be used only when

$$|h(\ell)| \leqslant |p_{c0}|\,|\Delta\rho p \tan\theta_0|^{-1}\left(\frac{\kappa(\ell)}{\kappa_w} - 1\right)$$

reflecting the fact that only as the oil stringer thickness, $h(\ell)$, heads toward zero (which occurs only at the leading and trailing edges of the stringer) is it necessary to include the wall effect. The factor p_{c0} is the scaling value of capillary pressure and may be taken either positive or negative to reflect water-wet or hydrocarbon-wet conditions, respectively, while the factor $\tan\psi$ (which can be independent of any sign and magnitude) can be used to reflect positive or negative angles of contact.

Other functional forms can be chosen for the effect of bulk and wall capillary pressure on the shaping of oil stringers under hydrodynamic conditions, and such choices should, presumably, be investigated to ascertain the total dynamic range of permeability variations in controlling both the bulk and wall capillary pressure effects. However, the form given above is more than adequate to reflect the variations being examined here.

8.1. Oil Accumulations

8.1.1. Permeability Increasing Up Dip

In the case of permeability increasing up dip, the low-permeability region down dip acts as a barrier to hydrocarbon development. The flow of water is kept at a fixed rate under steady-state flow conditions so that, in the low-permeability region, a higher fluid-pressure gradient is necessary to drive the steady water flow. In turn, this fluid-pressure gradient puts a much higher hydrodynamic force on a hydrocarbon stringer so that the length of the stringer will shorten in order that buoyancy can balance the stringer. In the high-permeability zone up dip, the stringer can be considerably longer than in the low-permeability region. This transition in the behavior of the shape of the areal mass of a hydrocarbon stringer as the leading edge of the stringer transits across the high/low permeability boundary and the change in areal mass allowed are the problems of interest here.

8.1.1.1. Increasing Aquifer Angle

Figure 8.3a–h exhibits the behavior of the areal mass of an oil stringer with position along the aquifer under fixed flow conditions as the angle of the aquifer increases from nearly horizontal to nearly vertical. To be noted here are the interesting changes in both shape and position of the areal mass as the angle increases.

8.1.1.2. Increasing Flow Speed

As the flow speed is increased in an aquifer at a fixed angle, it becomes more difficult for the hydrocarbon stringer to survive in the high-permeability region. The down-dip lower permeability region acts as a leaky barrier to the oil stringer so that there is a tendency for the stringer to start in the high-permeability region and be "squashed" up against the transition zone between the up-dip and down-dip regions of high and low permeability, respectively.

This particularly interesting behavior is exhibited in Fig. 8.4a–h.

8.1.1.3. Increasing Aquifer Width

Just as in previous situations investigated, as the width of the aquifer is increased, there is less need for the flow to speed up in transiting past an oil stringer.

Thus, the various shapings of an oil stringer with respect to starting position in the aquifer tend to an invariant pattern dependent dominantly on the flow speed as opposed to the hydrodynamic flux. These effects are exhibited most sharply in the sequence of Fig. 8.5a–h, which illustrates the variations in areal mass of oil that can be maintained at particular locations in the aquifer as the aquifer width increases.

8.1.2. Permeability Decreasing Up Dip

The situation in which permeability decreases up dip forms the fundamental basis for stratigraphic trapping of oil. The up-dip region of low permeability tends to block oil migration. Thus, a "plug" of low permeability has been inserted into the aquifer so that oil tends to accumulate on the down-dip side of the low-permeability plug, leading to a stratigraphically trapped oil stringer. The precise shape, position. and amount of oil trapped depend, of course, on physical conditions.

8.1.2.1. Increasing Aquifer Angle

As the aquifer angle increases, more of the buoyancy drive can be used to oppose the hydrodynamic force attempting to drive the oil down dip. Thus, the amount and location of the areal mass of oil vary substantively as aquifer angle changes.

This effect is clearly seen in the sequence of Fig. 8.6a–h for oil accumulations far down dip from the permeability plug interface and for those close to the transition region, and also for oil stringers fully contained in the upper medium of lower permeability.

8.1.2.2. Increasing Flow Speed

Once more, as the flow speed increases, the hydrodynamic force driving the oil down dip increases. As a consequence, the shapes and positions of the areal mass of oil stringers are changed under flow increases.

Representative variations are shown in Fig. 8.7a–h for oil stringers close to and far from the transition region between the high-permeability down-dip region and the low-permeability up-dip region.

8.1.2.3. Increasing Aquifer Width

Once more, we see from Fig. 8.8a–h that as the aquifer width increases, there is less need for the flow to be significantly distorted in passing by the stringer. Thus, the portion of the oil stringer again tends to an invariant form.

8.2. Gas Accumulations

To a large extent, the behaviors of gas stringers in a variable permeability aquifer mimic the trends uncovered above for oil stringers. Several technical factors make for differences in appearance between the detailed behaviors of gas and oil stringers: (i) the higher buoyancy pressure of gas compared to that of oil; (ii) the higher surface tension against water; (iii) the higher capillary pressure; and (iv) the lower viscosity. However, all factors operate in the same direction so that the basic physical patterns found for oil stringers are also obtained for gas stringers.

8.2.1. Permeability Increasing Up Dip

8.2.1.1. Increasing Aquifer Angle

Figure 8.3i–p illustrates the patterns of behavior for gas stringers. Direct comparison can be made with Fig. 8.3a–h for the oil stringer behavior.

8.2.1.2. Increasing Flow Speed

Figure 8.4i–p presents the gas stringer behaviors. Here, direct comparison can be made with Fig. 8.4a–h for the corresponding oil stringer results.

8.2.1.3. Increasing Aquifer Width

Figure 8.5m–p provides the computed gas stringer shapes. In this case, direct comparison can be made with the oil stringer shapes given in Fig. 8.5a–l.

8.2.2. Permeability Decreasing Up Dip

8.2.2.1. Increasing Aquifer Angle

Figure 8.6i–p illustrates the patterns of behavior for the gas stringers. Direct comparison can be made with Fig. 8.6a–h for the oil stringer behavior.

8.2.2.2. Increasing Flow Speed

Figure 8.7i–p presents the gas stringer behaviors. Here, direct comparison can be made with Fig. 8.7a–h for the corresponding oil stringer results.

8.2.2.3. Increasing Aquifer Width

Figure 8.8i–p provides the computed gas stringer shapes. In this case, direct comparison can be made with the oil stringer shapes given in Fig. 8.8a–h.

8.3. Oil or Gas Accumulations

The above patterns of behavior for oil and gas stringers provide a basis for understanding the variations in hydrocarbon masses that can be hydrodynamically maintained in constant-angle, constant-width aquifers. However, there are peculiarities of patterns of variation which are beholden to a combination of geometry variations in conjunction with permeability variations.

Here we examine three such cases to give an idea of the interesting patterns that can arise.

8.3.1. Permeability Oscillations

Consider a permeability variation along a planar, inclined, constant-width aquifer of the form

$$\kappa = \kappa_0 + \Delta\kappa \sin(2\pi\ell/\ell_0) \tag{8.1}$$

so that high and low values of permeability range from $\kappa_0 - \Delta\kappa$ to $\kappa_0 + \Delta\kappa$ over a spatial scale ℓ_0, as sketched in Fig. 8.9.

The corresponding shapes and positions of steady-state oil and gas stringers in such a hydrodynamically controlled aquifer are illustrated in Fig. 8.10a–f and 8.10g–l, respectively. The separation and distortion of the stringers into the high-permeability regions can be identified by inspection of the figures.

8.3.2. Anticline with Permeability Variations

Here two cases are considered: (i) when the anticline has a low permeability near its apex, and (ii) when the anticline has a high permeability near its apex (see Fig. 8.11).

Sketches of the positioning and shaping of oil and gas stringers are presented in Fig. 8.12a–l and show rather clearly the differences in characteristic response for oil and gas in both the relatively "impermeable" and relatively "permeable" anticline cases.

8.3.3. Syncline with Permeability Variations

For precisely the same parameters as in the anticlinal case, except with the geometry inverted (see Fig. 8.13), the corresponding patterns of oil and gas stringers are presented in Fig. 8.14a–l. Pictorial comparison of the syncline situation with the anticline patterns given immediately above provides an appreciation of the changes produced by a change in the geometric configuration.

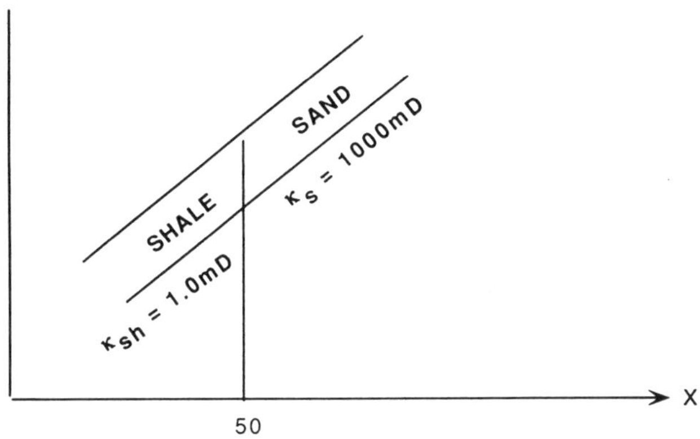

Figure 8.1. Sketch illustrating the shift from a high-permeability region (1000 mD) in the *upper* part of a constant-angle, constant-width aquifer to a low-permeability region (1 mD) in the *lower* part of the aquifer. The change in permeability occurs sharply at location 50—the same location as in Fig. 8.3a–p.

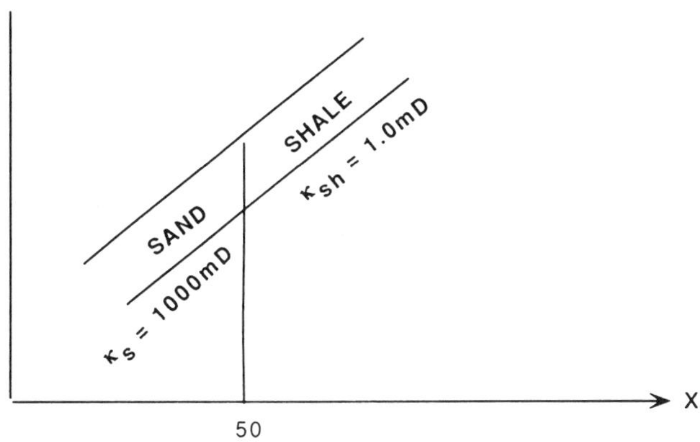

Figure 8.2. Sketch illustrating the shift from a high-permeability region (1000 mD) in the *lower* part of a constant-angle, constant-width aquifer to a low-permeability region (1 mD) in the *upper* part of the aquifer. The change in permeability occurs sharply at location 50—the same location as in Fig. 8.6a–p.

Table for Fig. 8.3. Aquifer Geometry: Fixed Width, Variable Angle with Shale Down Dip of Sand

Position along aquifer	Aquifer angle			
	Oil		Gas	
	40°	20°	40°	20°
75	8.3a	8.3e	8.3i	8.3m
60	8.3b	8.3f	8.3j	8.3n
52	8.3c	8.3g	8.3k	8.3o
35	8.3d	8.3h	8.3l	8.3p

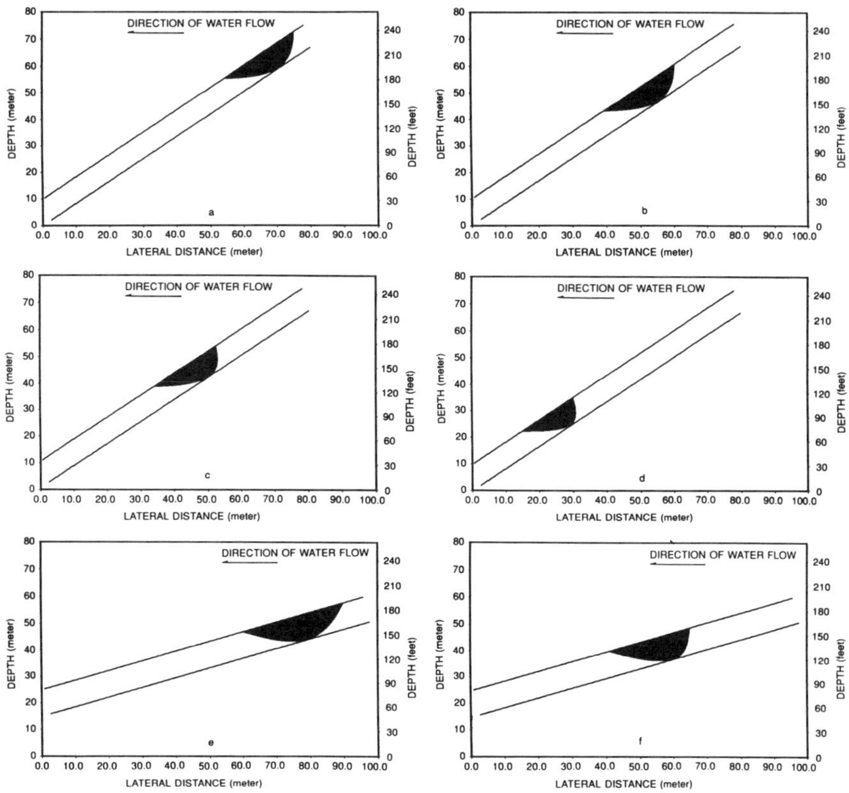

Figure 8.3. Hydrodynamic shapes of oil and gas stringers in the aquifer geometry and with the permeability variation of Fig. 8.1. The water flow speed is fixed at 100 cm/yr, and the oil (gas) density is 0.8 (0.1) g/cm^3. Both the angle of the aquifer and the location of the "nose" of the stringer are adjusted. The values of these parameters for each panel in the figure are provided in the accompanying table. Panels a–h refer to oil stringers, and panels i–p refer to gas stringers.

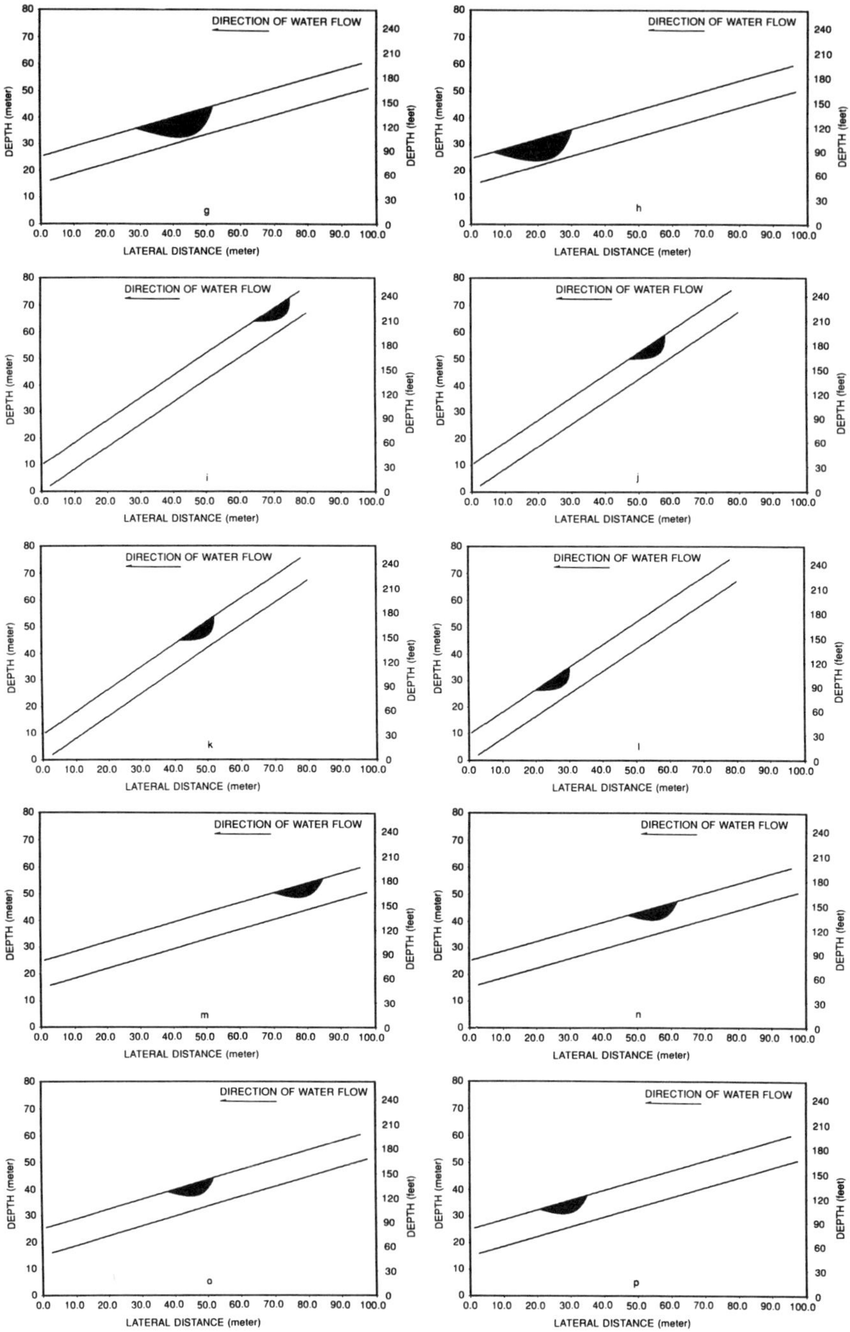

Figure 8.3. *(Continued)*

Table for Fig. 8.4. Aquifer Geometry: Fixed Width, Fixed Angle with Shale Down Dip of Sand

Position along aquifer	Water flow rate (cm/yr)			
	Oil		Gas	
	1.0	50	1.0	50
85	8.4a	8.4e	8.4i	8.4m
65	8.4b	8.4f	8.4j	8.4n
52	8.4c	8.4g	8.4k	8.4o
35	8.4d	8.4h	8.4l	8.4p

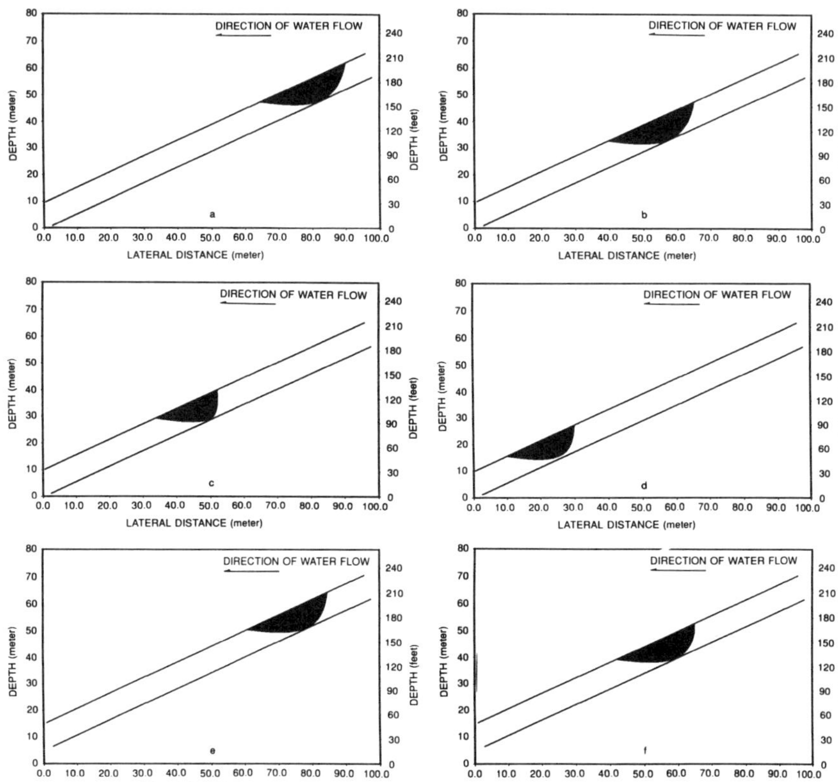

Figure 8.4. Hydrodynamic shapes of oil and gas stringers in the aquifer geometry and with the permeability variation of Fig. 8.1. The aquifer angle is fixed, and the oil (gas) density is 0.8 (0.1) g/cm^3. Both the water flow rate and the location of the "nose" of the stringer are adjusted. The values of these parameters for each panel in the figure are provided in the accompanying table. Panels a–h refer to oil stringers, and panels i–p refer to gas stringers.

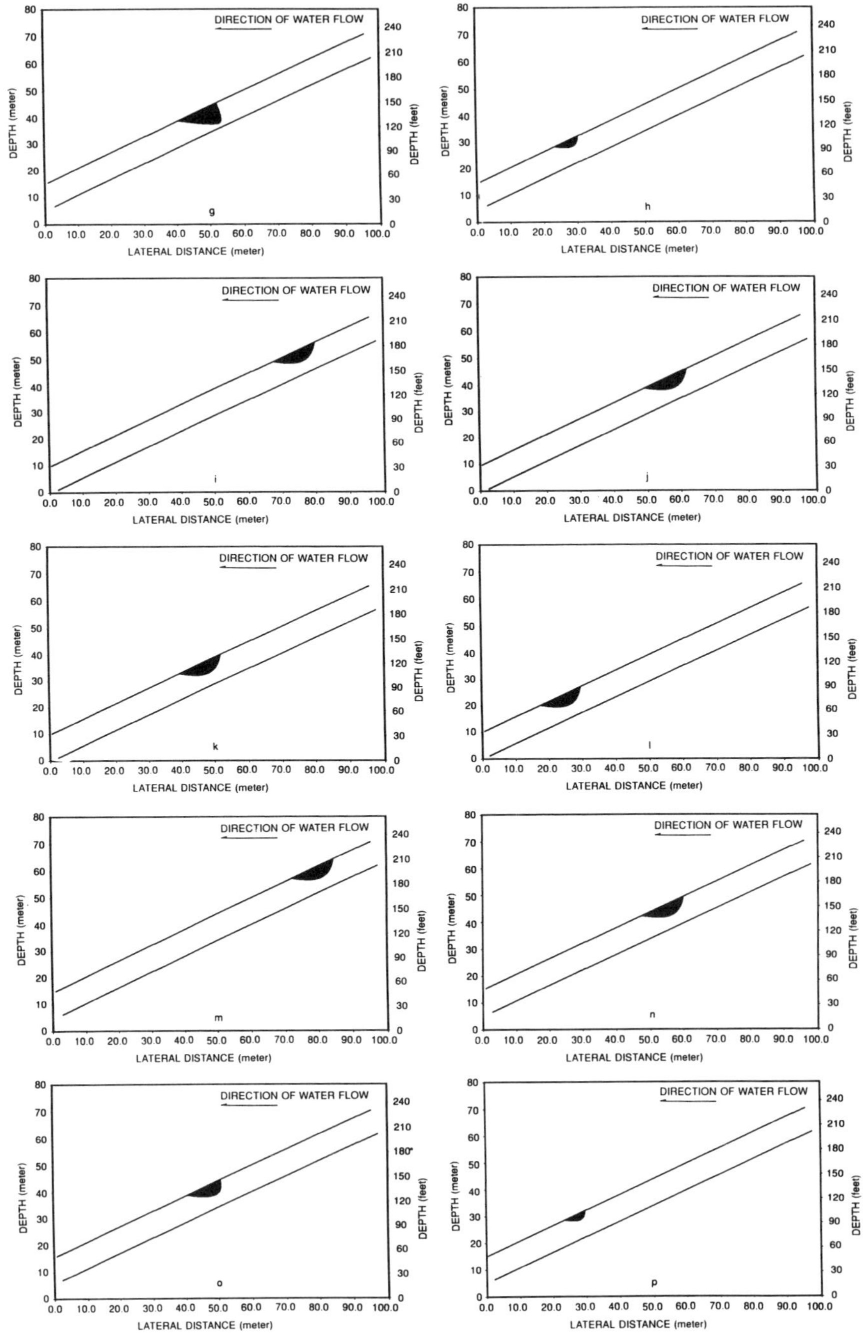

Figure 8.4. *(Continued)*

Table for Fig. 8.5. Aquifer Geometry: Fixed Angle, Variable Width with Shale Down Dip of Sand

Position along aquifer	Aquifer width (meters)			
	Oil		Gas	
	5	10	5	10
85	8.5a	8.5e	8.5i	8.5m
60	8.5b	8.5f	8.5j	8.5n
52	8.5c	8.5g	8.5k	8.5o
30	8.5d	8.5h	8.5l	8.5p

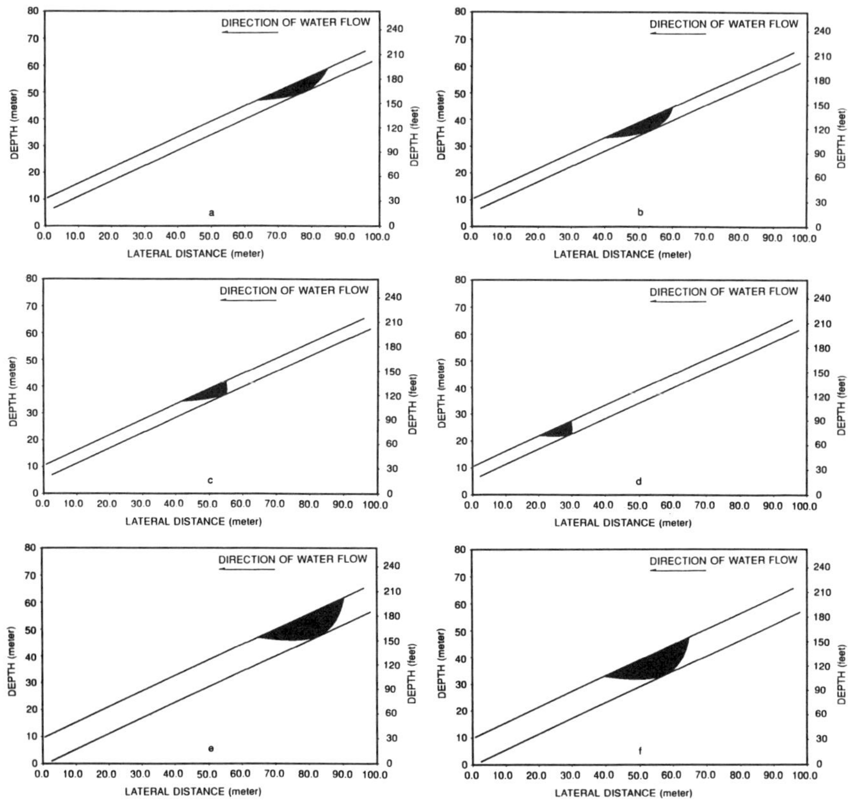

Figure 8.5. Hydrodynamic shapes of oil and gas stringers in the aquifer geometry and with the permeability variation of Fig. 8.1. The aquifer angle is fixed as is the flow rate (at 100 cm/yr), and the oil (gas) density is 0.8 (0.1) g/cm^3. Both the aquifer width and the location of the "nose" of the stringer are adjusted. The values of these parameters for each panel in the figure are provided in the accompanying table. Panels a–h refer to oil, and panels i–p refer to gas.

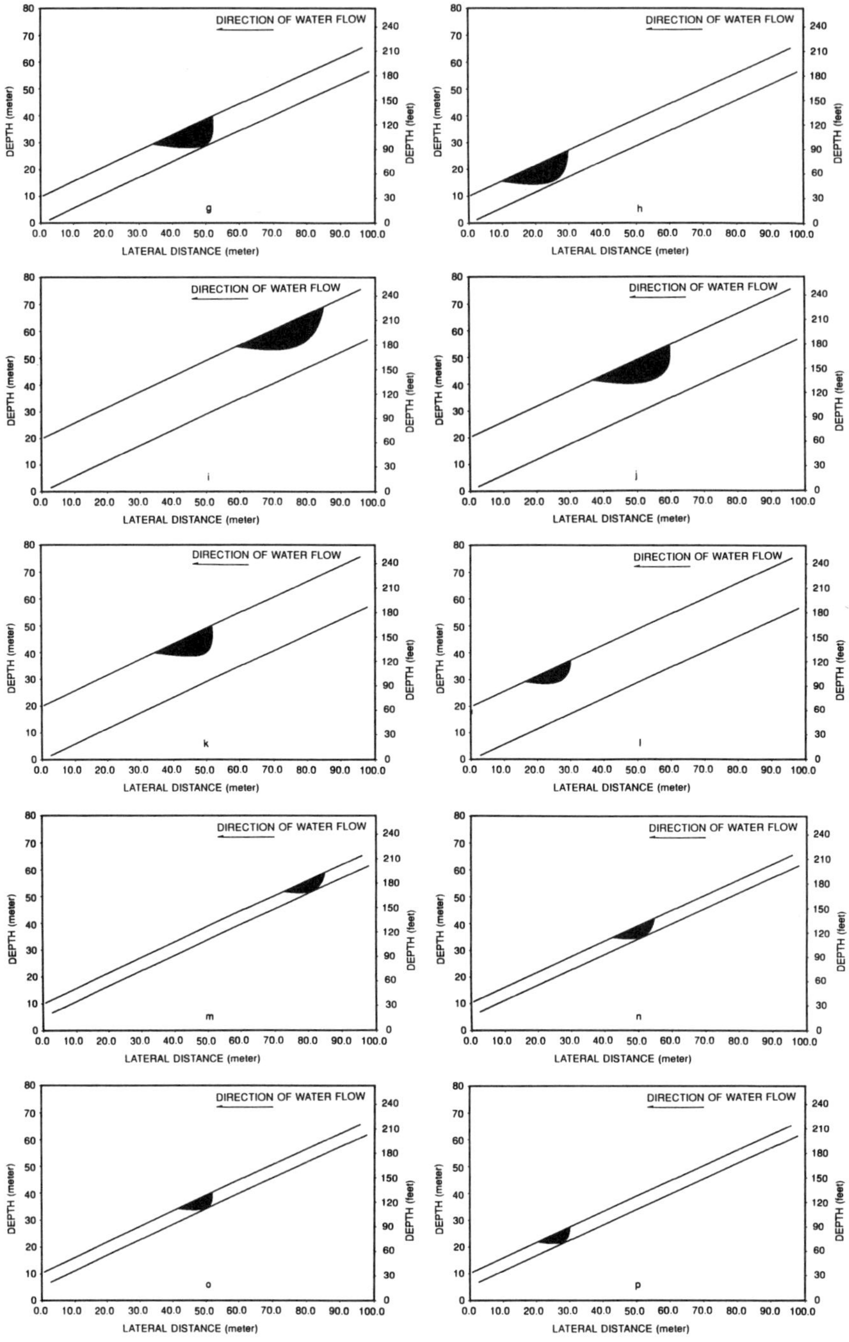

Figure 8.5. *(Continued)*

Table for Fig. 8.6. Aquifer Geometry: Fixed Width, Variable Angle with Sand Down Dip of Shale

Position along aquifer	Angle of aquifer			
	Oil		Gas	
	20°	40°	20°	40°
86 (oil), 80 (gas)	8.6a	8.6e	8.6i	8.6m
65 (oil), 60 (gas)	8.2b	8.6f	8.6j	8.6n
55 (oil), 51 (gas)	8.6c	8.6g	8.6k	8.6o
35 (oil), 30 (gas)	8.6d	8.6h	8.6l	8.6p

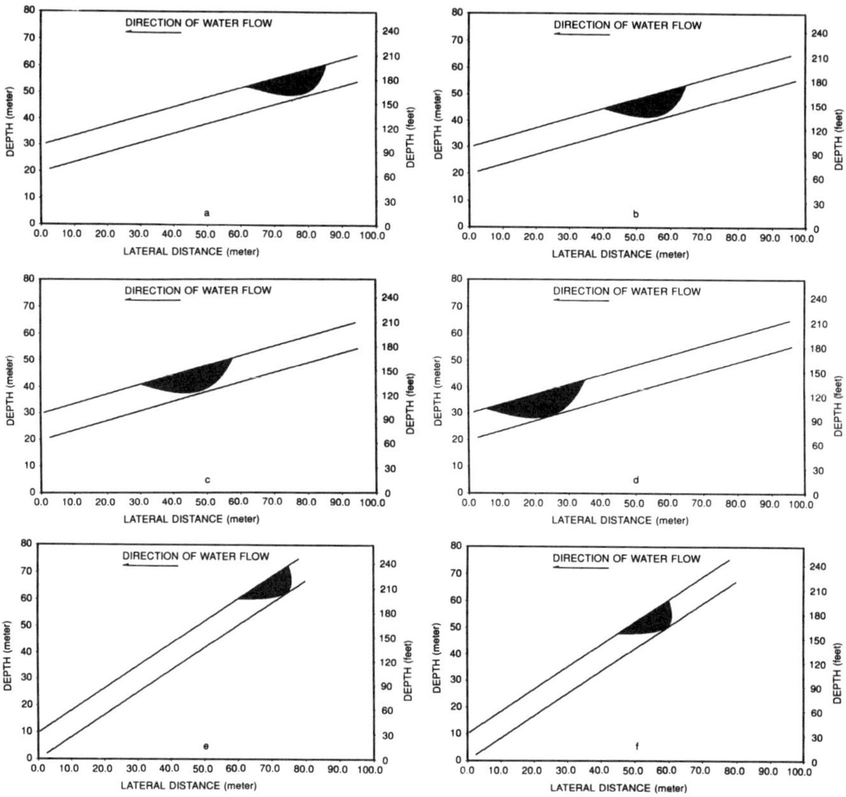

Figure 8.6. Hydrodynamic shapes of oil and gas stringers in the aquifer geometry and with the permeability variation of Fig. 8.2. The water flow speed is fixed at 100 cm/yr, and the oil (gas) density is 0.8 (0.1) g/cm^3. Both the angle of the aquifer and the location of the "nose" of the stringer are adjusted. The values of these parameters for each panel in the figure are provided in the accompanying table. Panels a–h refer to oil stringers, and panels i–p refer to gas stringers.

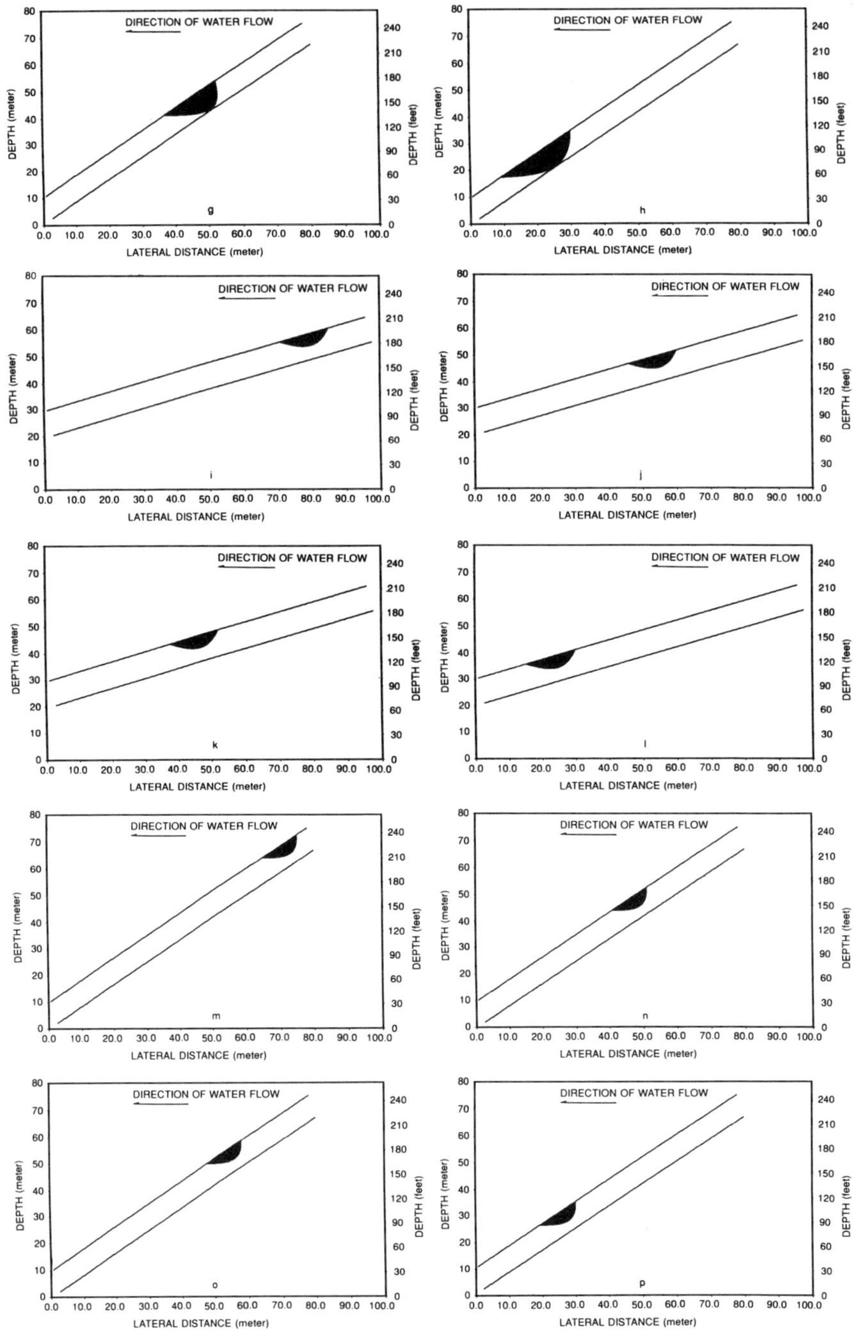

Figure 8.6. *(Continued)*

Table for Fig. 8.7. Aquifer Geometry: Fixed Width, Fixed Angle with Sand Down Dip of Shale

Position along aquifer	Water flow rate (cm/yr)			
	Oil		Gas	
	1.0	50	1.0	50
85	8.7a	8.7e	8.7i	8.7m
60	8.7b	8.7f	8.7j	8.7n
48	8.7c	8.7g	8.7k	8.7o
30	8.7d	8.7h	8.7l	8.7p

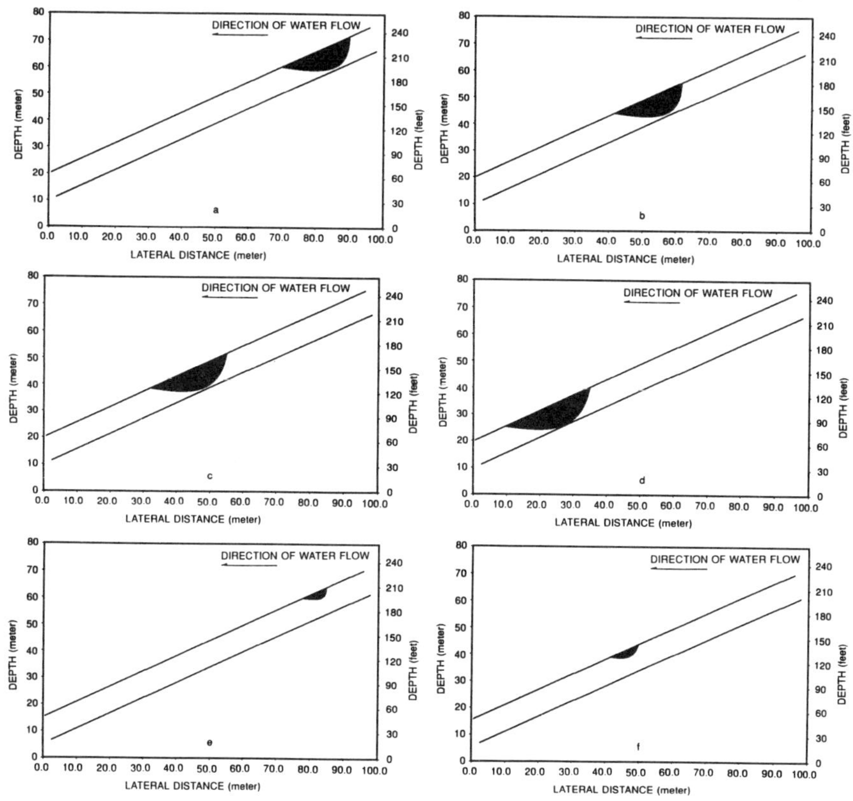

Figure 8.7. Hydrodynamic shapes of oil and gas stringers in the aquifer geometry and with the permeability variation of Fig. 8.2. The aquifer angle is fixed, and the oil (gas) density is 0.8 (0.1) g/cm³. Both the water flow rate and the location of the "nose" of the stringer are adjusted. The values of these parameters for each panel in the figure are provided in the accompanying table. Panels a–h refer to oil stringers, and panels i–p refer to gas stringers.

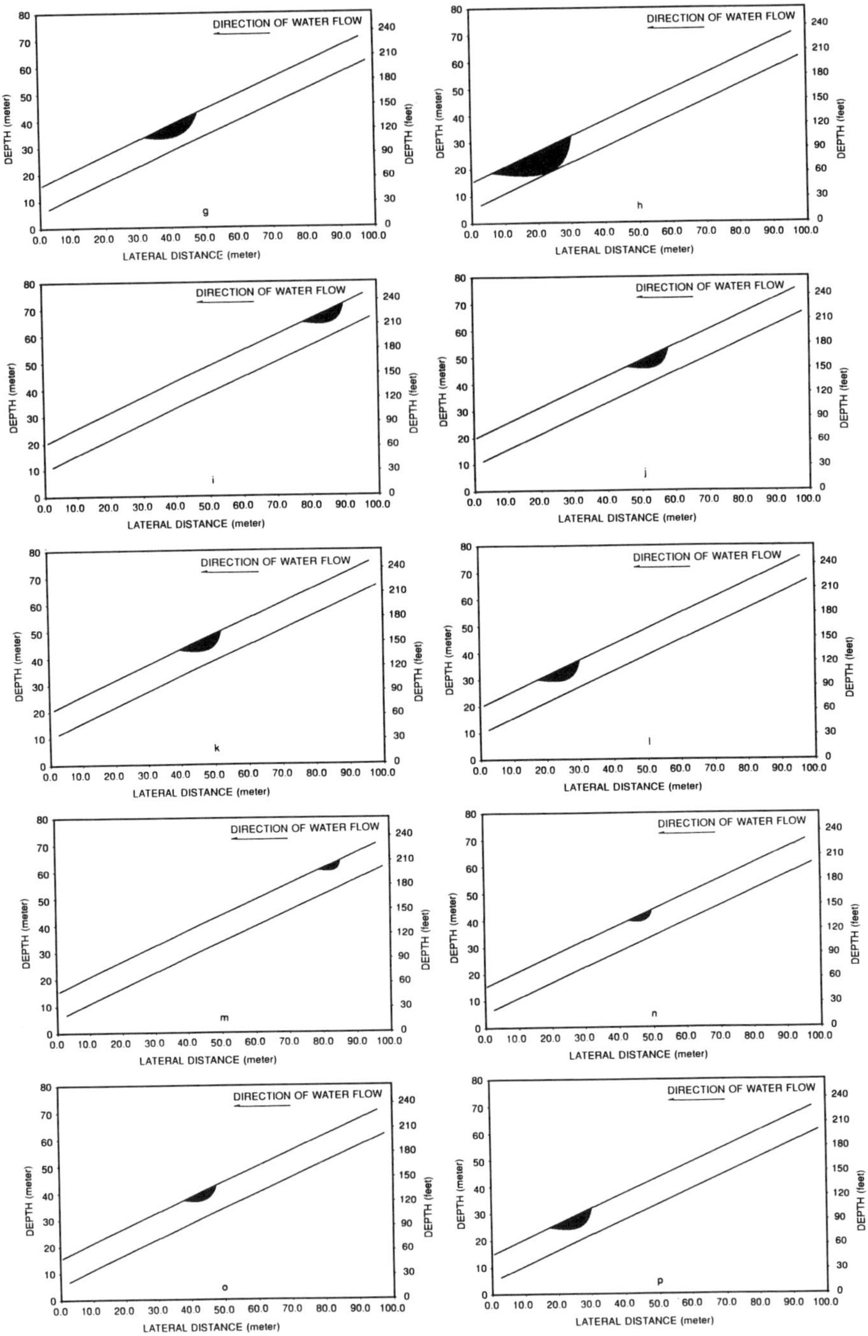

Figure 8.7. *(Continued)*

Table for Fig. 8.8. Aquifer Geometry: Fixed Angle, Variable Width with Sand Down Dip of Shale

	Aquifer width (meters)			
	Oil		Gas	
Position along aquifer	5	10	5	10
85	8.8a	8.8e	8.8i	8.8m
60	8.8b	8.8f	8.8j	8.8n
52	8.8c	8.8g	8.8k	8.8o
30	8.8d	8.8h	8.8l	8.8p

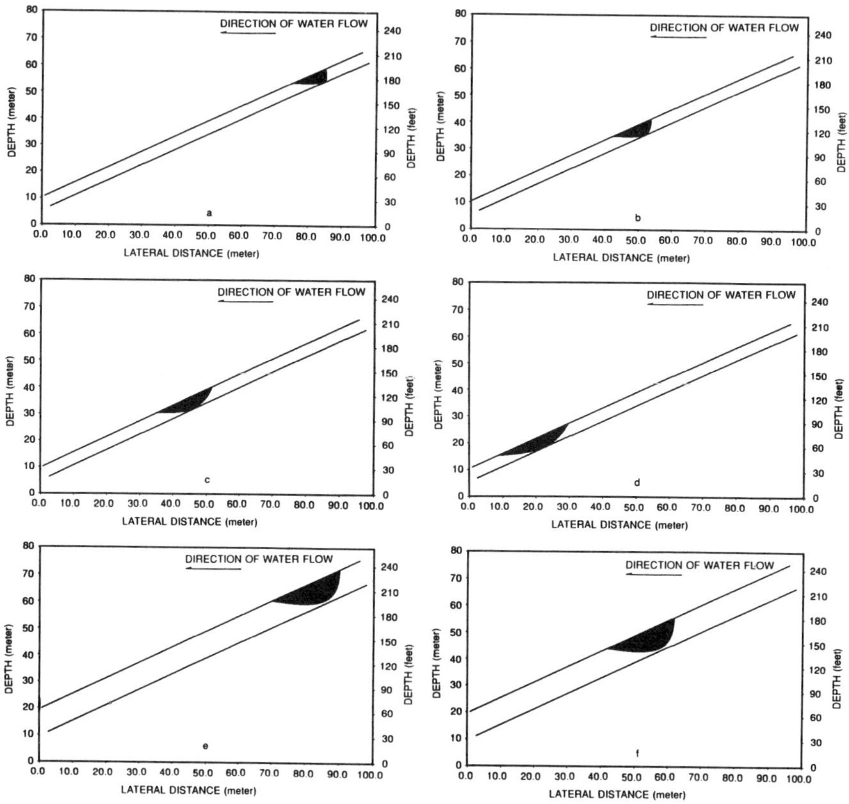

Figure 8.8. Hydrodynamic shapes of oil and gas stringers in the aquifer geometry and with the permeability variation of Fig. 8.2. The aquifer angle is fixed as is the flow rate (at 100 cm/yr), and the oil (gas) density is 0.8 (0.1) g/cm^3. Both the aquifer width and the location of the "nose" of the stringer are adjusted. The values of these parameters for each panel in the figure are provided in the accompanying table. Panels a–h refer to oil, and panels i–p refer to gas.

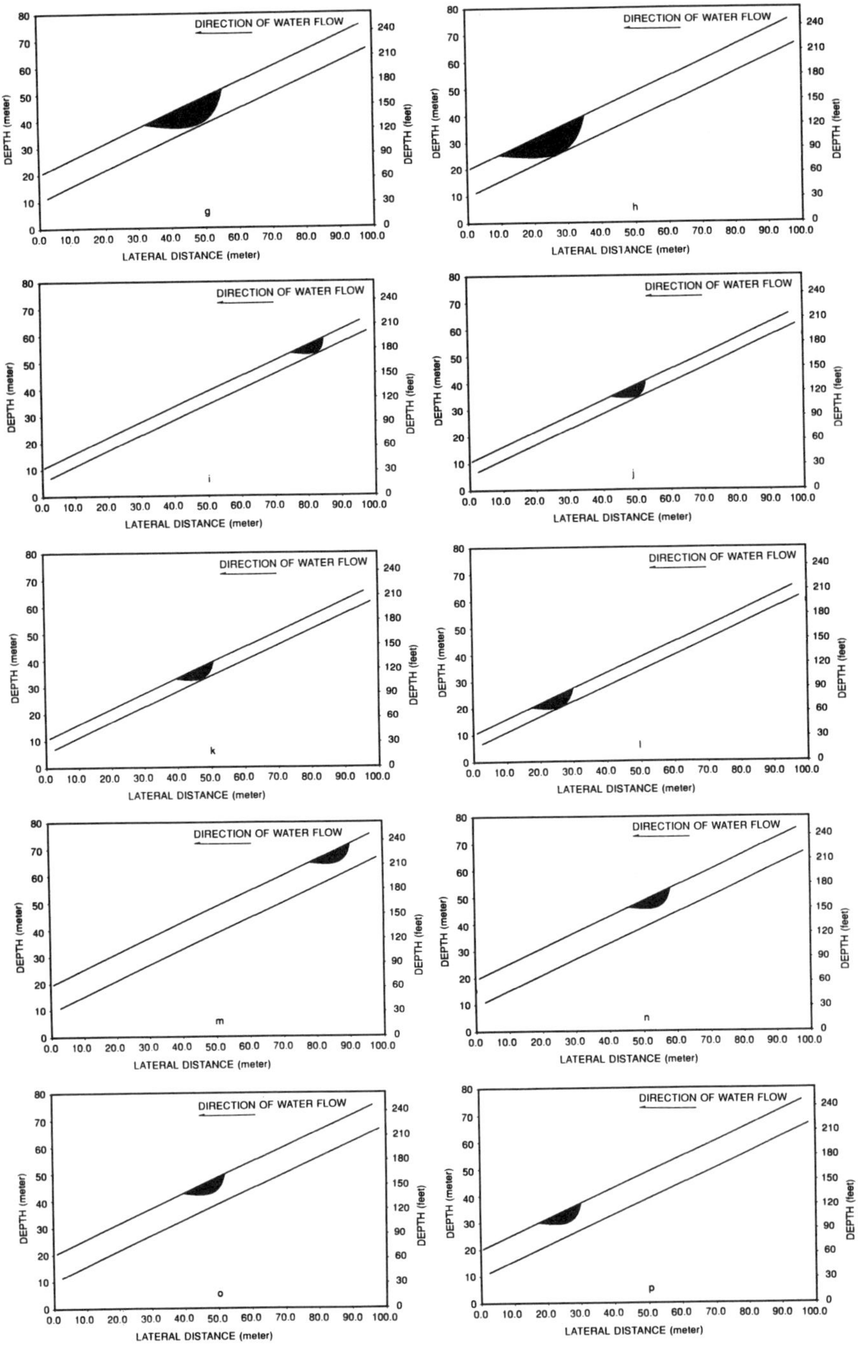

Figure 8.8. *(Continued)*

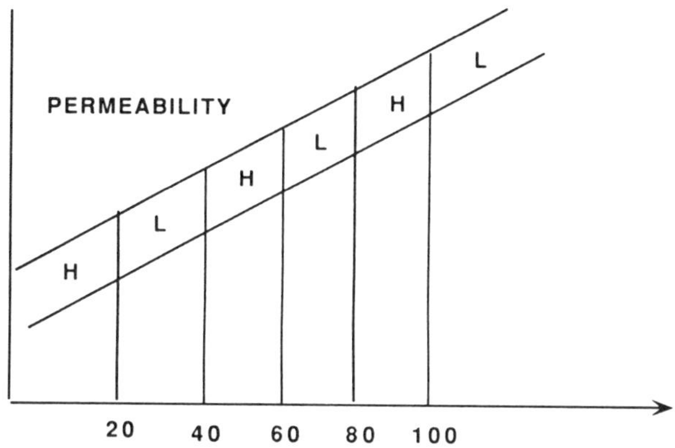

Figure 8.9. Sketch of a fixed-width, fixed-angle aquifer in which the permeability abruptly varies from 1000 mD to 1 mD at locations 20, 60, 100, . . . and varies from 1 mD to 1000 mD at locations, 40, 80, . . . , i.e., a periodic array sequence of sand–shale–sand–shale . . . along the aquifer. $\kappa_H = 1000$ mD, $\kappa_L = 1$ mD

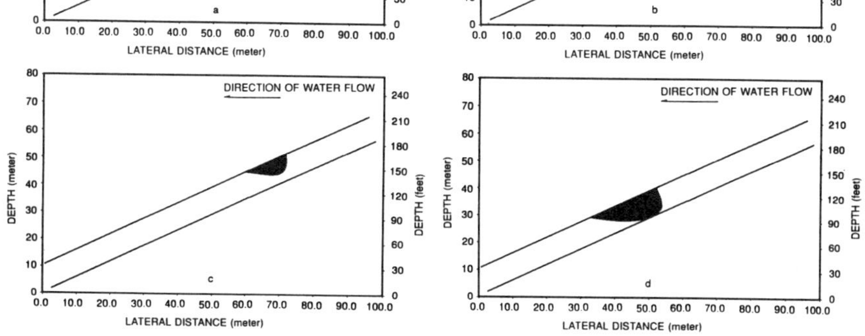

Figure 8.10. Hydrodynamic shapes of oil and gas stringers in the aquifer geometry and with the periodic permeability variation of Fig. 8.9. The water flow rate is maintained at 100 cm/yr, and the oil (gas) density is 0.8 (0.1) g/cm^3. Panels a–f show the shaping of oil stringers at positions 95, 78, 72, 52, 38, and 20, respectively, along the aquifer; panels g–l show the shaping of gas stringers at the same locations.

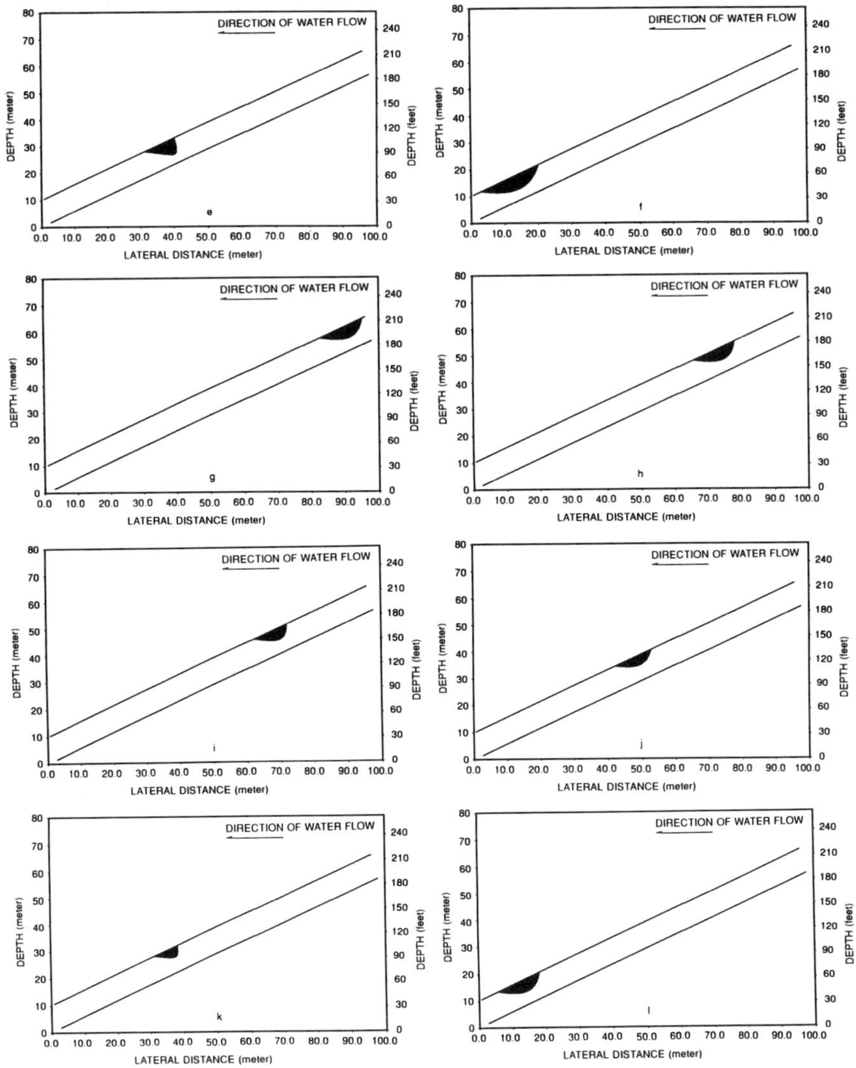

Figure 8.10. *(Continued)*

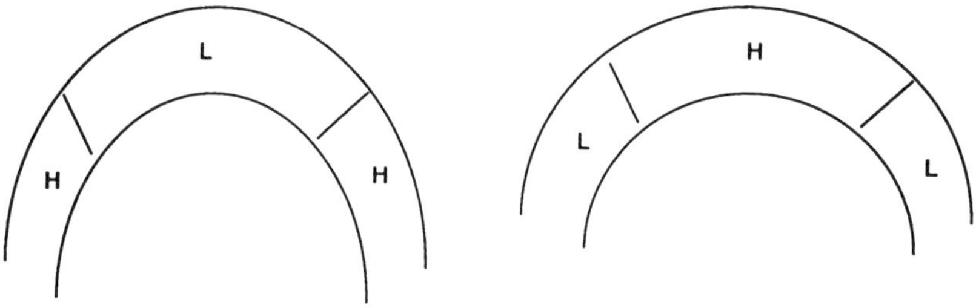

Figure 8.11. Sketch of an anticline with two different variations of permeability along the anticline: (a) with high-permeability regions to either side of the structural culmination and a low-permeability region in the center of the anticline; (b) with low-permeability regions to either side of the structural culmination and a high-permeability region in the center.

Table for Fig. 8.12. Aquifer Geometry: Single Anticline

Position along aquifer	Permeability sequence[a]			
	Oil		Gas	
	HLH	LHL	HLH	LHL
11	8.12a	8.12d	8.12g	8.12j
22	8.12b	8.12e	8.12h	8.12k
30	8.12c	8.12f	8.12i	8.12l

[a]Permeability sequence from flank through peak to flank: HLH, high–low–high; LHL, low–high–low.

Figure 8.12. Hydrodynamic shapes of steady-state oil and gas stringers in the anticline geometry and with the permeability variations of Fig. 8.11. The water flow speed is held fixed at 100 cm/yr; the high permeability is 1000 mD, and the low permeability is 1 mD. The position of the stringer is varied along the aquifer. The value of this parameter and the permeability sequence for each panel in the figure are provided in the accompanying table. Panels a–f refer to oil stringers (density 0.8 g/cm^3), and panels g–l refer to gas stringers (density 0.1 g/cm^3).

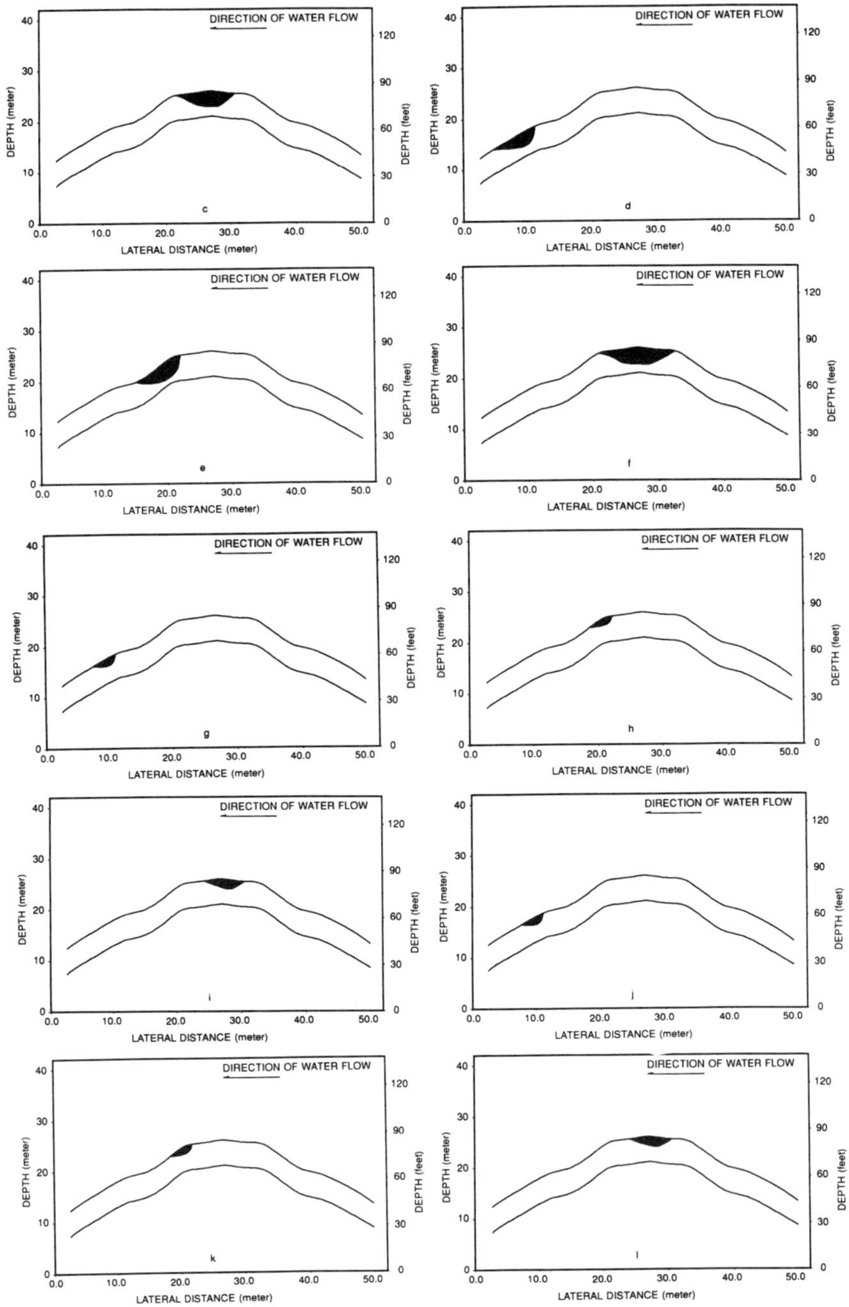

Figure 8.12. *(Continued)*

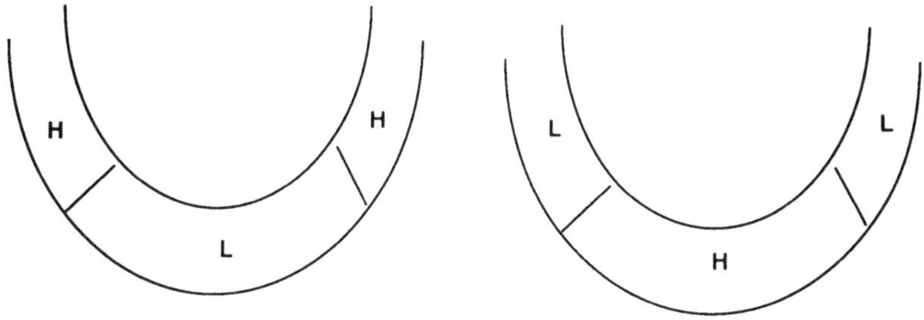

Figure 8.13. Sketch of a syncline with two different variations of permeability along the syncline: (a) with high-permeability regions to either side of the structural low and a low-permeability region in the center; (b) with low-permeability regions to either side of the structural low and a high permeability region in the center.

Table for Fig. 8.14. Aquifer Geometry: Single Syncline

Position along aquifer	Permeability sequence[a]			
	Oil		Gas	
	HLH	LHL	HLH	LHL
33	8.14a	8.14d	8.14g	8.14j
38	8.14b	8.14e	8.14h	8.14k
49	8.14c	8.14f	8.14i	8.14l

[a]Permeability sequence from flank through nadir to flank: HLH, high–low–high; LHL, low–high–low.

Figure 8.14. Hydrodynamic shapes of steady-state oil and gas stringers in the syncline geometry and with the permeability variations of Fig. 8.11. The water flow speed is held fixed at 100 cm/yr; the high permeability is 1000 mD, and the low permeability is 1 mD. The position of the stringer is varied along the aquifer. The value of this parameter and the permeability sequence for each panel in the figure are provided in the accompanying table. Panels a–f refer to oil stringers (density 0.8 g/cm^3) and panels g–l refer to gas stringers (density 0.1 g/cm^3).

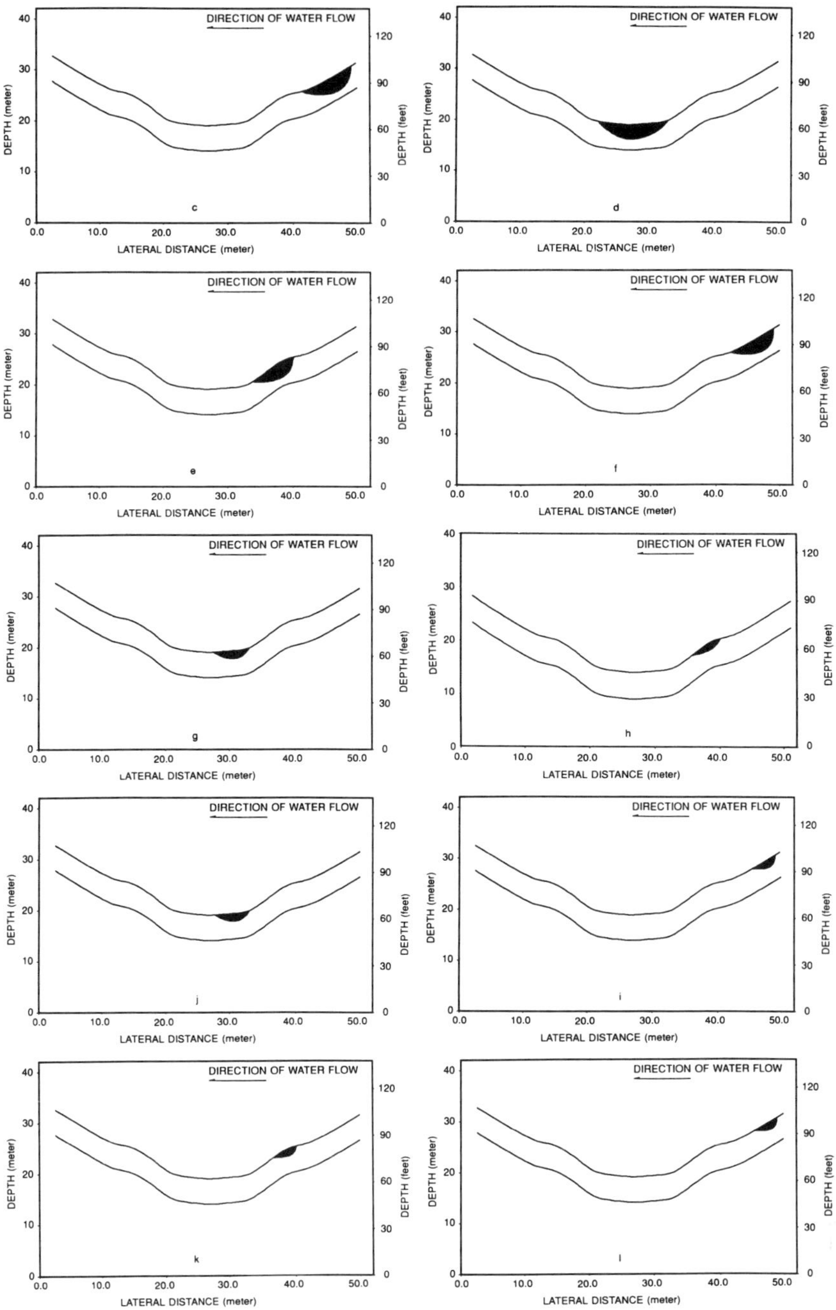

Figure 8.14. *(Continued)*

9

Saline and Fresh Water Flows

So far, the development of hydrodynamic shaping of buoyant masses has been concerned with steady-state situations in confined aquifers. However, several naturally occurring situations arise in which one of the aquifer boundaries is removed, normally the upper boundary. For instance, the flow of saline water through a fresh water regime and the flow of cold dense water into a warm, light ocean constitute situations, as sketched in Fig. 9.1, in which the "top" of the sealing aquifer is removed and the air–water interface under gravity acts as the unconfined marker position of the aquifer horizon.

The opposite situation—that of an aquifer with only a top sealing boundary—is relevant in the case of oil trapped on the underside of a competent, impermeable stratigraphic horizon and is also relevant to the spreading of buoyant salt masses pressed up against a competent anhydrite seal (Fig. 9.2).

We examine first the twin problems of cold river discharges into warm oceans and saline water flows into freshwater bodies, because these problems can be combined insofar as their quantitative mathematical development is concerned.

9.1. Saline or Cold Water Flows

Normal seawater contains about 35 parts per thousand (ppt) of salt with a corresponding density, ρ_s, of about 1.035 g/cm^3. In the subsurface, brines have been found to have up to 250–300 ppt of salt and a density of around 1.2–1.3 g/cm^3. Thus, saline waters are more dense than fresh waters at the same temperature.

In the case of cold waters being supplied to warm oceans, there is a density difference because of the thermal expansion of water, the highest

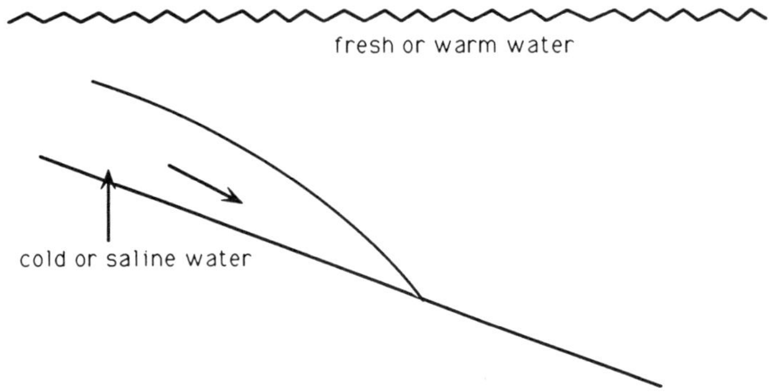

Figure 9.1. Sketch of cold or saline water entering a warm or fresh body of water and following the water floor contact as it moves.

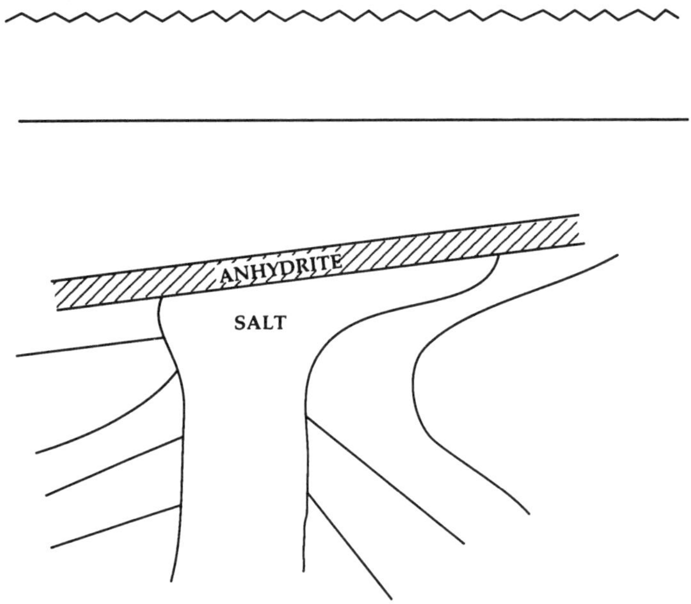

Figure 9.2. Sketch of buoyant salt motion and shaping produced when up-rising salt impacts against an impermeable, competent stratigraphic bed such as the layer marked anhydrite.

water density being at 4°C. Thus, cold water of density $\rho_c = \rho_4[1 - \alpha(T_c - 4)]$ (where α is the coefficient of thermal expansion, ρ_4 is water density at 4°C, and T_c is the cold water temperature, and warm water of density $\rho_w = \rho_4[1 - \alpha(T_w - 4)]$ (where T_w is the warm water temperature) have a density difference proportional to their temperature difference:

$$\Delta\rho = \rho_w - \rho_c = -\alpha\rho_4(T_w - T_c) > 0$$

Therefore, the steady-state flow of cold (warm) waters through a hydrodynamically warm (cold) water aquifer or of a saline brine through a freshwater aquifer (or vice versa, of course) can be treated mathematically on the same footing.

We now consider several numerical examples to illustrate typical response patterns.

9.1.1. Fresh, Warm Water Entering a Cold, Saline Ocean

The geometrical configuration envisaged is shown in Fig. 9.3a. The lower density of the fresh, warm water relative to the colder, more saline ocean implies that the fresh water stringer floats at the free surface of the water–air contact interface. We have also drawn the figure to correspond to a strand line to the left, representing the sourcing direction of the fresh water, with a basinal slope to deeper ocean water.

The numerical situations investigated show how an increasing mass of fresh water will "infill" the surficial region, depending on both the inflow velocity and inflow mass. For instance, Fig. 9.4a–c portrays the effect of an increasing flow speed on the shape of the fresh water stringer.

9.1.2. Saline, Cold Water Entering Fresh, Warm Water

The geometrical configuration under consideration is shown in Fig. 9.3b, where a cold, dense, saline flow is entering a fresh, warm water body of fluid through an aquifer with an unbounded top (air–water interface).

As the saline front advances, the shape and position of the saline stringer change. Numerical illustrations of saline water shapes for variable mass and increasing flux of fresh water across the saline stringer are given in Fig. 9.5a–c.

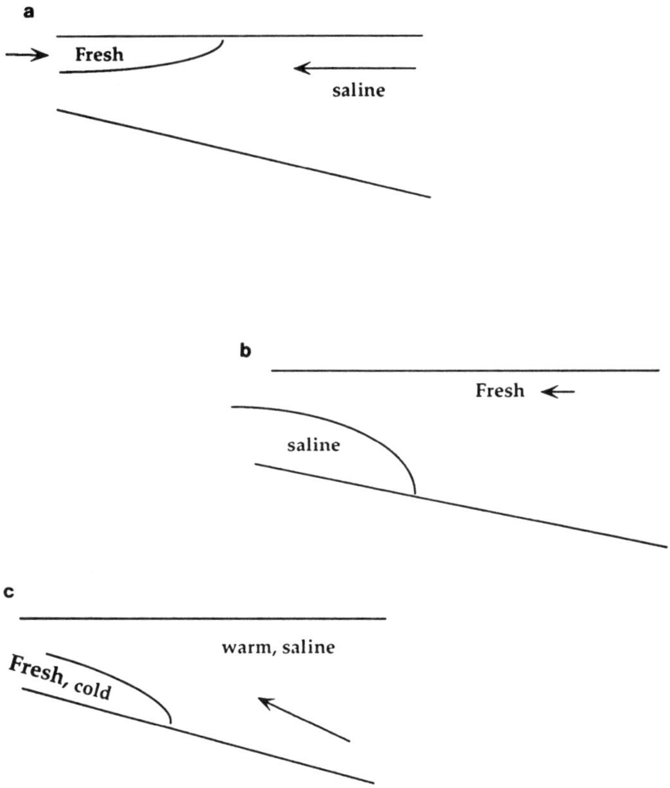

Figure 9.3. Sketch of various possible situations for nonmixing water masses: (a) Fresh, warm water entering a cold, saline ocean; (b) saline, cold water entering a fresh, warm body of water; (c) fresh, cold water entering a warm, saline ocean.

9.1.3. Fresh, Cold Water Entering a Warm, Saline Ocean

The geometrical configuration is sketched in Fig. 9.3c. Here we take the density of the fresh, cold water to be *greater* than that of the warm, saline water so that, as shown, the fresh water stringer now follows the bottom slope. Again, a strand line position is envisioned to the left of Fig. 9.3c, representing the sourcing of the fresh, cold water supply.

As in the previous situation, we could investigate numerically the shape and positioning of the stringer as the stringer mass and speed of flow change. However, an inspection of the previous situation (saline, dense, bottom stringer in a warm, fresh body of water) shows that the two problems are identical in mathematical, physical, and geometrical structure. Thus, with the same parameter values the patterns of behavior

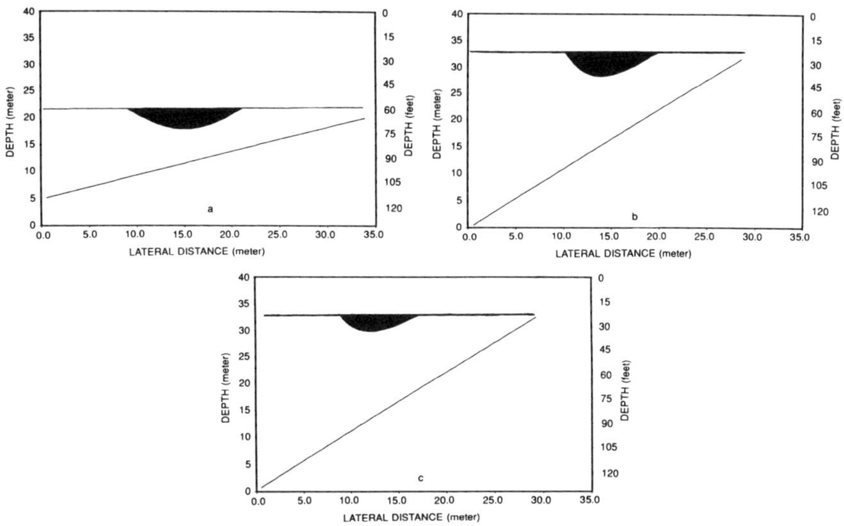

Figure 9.4. Numerical illustrations portraying the effect of increasing speed on a fresh, warm water stringer entering a cold, saline ocean. The calculations were done in a frame in which the stringer is at rest and the ocean flows from left to right. (a) Low-speed flow; (b) medium-speed flow; (c) high-speed flow.

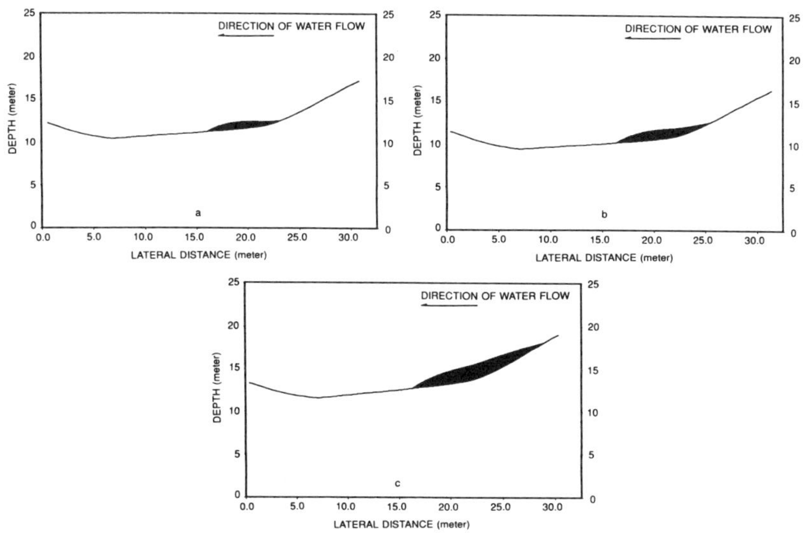

Figure 9.5. Numerical illustrations portraying the effect of increasing speed on a fresh, cold water stringer entering a warm, saline ocean (or a cold, saline water stringer entering a warm, fresh body of water). The calculations were done in a frame in which the stringer is at rest and the main body of water flows from left to right. (a) Low-speed flow; (b) medium-speed flow; (c) high-speed flow.

are identical. It is, then, of interest to perform a different set of numerical experiments in which the slope angle of the sediment–water interface line is increased while all other parameters are held constant.

Figure 9.5a–c (drawn for the saline case) is equally valid here and shows the shape and position of fresh, cold water stringers as the flux of warm, saline water increases. To be noted here is the increase in "anti-buoyancy" force of the dense stringer as the flow increases so that the fresh water stringer is more quickly emplaced in an ocean that is rapidly deepening away from the strand line.

9.2. Oil or Salt Flows

In the previous section we saw the effect of hydrodynamic drives in open-topped aquifers on the shaping and position of cold or saline waters. In this section we consider the case of open-bottom aquifers with a confining top. We have already seen (Chapter 4) examples of this type of behavior when we provided computer simulations of Hubbert's experiments. However, the problem is of greater interest. Diapiric salt, of constant density $\rho_s \approx 2.0\text{–}2.2\ \text{g/cm}^3$, rises through sedimentary cover once the sediment density exceeds that of salt (fully compacted sediments have a density of about 2.6–2.8 g/cm^3, whereas, at deposition, sediment density is around 1.6–1.9 g/cm^3). Emplacement of a competent stratigraphic unit, such as an anhydrite sheet, provides an impediment to the further rise of the salt. However, the buoyant force on the salt stock is still being exerted. If the competent layer does not yield, then the salt flows out along the underside of the competent horizon, displacing the sediments, as sketched in Fig. 9.6.

Equally, oil (of density ~0.8 g/cm^3) or gas (of density ~0.1 g/cm^3) will rise until trapped by a surface-sealing horizon such as an anhydrite layer, or even a previously emplaced salt sheet, as occurs in the Gulf of Mexico. Both situations can be modeled in a mathematically similar fashion.

9.3. Phreatic Island Lenses

The presence of a fresh water table below sandy islands in shallow marine environments is well established. Direct measurements of the fresh water–saline water contact often shows a lenslike shape as sketched in Fig. 9.7. The presence and maintenance of the fresh water lens is due to a variety of factors. Production of the phreatic lenses is due to rainfall settling through the island into the subsurface. Maintenance of a phreatic

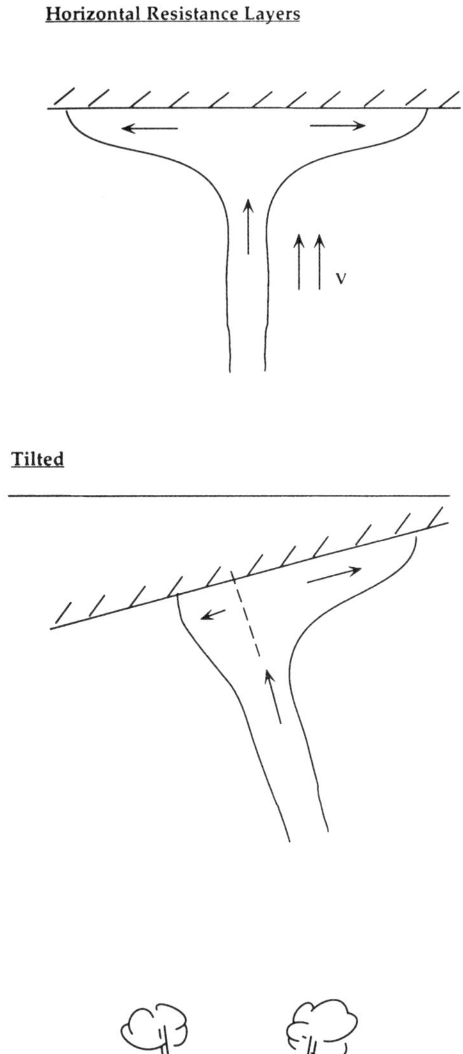

Figure 9.6. Sketch illustrating the difference in appearance of salt trapped under a competent seal for horizontal versus tilted sealing conditions when the salt is subject to a buoyancy drive.

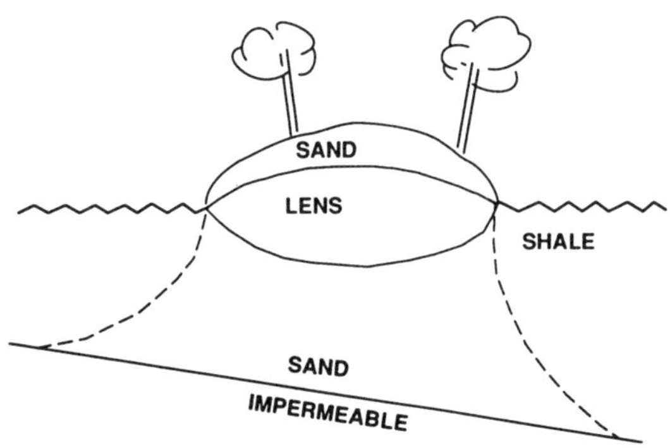

Figure 9.7. Sketch showing the lenslike shape of a fresh water accumulation in a saline ocean under an island subject to rainfall.

lens requires a stable balance of several factors. The first requirement is that solar evaporation and plant osmosis not exceed the rate of rainfall supply on a sustained basis, or else the phreatic lens will gradually dissipate. The second is that artesian well pumping also not exceed the recharge to the phreatic lens in order that diminution of the lens, and saline water in-filling replacement of the withdrawn fresh water, not compromise the integrity of the lens. The third, and most important for the purposes of this book, is that the effect of salinity variations, temperature variations, tides, and sustained water currents not be so large as to sweep the phreatic lens from its location under the sandy island. A fourth effect is that if the phreatic lens extends through the sandy island to the nearest impermeable lower-lying stratigraphic unit, then the shape and positioning of the phreatic lens may be significantly impacted by the dip variation of the substratum because of the differential buoyancy between the fresh water and the saline surroundings as sketched in Fig. 9.8a–d.

The effects of temperature and salinity variations are easily handled within the framework previously developed. The differential density, $\Delta\rho$, between fresh water and saline water can then be expressed as

$$\Delta\rho = \rho_{\text{saline}} - \rho_{\text{fresh}}$$

with

$$\rho_{\text{saline}} = \rho_{\text{fresh}}(1 + \alpha S)[1 - \beta(T - 4)] \cong \rho_{\text{fresh}}\,[1 + \alpha S - \beta(T - 4)]$$

where α is the salinity coefficient relating water density to salt absorbed [written as a fraction of the salinity, S (in units of per mill)], and β is the relative coefficient of thermal expansion for saline water versus fresh water. Then

$$\Delta\rho = \rho_{\text{fresh}}[\alpha S - \beta(T - 4)]$$

The spatial evolution of the phreatic lens under a steady current and in an ocean with a dipping seafloor, as sketched in Fig. 9.8, then depends on whether the current is directed up dip or down dip. An up-dip current aids the buoyant phreatic lens to move away from the supporting island recharge domain toward the up-dip direction, whereas a down-dip current tends to move the phreatic lens in the down-dip direction away from the neighborhood of the recharge island. However, if the down-dip current is weak, then the buoyancy drive will maintain the phreatic lens approximately positioned over the recharge island for a considerable time. In addition, the permeability of the recharge island, and of the surrounding

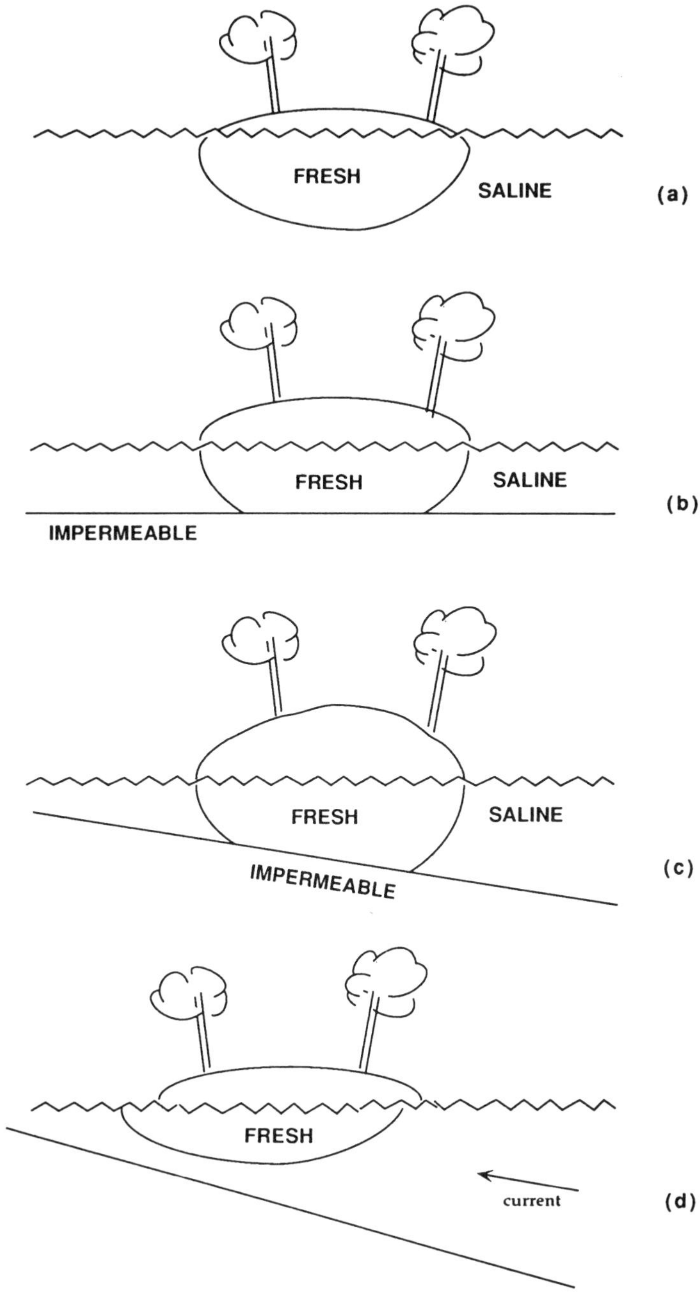

Figure 9.8. Sketch of a phreatic island lens of fresh water in a saline ocean when the lens base is well above any underlying impermeable base (a), when the lens impacts a horizontal base (b), when the lens impacts a tilted base (c), and when an ocean current directed onshore "pushes" the lens of fresh water away from the overlying island (d).

matrix of strata through which the saline waters flow in response to Darcy's law, is unlikely to be homogeneous with respect to position. Thus, the behavior sketched needs to be modified once again to allow for permeability variations along the up-dip and down-dip directions, as idealized in Fig. 9.8. By changing the permeability of the stratigraphic units surrounding the phreatic island lens to accommodate lithologic variations as a consequence of deposition, erosion, diagenesis, carbonate growth, and so on, a variety of case histories can be generalized to describe the general patterns of behavior that can arise.

The case of a light, fresh water stringer floating in a cold, saline ocean presented previously in Fig. 9.4 provides precisely the situation geometry and parameters to illustrate the effect. We refer the reader to that case study and to earlier chapters for an appreciation of the numerical details.

10

Dynamical Aspects of Permeable Flow

10.1. General Considerations

Throughout the preceding chapters, we have been concerned with the positioning and shaping of a stationary hydrocarbon mass under steady-state flow conditions. However, two factors indicate that the total story of hydrocarbon migration is considerably more complex. First, we showed that not all positions in an aquifer would admit of an arbitrary amount of hydrocarbons. Indeed, in some cases there was no steady-state solution available at all, implying that hydrocarbons could not occupy such sites indefinitely. If, therefore, a mass of hydrocarbons was placed at such a nonviable site, the mass would have to move, violating the assumption of a steady state. Second, physical, chemical, and geologic arguments strongly indicate that time-dependent effects all play roles in controlling the flow of hydrocarbons in aquifers.

For instance, at the risk of sounding trite, hydrocarbons in an aquifer must have arrived there from a source rock and must somehow be continuing on to a reservoir site. Thus, the contrapunctal interplay of source generation with time, aquifer transport, and site accumulation argues that time-dependent effects must play a role. Again, the changing solubilities of sulfates, carbonates, and silica as fluid composition, subsurface temperature, pressure, and lithology change with time argue for temporally and spatially varying permeability and porosity in carbonate, dolomite, and sandstone aquifers. Furthermore, the flux of water through an aquifer is unlikely to remain constant for all time, nor is an aquifer likely to remain of fixed geometrical proportions for all time; compaction and tectonic motion will alter the aquifer's characteristics.

In short, while the steady-state situations depicted in earlier chapters provide an excellent set of "snapshots" of the instantaneous response of a hydrocarbon stringer to prevailing conditions, as the conditions change and the hydrocarbons in the aquifer also change on a similar time scale (either in amount by supply and loss, or in type, for example, by oil converting to gas), it is necessary to investigate the long-term evolution patterns of unsteady flows in unsteady aquifers.

Two factors dictate that an accurate description of such flow patterns can be achieved. First, as we have seen in Chapter 1, the response of the buoyant hydrocarbons to prevailing conditions is rapid, so that a hydrocarbon stringer will almost always be close to the instantaneous pressure balance condition that we have used throughout this volume for the steady-state analysis. This rapid evolution, compared with much slower changes in hydrocarbon mass supply and loss and in the variation of the properties of the aquifer with time, leads immediately to the accurate adiabatic approximation developed extensively by Morse and Feshbach (1953). The essence of the approximation is to allow all parameters, which would be constant in a steady-state analysis, to be slowly varying with time. In turn, these slow variations cause slow changes in the positioning and shaping of a hydrocarbon accumulation that is otherwise in equilibrium with prevailing pressure forces.

The second factor is the use of the DuPuit approximation (Turcotte and Schubert, 1982), in which variations of parameters along the length of the aquifer are regarded as fundamentally more important than variations across the aquifer. The lateral variations are then picked up by global conservation arguments across a given cross section, as we have seen in Chapter 3.

These two factors can be gainfully intertwined to develop a quantitative equation describing the time- and space-dependent evolution of a variable hydrocarbon mass in an evolving aquifer.

Consider a small section of an aquifer as sketched in Fig. 10.1. At a time t, let the aquifer width and angle at position ℓ be $H(\ell, t$ and $\theta(\ell, t)$, respectively, and let the water flux across the aquifer be $Q(t)$. Here, ℓ is measured along the aquifer from the upper end of the hydrocarbon stringer. The width of the stringer is $h(\ell, t)$. Instantaneous motion, of velocity $V_H(\ell, t)$, of the hydrocarbon accumulation at position ℓ is represented by Darcy's law,

$$V_H(\ell, t) = -\kappa_H \mu_H^{-1} \frac{\partial p}{\partial \ell} \tag{10.1}$$

where κ_H is the permeability of the hydrocarbons in the aquifer, μ_H is their viscosity, and p is the total pressure acting on the hydrocarbons.

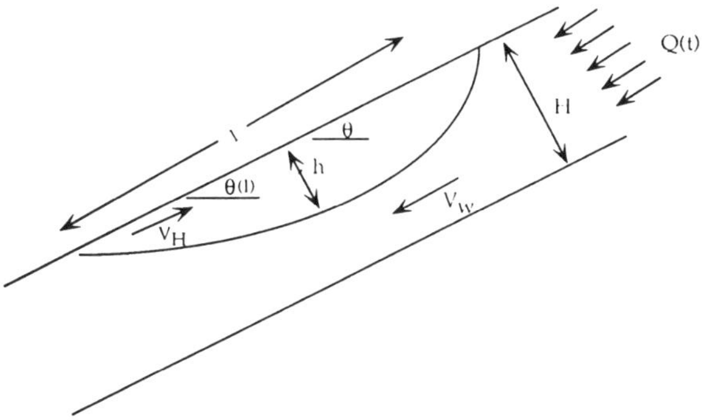

Figure 10.1. Sketch illustrating the supply upward and motion upward (at a speed V_H) to the hydrocarbon stringer against the downward drive of the water flux.

That pressure is made up of four components:

1. p_{buoyancy}: the buoyancy pressure of the hydrocarbons
2. p_{ST}: the surface tension pressure between the hydrocarbons and the prevailing fluid
3. p_c: the capillary pressure
4. p_{flow}: the fluid hydrodynamic pressure pushing on the surface of the hydrocarbon shape

At location ℓ the pressure gradient along the aquifer acting on the hydrocarbon shape is

$$\frac{\partial p_{\text{buoyancy}}}{\partial \ell} = \frac{\partial p}{\partial \ell} + 2T\frac{\partial}{\partial \ell}(h'') - \frac{\partial p_{\text{flow}}}{\partial \ell} + p_c\frac{\partial^2 h}{\partial \ell^2} \tag{10.2}$$

The gradient in buoyancy pressure at location ℓ at time t is

$$\frac{\partial p_{\text{buoyancy}}}{\partial \ell} = g\Delta\rho\frac{\partial h}{\partial \ell}\cos\theta(\ell, t) + g\Delta\rho\sin\theta(\ell, t) \tag{10.3a}$$

with $\Delta\rho = \rho_H - \rho_w$, where ρ_H is the hydrocarbon density, and ρ_w is the water density. The second term on the right-hand side of Eq. (10.2) is the surface tension gradient, on the DuPuit approximation being made, with $h'' = \partial^2 h(\ell, t)/\partial\ell^2$, and where the surface tension T can vary with ℓ and t. The third term, the flow pressure gradient, is given by (Chapter 3)

$$\frac{\partial p_{\text{flow}}}{\partial \ell} = \frac{\mu_w Q(t)}{\kappa_w[H(\ell, t) - h(\ell, t)]} \tag{10.3b}$$

with κ_w and μ_w the water permeability and viscosity, respectively, and $Q(t)$ is the flux of water through the aquifer. The fourth term is the capillary pressure term, $p_c(\partial^2 h/\partial \ell^2)$, where p_c can also vary with position ℓ and time t in the aquifer.

If the hydrocarbon accumulation is taken to be stationary ($V_H = 0$), then

$$\frac{\partial p}{\partial \ell} = 0 \tag{10.4}$$

which condition is the one used throughout the previous chapters.

However, as any of Q, H, p_c, T, μ_H, κ_H, μ_w, κ_w, θ, and $\Delta\rho$ slowly change with time, or as hydrocarbons are added to or subtracted from the accumulation, or as cementation/dissolution changes the porosity of the aquifer, then $\partial p/\partial \ell$ will not vanish. The hydrocarbons will then move until they are again in pressure balance.

Within the DuPuit approximation, the conservation of hydrocarbons in the accumulation, plus the supply/loss of hydrocarbons to the aquifer, is described by the mass conservation equation (Turcotte and Schubert, 1982)

$$\frac{\partial(\phi h \rho_H)}{\partial t} + \frac{\partial}{\partial \ell}(V_H h \rho_H) = S(\ell, t) - L(\ell, t) \tag{10.5}$$

where $S(\ell, t)$ and $L(\ell, t)$ are the rates of supply and loss (in grams per square centimeter per second) of hydrocarbons at position ℓ at time t; and $\phi(\ell, t)$ is the hydrocarbon-saturated porosity in the accumulation (for partial saturation, ϕ is replaced by ϕS_H, where S_H is the hydrocarbon saturation fraction).

Combining Eqs. (10.1) and (10.5), we can write

$$\frac{\partial}{\partial t}(\phi h \rho_H) = \frac{\partial}{\partial \ell}\left(\rho_H \frac{\kappa_H}{\mu_H}\frac{\partial p}{\partial \ell}\right) + \Delta M(\ell, t) \tag{10.6}$$

with $\Delta M(\ell, t)$ [$\equiv S(\ell, t) - L(\ell, t)$] as the net rate of increase in mass of hydrocarbons.

Use of Eq. (10.2) for the pressure gradient, $\partial p/\partial \ell$, in Eq. (10.6) then yields an equation describing the evolution of hydrocarbon shape once behaviors are prescribed for Q, κ_H, μ_H, κ_w, μ_w, $H(\ell, t)$, $\Delta M(\ell, t)$, $\Delta\rho$, $\theta(\ell, t)$, $T(\ell, t)$, p_c, and $\theta(\ell, t)$ as functions of position ℓ and time t.

The parameters and functional behaviors to be specified can be split into two major groups: Q, θ, H, and ϕ, which refer to properties of the aquifer, and κ_H, μ_H, κ_w, μ_w, ΔM, $\Delta\rho$, T, and p_c, which refer to properties of the fluids in the aquifer.

Thus, for a given set of fluid properties, the aquifer conditions can be changed to elucidate their impact on hydrocarbons; alternatively, for a given set of aquifer properties, the fluid masses and fluid properties can be varied to examine the different effects of oil, gas, temperature, and so on, on the shaping, positioning, and dynamical evolution. More generally, both the aquifer and fluid properties can be jointly interactive in a dynamical sense, depending upon the particular situation to be investigated. An exhaustive analysis of such situations is quite complex and involved and, in any event, is well beyond the scope of this book. However, as a guide to those who wish to pursue the time-dependent problem further, it seems appropriate to present here models for several processes that can be of importance.

10.2. A Kinematic Model of Cementation/Dissolution

The hydrodynamic water flow through the aquifer will not be fresh water in general. The solubilities of minerals dissolved in the water will ensure that concentrations of minerals will vary depending on prevailing conditions. For instance, the solubilities of calcium and magnesium carbonates and sulfates *decrease* with *increasing* temperature (Barnes, 1979). Thus, water initially saturated in carbonates flowing through an aquifer from shallow to deep positions will tend to precipitate carbonates. If the structure of the aquifer is synclinal, as sketched in Fig. 10.2, then carbonate precipitation will occur on the up-dip side of the aquifer, with carbonate rock dissolution on the down-dip side. Thus, the porosity and permeability will both decrease with time toward the up-dip side of the synclinal base, whereas both will *increase* with time to the down-dip side.

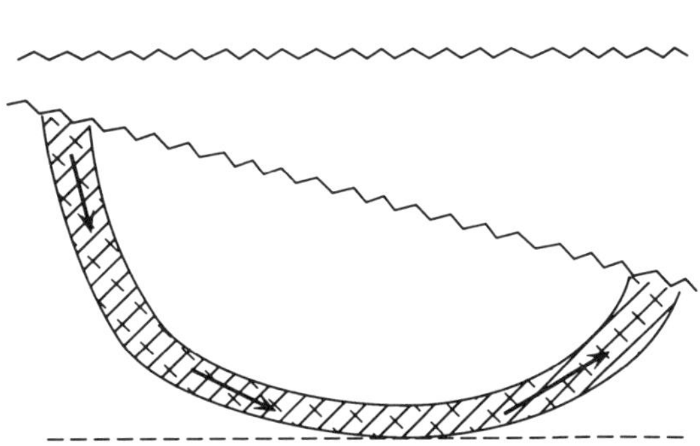

Figure 10.2. Sketch of meteoric or marine recharge through an aquifer causing flushing.

For an anticline, sourced by deep hot waters, the converse is true. The flow will dissolve carbonate rocks on its rise, because of the lower temperature near the anticlinal top, and will deposit carbonates on its descent.

For silica precipitation in sandstone reservoirs and aquifers, the solubility of silica *decreases* with *decreasing* temperature (Barnes, 1979). A quantitative procedure for handling this problem has already been provided by Leder and Park (1986) for silica overgrowth, and by Garrels and Christ (1965) and Barnes (1979) for carbonates and sulfates. We refer the interested reader to the above references for an appreciation of the detailed technical development.

10.3. Leakage and Fracturing

Apart from changes in permeability along an aquifer as a consequence of either lithology shifts (stratigraphic trapping) or cementation/dissolution effects (diagenetic trapping), there is also the possibility of permeability increases being caused by excess fluid or buoyancy pressure. It is an empirical fact that sedimentary rocks fracture when the *total* fluid pressure gradient (hydrostatic plus excess) acting on the rock is around 0.7–0.85 of the lithologic overburden pressure gradient (Dutta, 1987; Lerche, 1990a). Accordingly, as hydrocarbons push on the aquifer walls, they provide a buoyancy pressure, usually acting on the upper boundary of the aquifer. As the mass of hydrocarbons in the aquifer is increased, or as the buoyancy pressure increases (perhaps by making lower specific gravity hydrocarbons), the fluid pressure pushing on the aquifer wall also increases. If this pressure is large enough, it can cause fracturing. In addition, it is possible that, at particular locations along the aquifer, the encompassing rock may already be fractured. In either event, the "fractured" side of the aquifer then provides a leakage domain for hydrocarbon loss from the aquifer.

In that case the dominant buoyancy causes a vertical pressure gradient of $g\Delta\rho\,[\cos\theta(\ell, t)]\,[\partial h(\ell, t)/\partial\ell]$ at position ℓ. The mass flux loss of hydrocarbons through the wall of the aquifer is then

$$F = \int_{\ell_1}^{\ell_2} \rho_H \kappa_H(\text{wall}, t)\mu_H^{-1} g\Delta\rho \cos\theta(\ell, t)\,\frac{\partial h(\ell, t)}{\partial\ell}\,d\ell \qquad (10.7)$$

where ℓ_1 and ℓ_2 mark the beginning and end positions of the fractured domain, and $\kappa_H(\text{wall}, t)$ is the fracture permeability, as sketched in Fig. 10.3. Thus, the mass loss rate $L(\ell, t)$ is given by F, and the evolution

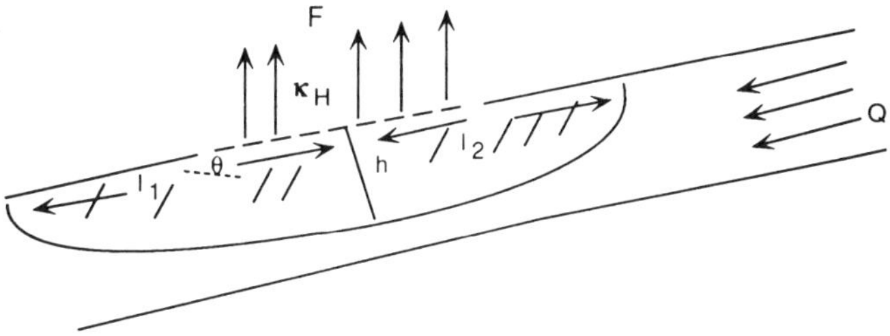

Figure 10.3. Sketch illustrating the effects on hydrocarbons of leakage through the top of an aquifer when there is also a hydrodynamic force acting to drive the hydrocarbon accumulation down dip against buoyancy.

equation (10.5) can then be solved for a given model of the domain of fracturing and wall permeability.

10.4. Oil to Gas Conversion

One of the difficulties that arises in attempts to handle unsteady gas flow versus oil flow in a hydrodynamic aquifer is that whereas oil is virtually incompressible, gas is highly compressible. The equation of state for oil is then approximately that the density, ρ_o, is constant. The equation of state for gas, well above its critical point, is accurately approximated by the perfect gas law

$$p = R\rho_g T \tag{10.8}$$

where R is the gas constant, T is the absolute temperature, ρ_g is the gas density, and p is the gas pressure.

Under the adiabatic approximation of slow changes in conditions, the gas pressure, p, must then be in balance with external pressures exerted on the gas. These external pressures are the hydrodynamic pressure, the buoyancy pressure, the capillary pressure, and the surface tension pressure. Thus, the flow speed of gas along the aquifer is

$$V_g = -\frac{\kappa_g}{\mu_g}\frac{\partial p}{\partial \ell} = -\frac{\kappa_g}{\mu_g} R \frac{\partial}{\partial \ell}(\rho_g T) \tag{10.9}$$

which can be used in Eq. (10.5).

We must also have

$$R\frac{\partial}{\partial \ell}(\phi\rho_g T) = \frac{\partial p_{\text{buoyancy}}}{\partial \ell} + 2\frac{\partial}{\partial \ell}(\tau h'') + p_c\frac{\partial^2 h}{\partial \ell^2} - \frac{\partial p_{\text{fluid}}}{\partial \ell} \tag{10.10}$$

where we have written the surface tension as τ to avoid confusion with absolute temperature, T. However, the buoyancy pressure is dependent on the density contrast, $\rho_g - \rho_w$. Hence, we now have a pair of highly nonlinear coupled equations to solve for both the height, h, of an unsteady gas accumulation and the density of the gas, ρ_g.

The unsteady evolution of a gas accumulation is, therefore, considerably more complex than that of an incompressible oil accumulation. Even more difficulties arise if there is significant conversion of oil to gas in the aquifer because then the shift from one equation of state (ρ = constant) to another ($p = R\rho T$) must be made for that fraction of oil converted to gas. Thus, a multiphasic set of flow equations is needed.

In the steady-state case, none of these problems arise because the steady-state shape ($\partial p/\partial \ell = 0$) is computed for a given density. The equation of state for gas can then be used to compute the temperature variation required to conform to the shape. This degree of freedom is removed in the case where the variation of temperature is specified along the aquifer *a priori*.

10.5. Mass Addition and Solubility Flushing

10.5.1. Mass Addition

In order to place hydrocarbons in an aquifer, they must be generated. If the mass rate of generation of hydrocarbons of density ρ_H is $r(\ell, t)$ g/cm^2 per second at location ℓ at time t in the aquifer, then we have provided an expression for the source term $S(\ell, t)$ in the dynamic equation (10.5). This generation rate term can be obtained either from kinetic models (Tissot and Welte, 1978; Braun and Burnham, 1987) that describe the breakdown of solid kerogen to mobile hydrocarbons or by positing a behavior at, say, the lower "end" of an aquifer where a hydrocarbon source rock is located. In either event, such models of hydrocarbon generation provide precisely the terms needed to describe the accumulating and time-dependent behavior of hydrocarbon stringers in hydrodynamic aquifers.

10.5.2. Solubility Flushing

Apart from hydrocarbon mass loss effects caused by leakage out of an aquifer by physical transport processes, there is also the problem of mass loss from a hydrocarbon accumulation because of solubility. Solubilities of some types of hydrocarbons (e.g., isoalcohols) in water are extremely high whereas those of others (e.g., coal tar) are quite low (parts per million), and solubilities appear to be highly temperature dependent (Hunt, 1979). Consider then the effect of water flux at speed V_w past the hydrocarbon shape, moving at speed V_H. The water is initially devoid of hydrocarbons and so can absorb them up to the solubility limit, S. In passing by the hydrocarbon stringer, the water will then absorb hydrocarbons at the mass rate

$$\frac{dM}{dt} = \rho_H S \left(1 + \left(\frac{\partial h}{\partial \ell}\right)^2\right)^{1/2} (V_w - V_H) \tag{10.11}$$

at location ℓ at time t. The hydrocarbon stringer must lose mass at the same rate, thereby decreasing its overall size. Clearly, the problem of determining the shape of an evolving hydrocarbon accumulation under such conditions is a highly complex Stefan type of problem (Crank, 1987) and is well beyond the scope of the present modest volume.

The fact that we have not investigated this set of time-dependent problems here should not be construed as a statement that we consider them to be unimportant. Indeed, quite the opposite is true. In any attempt to come to grips with the supply, migration, and accumulation of hydrocarbons in the subsurface, the time-dependent nature is likely to be of paramount importance.

What is clearly needed is an investigation, much more extensive than we have given, of the orders of magnitudes of the above influences so that we gain an appreciation of the subsurface conditions under which we can expect a dominant influence to be exerted by each factor. In this way we improve our understanding of the likely sites of hydrocarbon accumulations and decrease the risk factor for obtaining dry holes.

11

The Inverse Oil/Gas Stringer Problem

11.1. General Considerations

Given the shape $h(l)$ of an oil/gas stringer extending in $0 \leqslant l \leqslant L_{max}$ with $h(l) = 0$ on $l = 0$ and again on $l = L_{max}$, the question is: Is it possible to reconstruct the steady-state flow, shape, and position of the aquifer and the permeability along the aquifer and determine the average (constant) density difference between oil/gas and water from information on $h(l)$? This chapter shows how to obtain the required information.

The equation governing hydrocarbon shaping (from Chapter 3 in this volume) can be written in the form

$$2T \frac{d^2h}{dl^2} = \frac{Q}{\mu} \int_0^l \frac{dl'}{\kappa(l')[H(l') - h(l')]} - \Delta\rho[\text{Stuff}(l)] \tag{11.1}$$

where Stuff(l) is the buoyancy factor, dependent on aquifer length, dip angle θ, etc.

The problem of obtaining the desired information splits conveniently into two parts. Note from Eq. (11.1) that *linearity* in the constants T, Q/μ, and $\Delta\rho$ obtains, because if all are doubled, Eq. (11.1) stays the same. Note also from Eq. (11.1) that the response is nonlinear in κ, H, and aquifer geometry.

We can capitalize on this difference as follows. First, suppose that an aquifer shape, width, and dip and a permeability along the aquifer have been assigned (we return in Section 11.3 to this nonlinear part of the problem). Then Eq. (11.1) provides a least-squares procedure for obtaining T, Q/μ, and $\Delta\rho$, as follows.

11.2. Least-Squares Method

Rewrite equation (11.1) in the form

$$Tf(l) - (Q/\mu)I(l) + \Delta\rho[\text{Stuff}(l)] = 0 \tag{11.2}$$

$$\text{where } f(l) = 2d^2h/dl^2, \text{ and} \tag{11.3a}$$

$$I(l) = \int_0^l \frac{dl'}{\kappa(l')[H(l') - h(l')]} \tag{11.3b}$$

Introduce the least-squares control integral (over $0 \leqslant l \leqslant L_{\max}$) given by

$$X^2(T, Q/\mu, \Delta\rho) = \frac{1}{L_{\max}} \int_0^{L_{\max}} \{Tf(l') - \frac{Q}{\mu} I(l') + \Delta\rho[\text{Stuff}(l')]\}^2 \, dl' \tag{11.4}$$

A minimum of X^2 with respect to T, Q/μ, and $\Delta\rho$ occurs when

$$\frac{\partial X^2}{\partial T} = 0 \equiv T \int_0^{L_{\max}} [f(l')]^2 \, dl' - \frac{Q}{\mu} \int_0^{L_{\max}} I(l')f(l') \, dl' + \Delta\rho \int_0^{L_{\max}} \text{Stuff}(l')f(l') \, dl' \tag{11.5a}$$

and

$$\frac{\partial X^2}{\partial (Q/\mu)} = 0 \equiv T \int_0^{L_{\max}} f(l')I(l') \, dl' - \frac{Q}{\mu} \int_0^{L_{\max}} I(l')^2 \, dl' + \Delta\rho \int_0^{L_{\max}} \text{Stuff}(l')I(l') \, dl' \tag{11.5b}$$

and

$$\frac{\partial X^2}{\partial \Delta\rho} = 0 \equiv T \int_0^{L_{\max}} f(l')\text{Stuff}(l') \, dl' - \frac{Q}{\mu} \int_0^{L_{\max}} I(l')\text{Stuff}(l') \, dl' + \Delta\rho \int_0^{L_{\max}} \text{Stuff}(l')^2 \, dl' \tag{11.5c}$$

Introduce matrix elements

$$
\begin{aligned}
a_{11} &= \textstyle\int_0^{L_{\max}} f(l')^2\, dl', \qquad a_{12} = \int_0^{L_{\max}} I(l')f(l')\, dl' \equiv a_{21} \\
a_{13} &= \textstyle\int_0^{L_{\max}} \mathrm{Stuff}(l')f(l')\, dl' \equiv a_{31} \\
a_{22} &= \textstyle\int_0^{L_{\max}} I(l')^2\, dl', \qquad a_{33} = \int_0^{L_{\max}} \mathrm{Stuff}(l')^2\, dl' \\
a_{23} &= \textstyle\int_0^{L_{\max}} \mathrm{Stuff}(l')I(l')\, dl' \equiv a_{32}
\end{aligned}
$$

so that Eqs. (11.5) can be written

$$Ta_{11} - \frac{Q}{\mu} a_{12} + \Delta\rho a_{13} = 0 \tag{11.6a}$$

$$Ta_{12} - \frac{Q}{\mu} a_{22} + \Delta\rho a_{23} = 0 \tag{11.6b}$$

$$Ta_{13} - \frac{Q}{\mu} a_{23} + \Delta\rho a_{33} = 0 \tag{11.6c}$$

Equations (11.6) have a solution if and only if the determinant $(a_{ij}) = 0$, i.e.,

$$\begin{vmatrix} a_{11} & a_{12} & a_{13} \\ a_{12} & a_{22} & a_{23} \\ a_{13} & a_{23} & a_{33} \end{vmatrix} = 0 = \det(a_{ij}) \tag{11.7}$$

which provides a constraint to be used later. When the constraint equation (11.7) is satisfied, it follows that

$$\frac{Q}{\mu} = \Delta\rho(\tau a_{21} + a_{23})/a_{22} \tag{11.8a}$$

with

$$\tau = (a_{12}a_{23} - a_{13}a_{22})/(a_{11}a_{22} - a_{12}{}^2) \tag{11.8b}$$

and

$$T = \Delta\rho\tau \tag{11.8c}$$

When the values given in Eqs. (11.8) are substituted into the control function X^2 (Eq. 11.4), we obtain

$$X^2 = \frac{(\Delta\rho)^2}{L_{\max}} \int_0^{L_{\max}} \left[f(l')\frac{(a_{12}a_{23} - a_{13}a_{22})}{(a_{11}a_{22} - a_{12}^2)} - I(l')\frac{(a_{23}a_{11} - a_{21}a_{13})}{(a_{11}a_{22} - a_{12}^2)} + \mathrm{Stuff}(l') \right]^2 dl' \tag{11.9}$$

For given functional values of $\theta(l)$, $H(l)$, and $\kappa(l)$, Eqs. (11.8) and (11.9) represent a minimum value of X^2. However, the freedom exists to vary θ, H, and κ in such a way as to obtain the minimum of all minima of X^2, thereby producing the best fit aquifer geometry and physical parameters for the aquifer, consistent with knowledge of the hydrocarbon stringer behavior. We now consider this nonlinear aspect of the problem.

11.3. Nonlinear Determination of Aquifer Properties

We note from Eq. (11.9) that the matrix entries a_{ij} depend on $f(l)$, $I(l)$, and Stuff(l), and these three functions depend, in turn, on the behavior of κ, H, θ, etc., that is, on the geometrical and physical aquifer properties. To determine the geometry parameters, we proceed in two stages. First, we assign generalized functional shapes to relevant quantities, and then we examine how to determine the variables in the generalized functions using the control function given by Eq. (11.9) and the constraint condition given by Eq. (11.7).

11.3.1. Generalized Functional Forms

Let the angle θ of the top surface of the aquifer be given through

$$\theta = \theta_0 + \theta_1 \frac{l}{L_{\max}} + \sum_{n=1}^{N} \phi_n \sin\left(\frac{n\pi l}{L_{\max}}\right) \tag{11.10}$$

with $\theta = \theta_0$ on $l = 0$, where $h = 0$, and where θ_0, θ_1, and $\{\phi_n\}$ are parameters to be determined.

Then the top of the aquifer is described by the x,y coordinates

$$y_{\text{top}} = y_0 + \int_0^l \sin\theta(l')\, dl' \tag{11.11a}$$

$$x_{\text{top}} = x_0 + \int_0^l \cos \theta(l')\, dl' \tag{11.11b}$$

where y_0, x_0 are the coordinates on $l = 0$. The bottom of the aquifer is described by

$$y_{\text{bottom}}(l) = y_{\text{top}}(l) - H(l) \cos \theta(l) \tag{11.12a}$$

$$x_{\text{bottom}}(l) = x_{\text{top}}(l) + H(l) \sin \theta(l) \tag{11.12b}$$

The width $H(l)$ of the aquifer must exceed the oil stringer width $h(l)$. So write

$$H(l) = h(l) + \Delta_0 \exp\left[\frac{\Delta_1 l}{L_{\max}} + \sum_{n=1}^{N} b_n \sin\left(\frac{n\pi l}{L_{\max}}\right)\right] \tag{11.13}$$

with $\Delta_0 > 0$ ($H = \Delta_0$ on $l = 0$, where $h = 0$), and where Δ_0, Δ_1, and $\{b_n\}$ are parameters to be determined.

Equally, the permeability (which is positive) may vary along the aquifer. So introduce

$$\kappa(l) = \kappa_0 \exp\left[\frac{\kappa_1 l}{L_{\max}} + \sum_{n=1}^{N} k_n \sin\left(\frac{n\pi l}{L_{\max}}\right)\right] \tag{11.14}$$

with $\kappa_0 > 0$ and $\kappa(l) = \kappa_0$ on $l = 0$, and where κ_0, κ_1, and $\{k_n\}$ are parameters to be determined.

The total set of unknown parameters to be determined are represented by the set

$$\theta_0,\ \theta_1,\ \{\phi_n\};\ \Delta_0,\ \Delta_l,\ \{b_n\};\ \kappa_0,\ \kappa_1,\ \{k_n\}$$

Introduce the parameter vector $\boldsymbol{P}$ which is constructed as

$$\boldsymbol{P} = \{\theta_0, \theta_1, \phi_1, \phi_2 \ldots \phi_{\text{n}}; \Delta_0, \Delta_1, \text{b}_1, \ldots \text{b}_{\text{n}}; \kappa_0, \kappa_1, \ldots \text{k}_{\text{n}}\}$$

Note that Δ_0 has the dimensions of length, and κ_0 has the dimensions of darcies; the remaining unknowns in the parameter vector $\boldsymbol{P}$ are dimensionless.

11.3.2. Determination of the Parameter Vector, *P*

For any given choice of the components of the parameter vector, ***P***, two factors have to be considered.

First, the single constraint condition (Eq. 11.7)

$$\det(a_{ij}) = 0 \tag{11.15}$$

must be satisfied, therefore providing one relation between the parameter vector components. To be noted from the structural functional form of the individual matrix elements a_{ij} is that the product $\kappa_0\Delta_0$ is not an integral component of $I(l)$; i.e.,

$$I(l) \equiv \int_0^{L_{\max}} \frac{dl'}{\kappa(l')[H(l') - h(l')]} \equiv \frac{1}{\kappa_0\Delta_0} \int_0^{L_{\max}} dl'$$

$$\exp\left[-(l'/L_{\max})(\kappa_1 + \Delta_1) - \sum_{n=1}^{N} (k_n + b_n) \sin(n\pi l'/L_{\max})\right] \tag{11.16}$$

$$\equiv (\kappa_0\Delta_0)^{-1}\,\mathcal{F}, \qquad \text{with } \mathcal{F} > 0.$$

Thus, the product $\kappa_0\Delta_0$ is a multiplicative factor for $I(l)$.

It is, therefore, of use to regard the constraint determinant condition as providing a value for the product $\kappa_0\Delta_0$ in terms of all other parameters of the system. Write this constraint result in the form

$$\kappa_0\Delta_0 = F(\psi) \tag{11.17}$$

where ψ is the reduced parameter vector

$$\psi = (\theta_0, \theta_1, \phi_1, \phi_2, \ldots, \phi_n; \Delta_0, \Delta_1, b_1, b_2, \ldots, b_n; \kappa_1, k_1, k_2, \ldots, k_n)$$

Note the absence of κ_0 in ψ, so that we regard κ_0 as a derived parameter from the constraint condition, with Δ_0 being taken as an independent variable.

Then X^2 is a functional of the reduced parameter vector ψ, with all components of ψ being freely available for adjustment (with the isolated restriction $\Delta_0 \geqslant 0$, $\kappa_0 > 0$), i.e., $F(\psi) > 0$.

Direct insertion of the functional forms for H, θ, and κ into Eq. (11.9) enables us to write this equation in the form

$$X^2 = \frac{(\Delta\rho)^2}{L_{\max}}\, C(\psi) > 0 \tag{11.18}$$

The problem then is to obtain a procedure to minimize $C(\psi)$ with respect to the reduced parameter vector ψ so that the best solution parameters are obtained.

A procedure for obtaining such values is as follows. Choose an arbitrary value, $\psi(0)$. Then construct the nth nonlinear iteration for the jth component of ψ from

$$\psi_j(n + 1) = \psi_j(n) \exp\left(- \alpha_j \frac{\partial C[\psi(n)]}{\partial \psi_j(n)} \right) \tag{11.19}$$

with

$$\alpha_j = \left| \frac{\partial C[\psi(0)]}{\partial \psi_j(0)} \right|^{-1} \ln\left(1 + \frac{1}{R\psi_j(0)} \right)$$

where R is the number of times the nonlinear iteration is to be undertaken.

The form (11.19) of the iteration scheme guarantees that $C(\psi)$ will always head toward a minimum in the multidimensional space of the reduced parameter vector ψ.

In the event that a restricted search is required over a finite parameter domain (possibly because there is sufficient information available to provide constraints), such a restrictive search procedure can be handled generally. Let each ψ_j be bounded by

$$\psi_j(\min) < \psi_j < \psi_j(\max)$$

Introduce u_j through

$$\psi_j = [\psi_j(\max) - \psi_j(\min)]\, u_j \tag{11.20}$$

so that $0 \leqslant u_j \leqslant 1$.

Then regard the reduced vector $\mathbf{u}$ as a basic quantity with the vector ψ being given parametrically through Eq. (11.20).

In place of Eq. (11.19) we can then write

$$u_j(n + 1) = u_j(n) \exp\left(- \beta_j \frac{\partial C[\mathbf{u}(n)]}{\partial u_j(n)} \right) \tag{11.21}$$

where

$$\beta_j = \left|\frac{\partial C[\mathbf{u}(0)]}{\partial u_j(0)}\right|^{-1} \ln\left(1 + \frac{1}{Ru_j(0)}\right) \tag{11.22}$$

An initial estimate, in the absence of further knowledge, might be $u_j(0) = \frac{1}{2}$, representing a midpoint estimate within the declared search range.

Once the iteration procedure has been carried through to completion, then X^2 is at a minimum. The final parameter vector value, ψ, can then be used to construct best estimates of $y_{\text{top}}(l)$, $x_{\text{top}}(l)$, $H(l)$, $\kappa(l)$, and $\sin\theta(l)$. In turn, the best possible relative values of $T/\Delta\rho$ and $Q/(\mu\Delta\rho)$ can then be written down, and the inverse problem is essentially complete. Problems of resolution, uniqueness, precision, and sensitivity can be examined with the use of conventional methods (Lerche, 1990a,b).

12

Migration and Diffusion of Hydrocarbons in Complex Stratigraphic Sequences

In this chapter a mean-field method is used to describe the variations in both average and fluctuating flow properties for water and oil in geometrical configurations of juxtapositional lithologies. The mean vertical flow properties (speed, permeability, and pressure gradient), as well as the root mean square lateral variations in diffusive flow as a consequence of spatial variability in lithologic units, will be shown to be related directly to the autocorrelation tensor of permeability fluctuations. Comparison of oil and water flow characteristics provides an indication of the relative strengths of capillarity and buoyancy in controlling the mean and fluctuating components of oil flow in relation to water flow. Effects due to both sandiness variations and water wetness fluctuations are also incorporated, as is the effect of decreasing porosity with increasing subsurface burial. Numerical estimates suggest that complex stratigraphic sequences have a major impact on both mean and fluctuating components of flow for sandiness percentages between about 20% and 80%. Outside of this regime the sequence is sufficiently shalelike, for sandiness less than 20%, or sandlike, for sandiness greater than 80%, that the dominant behavior is controlled by the end-member situations of shale and sand, respectively.

In general, the extra effects of vertical buoyancy and capillarity in oil flow tend to make the lateral diffusion effects less than for water flow, although these extra effects can be overridden in exceptional circumstances by the correlated behavior (lateral or vertical) of lithologic variations.

12.1. Rationale

After leaving the source bed, the flow of hydrocarbons in the subsurface is described as secondary migration. Two dominant forces act to control the migration speed and direction. The first is the effect of any fluid pressure, ∇p, in excess of hydrostatic, and the second is buoyancy as a consequence of the lower density of hydrocarbons relative to that of water. In the context of Darcy's law, which is the appropriate vehicle (Hubbert, 1953) for describing flow in the subsurface, we can write the hydrocarbon vector flow velocity, $\mathbf{v}$, as

$$\mathbf{v} = -\frac{\kappa}{\mu}(\nabla p + g\Delta\rho\hat{z} + \nabla p_c) \tag{12.1}$$

where κ is the formational permeability to hydrocarbon flow, μ is the oil kinematic viscosity, ρ_o is hydrocarbon density, g is acceleration due to gravity, $\Delta\rho$ is the difference in density between water, ρ_w, and hydrocarbons ($\Delta\rho = \rho_w - \rho_o$), with the vertical direction z measured upward toward the sediment mudline. Also, p_c measures the capillary pressure acting on the hydrocarbons at a given location, influencing upward mobility of the hydrocarbons (Hubbert, 1953) as a consequence of juxtaposition of less permeable, lithologic layers overriding more permeable layers.

Three major effects confound the deterministic problem of attempting to assess the flow of hydrocarbons from source rock to reservoir accumulation sites. First, even if the geometrical configuration of stratigraphic sequences and the dynamical states (permeability, overpressure) of the rocks do not alter with time or spatial location, it is certain that the source supply of hydrocarbons varies both with time and with spatial location. Second, even if the source supply were steady and the geometrical configuration of stratigraphic sequences remained unaltered, nevertheless both the permeability and the fluid pressure surely change with time and spatial position (Lerche, 1990a,b). Third, even if the source supply were steady and the dynamical state of the rocks did not alter, nevertheless the effects of later sedimentation, of erosion, and of tectonic forces will alter the geometrical configuration of stratigraphic sequences. When taken together, these complex and interacting behaviors suggest that it is unlikely that an absolutely deterministic model of hydrocarbon flow in the subsurface can be constructed for some considerable time.

In addition, the identification of a particular lithologic unit as, say, sand does not preclude the sand containing shale stringers, which may themselves vary spatially in number, thickness, and intrinsic physical properties. Thus, the broad recognition of basic types of lithologic units is compromised by the fine-scale detail of each such unit. Even on some broad scale, lithologic units vary, because sedimentary formations pinch out, thin, are being deposited discontinuously as isolated bodies, are laid down in low-, medium-, and high-energy environments, and so on.

It is unlikely that finite sampling by drill hole, or coarse-grained seismic resolution, will ever provide sufficiently accurate information for a precise determination of the evolving dynamical milieu through which generated hydrocarbons pass on their way to accumulation sites. This problem is made even more acute by the fact that models of basin evolution, even when consistent with present-day data, nevertheless have problems of resolution due to intrinsic assumptions, parameters, and uncertainties in the data that go into such models (Yukler and Kokesh, 1984; Cao 1987).

For all of the above reasons, it seems appropriate to replace the precise, but inaccurate, deterministic modeling of hydrocarbon migration with a statistical treatment in which the uncertain nature of the spatial and dynamical factors is incorporated. In this way it should, at the least, be possible to assess the likelihood of mean flow directions and variations around the mean (diffusion) in terms of statistically average properties of the subsurface.

Success of such an endeavor carries with it the possibility of allowing, in an average sense, for variable emission of hydrocarbons from a source rock, depending on both spatial location and time, so that a composite picture can be built of the general pattern of evolution of hydrocarbon pathways and accumulation sites in relation to the ongoing dynamical and geometrical evolution of a sedimentary basin.

The vehicle that we have chosen to categorize the statistical behavior is the mean-field method (Frisch, 1968), which has been used with considerable success in a wide range of disciplines covering a wide range of problems. Effectively, the method is based on the argument that the complexities are such that the forces involved are best described by mean values, which may vary deterministically with time and space, and fluctuations (spatial or temporal) around the mean values, which are best handled as random variables.

The purpose of this chapter is to attempt to account quantitatively for the gross behavior of hydrocarbon transport and to discuss the diffusion characteristics of the motion.

The majority of stochastic models for subsurface flow fall into two categories: (1) Monte Carlo simulations, which, although quite powerful, consume much computer time and do not yield a general theory that can be used to provide insight into long-standing problems or suggest new lines of approach to new problems; and (2) stochastic differential equations and analytic solution by small-perturbation approximations. Most of the studies in the latter category, with few exceptions, have been restricted to one-dimensional, steady-state fluid flow with the logarithm of permeability regarded as randomly varying.

Our approach here is to follow the second line of attack but not to restrict calculations to the one-dimensional situation, and (a) to relate bulk flow to mean properties of the system and (b) to define, and then relate, an effective diffusion coefficient for the disordered component of the hydrocarbon flow (which is randomly plus or minus and so has no mean value) to root mean square lateral flow and thus to mean properties of the system. As we shall see, this lateral randomly fluctuating component of flow can be dominant.

12.2. Basic Results

The technical development of the statistical properties of the flow, both for the mean and for the diffusive components, is quite involved. For that reason, and also to maintain the thread of continuity in the body of the chapter, we have relegated the main details of the calculations to an appendix at the end of this chapter. Here we provide a précis of the main results that will be used later in the chapter.

Two main classes of behaviors are available: (1) mean properties associated with flow, which would persist in the absence of any spatial fluctuations in the medium, but which are modified by the spatial fluctuations; and (2) properties which owe their existence to the fact that the medium does have spatial fluctuations. The latter class includes such properties as diffusion of flow around the mean direction and pressure fluctuations as a consequence of juxtaposition of spatially fluctuating permeability associated with abrupt shifts in lithology.

To provide a convenient framework, the mean properties are compared to the corresponding values that would obtain in a medium with the same average value for log permeability but without any spatially fluctuating components. In this way a relatively simple standard of comparison is provided against which changes produced by fluctuations in permeability are maximally contrasted.

12.2.1. Nonfluctuating Medium Results

12.2.1.1. Water Flow

In a transversely isotropic medium, in which the average water permeability $\kappa_w(z)$ is a smoothly continuous function solely of depth, z, measured from the sediment–water interface, the steady-state flow of water through the sediments is vertically directed and described through a mean *excess fluid* pressure gradient, $f_*(z)$, which varies as

$$f_*(z) = f_*(0)[k_w(0)/k_w(z)] \tag{12.2}$$

where $k_w(z)$ is the hydraulic conductivity of water, given by

$$k_w(z) = \kappa_w(z)/\rho_w\mu_w \tag{12.3}$$

where ρ_w and μ_w are, respectively, the density and the viscosity of water.

The vertical flow speed V_{w*} of water is then given by

$$V_{w*} = -k_w(z)f_*(z) = -k_w(0)f_*(0) = \text{constant} \tag{12.4}$$

12.2.1.2. Oil Flow

Under the same conditions of medium properties as for water flow, the steady-state oil flow in a transversely isotropic medium is described through a mean *total* driving vertical pressure gradient, $G_*(z)$, which varies as

$$G_*(z) = G_*(0)[k_o(z = 0)/k_o(z)] \tag{12.5}$$

where $k_o(z)$ is the hydraulic conductivity of oil, related to the smoothly continuous oil permeability, $\kappa_o(z)$, through

$$k_o(z) = \kappa_o(z)/\rho_o\mu_o \tag{12.6}$$

where ρ_o and μ_o are, respectively, the density and the viscosity of oil.

The vertical flow speed, V_{o*}, of the oil is then given by

$$V_{o*} = -k_o(z)G_*(z) = -k_o(0)G_*(0) = \text{constant} \tag{12.7}$$

The total driving pressure gradient on the oil is made up of components due to excess fluid pressure, to buoyancy, and to capillary pressure so that we can write

$$G_*(z) = f_*(z) + \Delta\rho g + p_c'(z) \tag{12.8}$$

where $\Delta\rho = \rho_w - \rho_o$, and $p_c'(z)$ is the mean vertical capillary pressure gradient in the smoothly continuous medium.

12.2.2. Fluctuating Medium Results

12.2.2.1. Water Flow

In the presence of fluctuations in permeability in the medium, with the same average vertically dependent permeability as in the nonfluctuating case, the water flow will now seek out, to the extent it can, those pathways for flow that are continuously connected through the highest permeability regions, or, if there is no continuously connected path, the pathways will be across the regions of least low permeability. Thus, both the mean vertical excess fluid pressure gradient and the mean vertical flow speed will be influenced by the spatial fluctuations in permeability. In addition, because of the spatial fluctuations in permeability, the flow pathways will no longer be exactly vertical so that there will be fluctuations in velocity around the mean flow speed, and the fluctuations will have both transverse (to z) and parallel (to z) components. The excess fluid pressure will also no longer be solely vertically directed, but will have fluctuations around the vertically directed mean excess fluid pressure gradient (which is itself modified from the nonfluctuating medium value by the presence of the permeability fluctuations) that are related to the spatial fluctuations in permeability. Consider each in turn.

a. Mean Excess Fluid Pressure Gradient. From the appendix at the end of this chapter (Eq. A50), the mean vertical excess fluid pressure gradient, $f(z)$, is given by

$$f(z) = f_*(z)[f_*(z)/f_*(0)]^{-n} \tag{12.9}$$

where

$$n = \delta_w^2 \int d^3k\, k_z^2\, k^{-2} r(\mathbf{k}) \geqslant 0 \tag{12.10}$$

with $r(\mathbf{k})$ the normalized power spectrum of spatial fluctuations in $\ln k_w(\mathbf{x})$, the logarithmic variation in permeability. The normalization is

$$\int d^3\mathbf{k}\ r(\mathbf{k}) = 1 \tag{12.11}$$

Here $\delta_w^2 = \langle(\delta \ln k_w(\mathbf{x}))^2\rangle$ is the variance in the fluctuations in $\ln k_w(\mathbf{x})$.

Note that relative to an increasing (with depth) vertical excess pressure gradient, $f_*(z)$, in the *absence* of spatial fluctuations, the vertical excess pressure gradient in the *presence* of fluctuations will be increasing (decreasing) with increasing depth according as $1 - n > 0$ $(1 - n < 0)$.

b. Mean Vertical Water Flow. From the appendix at the end of this chapter (Eq. A53), the mean vertical water flow, V_w, in the *presence* of spatial fluctuations in the permeability is given by

$$V_w = V_{w*}[f_*(z)/f_*(0)]^{-n}\left(1 + n + \frac{1}{2}\delta_w^2\right) \tag{12.12}$$

so that the water flow is increased at the sediment–water interface by the amount $1 + n + \frac{1}{2}\delta_w^2$ relative to the mean flow, V_{w*}, obtaining in the absence of fluctuations, but, as depth increases, the flow speed V_w eventually drops below the value in the the nonfluctuating medium for an increasing excess fluid pressure gradient, $f_*(z)$.

c. Excess Fluid Pressure Gradient Fluctuations. The spatial fluctuations, $\delta\mathbf{f}$, in excess fluid pressure *gradient* are characterized by the variance matrix $\langle\delta f_i(\mathbf{x})\ \delta f_j(\mathbf{x})\rangle$ given by (Eq. A54)

$$\langle\delta f_i(\mathbf{x})\ \delta f_j(\mathbf{x})\rangle = f(z)^2\delta_w^2\ q_{ij} \tag{12.13}$$

where the angular, dimensionless, correlation matrix is given through

$$q_{ij} = \int d^3k\ k_z^2 k_i k_j k^{-4} r(\mathbf{k}) \tag{12.14}$$

The spatial extent of the excess pressure fluctuations in different directions (i.e., the "compartmentalization" of pressure gradient fluctuations) is measured by the ratio of the variance of pressure fluctuations to that of pressure gradient fluctuations. An equivalent correlation length matrix,

l_{ij}, for pressure compartments is then given through

$$l_{ij} = \left(\frac{|q_{ij}|}{\int d^3k\ k_z^2 k^{-4} r(\mathbf{k})} \right)^{1/2} \tag{12.15}$$

The variations in excess fluid pressure δp are characterized by the variance

$$\langle \delta p(\mathbf{x}) \delta p(\mathbf{x}) \rangle = f(z)^2\ \delta_w^2 \int d^3k\ k_z^2\ k^{-4} r(\mathbf{k}) \tag{12.16}$$

d. Water Velocity Fluctuations. The variance matrix in water flow speed, $\delta \mathbf{w}$, describes the disordered components of flow as a consequence of water flow no longer being rigorously vertical. From the Appendix (Eqs. A60c and A60d), we have the transverse (to z) components characterized by

$$\langle \delta w_i(\mathbf{x})\ \delta w_j(\mathbf{x}) \rangle = V_{w*}^2\ \delta_w^2\ \exp(\delta_w^2)[k_w(z)/k_w(0)]^{n+1} q_{ij} \qquad \text{for } i \neq z \neq j \tag{12.17}$$

while for the vertical component we have

$$\langle \delta w_z(\mathbf{x})\ \delta w_z(\mathbf{x}) \rangle = V_{w*}^2\ \delta_w^2\ \exp(\delta_w^2)[k_w(z)/k_w(0)]^{n+1}(q_{xx} + q_{yy}) \tag{12.18}$$

Note that for a decreasing mean permeability $\kappa_w(z)$ with increasing depth, the fluctuations in flow velocity decrease.

12.2.2.2. Oil Flow

As shown in the appendix at the end of this chapter, the development of the mean and fluctuating components of physical flow properties of oil closely parallels the development for water. The main difference in results arises from two causes. First, the permeability of lithofacies to oil penetration differs from that for water. Second, the oil flow feels the impact not only of the excess fluid pressure gradient, but also of differential buoyancy and of capillary pressure. The combination of the effects is to provide a total pressure gradient acting on the oil and to make for a more complex development than in the case of water flow.

a. Mean Total Pressure Gradient. From Eq. (A39), we have a vertical total mean pressure gradient, $G(z)$, given through

$$G(z) = G_*(z)[G_*(z)/G_*(0)]^{-m} \tag{12.19}$$

where

$$m = \delta_o^2 \int d^3k\, k_z^2\, k^{-2} R(\mathbf{k}) \tag{12.20}$$

with $R(\mathbf{k})$ the normalized power spectrum of spatial fluctuations in ln $k_o(\mathbf{x})$, the logarithmic variation of permeability. The normalization is

$$\int d^3\mathbf{k}\, R(\mathbf{k}) = 1 \tag{12.21}$$

Here $\delta_o^2 = \langle \delta \ln k_o(\mathbf{x}))^2 \rangle$ is the variance in the fluctuations in ln $k_o(\mathbf{x})$.

Note that relative to an increasing vertical total pressure gradient, $G_*(z)$, with increasing depth in the *absence* of spatial fluctuations, the vertical total pressure gradient in the *presence* of fluctuations will increase (decrease) with increasing depth according as $1 - m > 0$ $(1 - m < 0)$.

b. Mean Vertical Oil Flow. From Eq. (A42), we have the mean vertical oil velocity, V_o, in the presence of fluctuations in the permeability given by

$$V_o = V_{o*} [G_*(z)/G_*(0)]^{-m} \left(1 + m + \frac{1}{2}\delta_o^2\right) \tag{12.22}$$

so that the oil flow is increased at the sediment–water interface by the factor $1 + m + \frac{1}{2}\delta_o^2$ relative to the value for the nonfluctuating medium but decreases with increasing depth to a value below that in the nonfluctuating case provided the total pressure gradient $G_*(z)$ increases with increasing depth.

c. Total Pressure Fluctuations. The spatial fluctuations, δG, in *total* pressure gradient are characterized by the variance matrix $\langle \delta G_i(\mathbf{x})\, \delta G_j(\mathbf{x}) \rangle$ given by (Eq. A44)

$$\langle \delta G_i(\mathbf{x})\, \delta G_j(\mathbf{x}) \rangle = G(z)^2 \delta_o^2 s_{ij} \tag{12.23}$$

where the angular, dimensionless correlation matrix is given through

$$s_{ij} = \int d^3k\, k_z^2\, k^{-4} k_i k_j R(\mathbf{k}) \tag{12.24}$$

The spatial extent of the total pressure fluctuations in different directions is characterized by a correlation length matrix, L_{ij}, given through

$$L_{ij} = \left(|s_{ij}| \Big/ \int d^3k\, k_z^2\, k^{-4} R(\mathbf{k}) \right)^{-1/2} \tag{12.25}$$

The variations in total pressure, δM, are characterized by the variance

$$\langle \delta M(\mathbf{x})\, \delta M(\mathbf{x}) \rangle = G(z)^2 \int d^3k\, k_z^2\, k^{-4} R(\mathbf{k}) \tag{12.26}$$

d. Oil Velocity Fluctuations. The variance matrix characterizing the fluctuations, $\delta\mathbf{v}$, in the oil flow is described by

$$\langle \delta v_i(\mathbf{x})\, \delta v_j(\mathbf{x}) \rangle = V_{o*}^2 \delta_o^2 \exp(\delta_o^2) [k_o(z)/k_o(0)]^{m+1}\, s_{ij} \tag{12.27}$$

for transverse (to z) fluctuations ($i \neq z \neq j$), while for fluctuations parallel to the mean flow of oil we have

$$\langle \delta v_z(\mathbf{x})\, \delta v_z(\mathbf{x}) \rangle = V_{o*}^2 \delta_o^2 \exp(\delta_o^2) [k_o(z)/k_o(0)]^{m+1} (s_{xx} + s_{yy}) \tag{12.28}$$

Note that for an oil permeability decreasing with increasing depth, it follows that the fluctuations in oil flow velocity also decrease.

12.2.3. Comparisons of Mean and Fluctuating Quantities

Of interest is the question of whether fluctuations are large or small compared to mean values, so that an assessment can be made of the conditions under which it is necessary to pay attention to fluctuations as they would then seriously modify otherwise preconceived ideas of flow patterns. Three comparisons are particularly relevant: (i) comparison of mean values in the presence of fluctuations with the corresponding values in the absence of fluctuations; (ii) comparison of fluctuating components with mean values to determine the relative level of impact of fluctuations; and (iii) intercomparison of oil and water mean and fluctuating components, so that the relative impact of fluctuations in more dominantly altering one phase of the flow relative to another can be evaluated.

12.2.3.1. Mean Quantities

a. Excess Fluid Pressure and Mean Water Flow Speed. From Eqs. (12.9) and (12.12) we can write the ratio of excess fluid pressure and mean water flow speed relative to the corresponding values obtaining in the absence of fluctuations in terms of the mean permeability of the nonfluctuating medium. We obtain:

$$f(z)/f_*(z) = [k_w(z)/k_w(0)]^n \tag{12.29a}$$

and

$$V_w(z)/V_{w*} = [k_w(z)/k_w(0)]^n \left(1 + n + \frac{1}{2}\delta_w^2\right) \tag{12.29b}$$

for the mean excess pressure gradient and mean water flow speed, respectively.

We also have V_{w*} = constant, and

$$f_*(z) = f_*(0)[k_w(0)/k_w(z)] \tag{12.29c}$$

b. Total Mean Pressure and Mean Oil Flow Speed. From Eqs. (12.19) and (12.22) we can write the ratios

$$G(z)/G_*(z) = [k_o(z)/k_o(0)]^m \tag{12.30a}$$

and

$$V_o(z)/V_{o*} = [k_o(z)/k_o(0)]^m \left(1 + m + \frac{1}{2}\delta_o^2\right) \tag{12.30b}$$

We also have V_{o*} = constant, and

$$G_*(z) = G_*(0)[k_o(0)/k_o(z)] \tag{12.30c}$$

12.2.3.2. Fluctuating and Mean Quantities

The fluctuations in pressure gradient and flow speed for both water and oil are related not only to the magnitude of the fluctuations in permeability for each phase but also to the statistical directions of correlated behavior.

For water flow we have the relative strength measures

$$\langle \delta f_i \, \delta f_j \rangle / f(z)^2 = \delta_w^2 \, q_{ij} \tag{12.31a}$$

and (for $i \neq z \neq j$)

$$\langle \delta w_i \, \delta w_j \rangle / V_w(z)^2 = \delta_w^2 \exp(\delta_w^2)[k_w(z)/k_w(0)]^{1-n}\left(1 + n + \frac{1}{2}\delta_w^2\right)^{-2} q_{ij} \tag{12.31b}$$

together with

$$\langle \delta w_z \, \delta w_z \rangle / V_w(z)^2 = \delta_w^2 \exp(\delta_w^2)[k_w(z)/k_w(0)]^{1-n}\left(1 + n + \frac{1}{2}\delta_w^2\right)^{-2}(q_{xx} + q_{yy}) \tag{12.31c}$$

Thus, fluctuations in pressure gradient tend to maintain a constant ratio to the mean pressure gradient, while the fluctuations in water flow speed relative to the mean vertical water flow speed decrease (increase) with increasing depth, provided the mean water permeability decreases with increasing depth, according as $1 > n$ $(1 < n)$. For oil flow we have the relative strength measures

$$\langle \delta G_i \, \delta G_j \rangle / G(z)^2 = \delta_o^2 s_{ij} \tag{12.32a}$$

and (for $i \neq z \neq j$)

$$\langle \delta v_i \, \delta v_j \rangle / V_o(z)^2 = \delta_o^2 \exp(\delta_o^2)\left(1 + m + \frac{1}{2}\delta_o^2\right)^{-2}[k_o(z)/k_o(0)]^{1-m} s_{ij} \tag{12.32b}$$

together with

$$\langle \delta v_z \, \delta v_z \rangle / V_o(z)^2 = \delta_o^2 \exp(\delta_o^2)\left(1 + m + \frac{1}{2}\delta_o^2\right)^{-2}[k_o(z)/k_o(0)]^{1-m}(s_{xx} + s_{yy}) \tag{12.32c}$$

so that, just as for water flow, the relative strength of total pressure gradient fluctuations remains a sensibly constant fraction of the mean pressure gradient, while the fluctuations in velocity, both transverse and parallel to the mean flow, become smaller (larger) as depth increases provided the mean oil permeability, $k_o(z)$, decreases with increasing depth, according as $1 > m$ $(1 < m)$.

12.2.3.3. Intercomparison of Oil and Water Properties

a. Mean and Fluctuating Flow Components. A question of some interest is whether the impact of fluctuations in permeability is more dominant on oil or water. Estimates of the relative impact can be obtained by comparing the ratio of oil-related flow properties to the corresponding water properties. Thus, for mean flow we have the ratio of oil to water effects of

$$[V_o(z)/V_w(z)]/[V_{o*}/V_{w*}] = [k_o(z)/k_o(0)]^m[k_w(z)/k_w(0)]^{-n}\left(1 + m + \frac{1}{2}\delta_o^2\right)\left(1 + n + \frac{1}{2}\delta_w^2\right)^{-1} \tag{12.33a}$$

For fluctuations in flow speed we have the ratio of oil to water properties for vertical fluctuations

$$\langle\delta v_z^2\rangle/\langle\delta w_z^2\rangle = R_{ow}(s_{xx} + s_{yy})/(q_{xx} + q_{yy}) \tag{12.33b}$$

and for transverse (to z) fluctuations

$$\langle\delta v_i\, \delta v_j\rangle/\langle\delta w_i\, \delta w_j\rangle = R_{ow}\, s_{ij}/q_{ij} \tag{12.33c}$$

where

$$R_{ow} = (\delta_o/\delta_w)^2 \exp(\delta_o^2 - \delta_w^2)(V_{o*}/V_{w*})^2\, [k_o(z)/k_o(0)]^{m+1}[k_w(z)/k_w(0)]^{-(n+1)} \tag{12.33d}$$

We see from Eqs. (12.33) that the relative magnitude of the impact of fluctuations is, apart from the geometric factors, related to the relative strength of logarithmic fluctuations in permeability, δ_w^2 and δ_o^2, for water and oil, respectively, as well as to the mean values of permeabilities, k_w and k_o, in the corresponding nonfluctuating medium. Clearly, both mean values, mean square fluctuations, and geometrical factors (s_{ij} and q_{ij}) are specifically dependent on each geological situation so that it is not possible to generalize whether oil or water is impacted more by the permeability fluctuations. Specific examples are presented in Section 12.3.

b. Mean Capillary Pressure Gradient. From Eq. (A57c) and using the definitions of mean steady-state flow V_{o*} and V_{w*}, we see that the fluctuations in permeability require the supporting existence of a mean capillary pressure gradient for oil flow, $p_c'(z)$, given by

$$p_c'(z) = V_{w*}\, k_w(0)^{-1}\{[k_w(z)/k_w(0)]^{n-1} - 1\} - V_{o*}k_o(0)^{-1}\{[k_o(z)/k_o(0)]^{m-1} - 1\} + p_c'(0) \tag{12.34}$$

where $p_c'(0)$ is the scaling value on the sediment–water interface.

12.2.4. Spatial Diffusion of Oil

In addition to investigating the dependence of the velocity components of oil flow on the spatial fluctuations in permeability, it is interesting to track the motion of a single element of oil after it leaves a source (located at $z = z_0$, $x = x_0$, $y = y_0$) at time $t = t_0$. The deterministic motion of an individual element of oil is sufficiently complex that an appropriate representation is as given by a statistical Markov probability description (Feller, 1966; Frisch, 1968; Lumley, 1970). In such a representation of oil flow, the deterministic picture of oil flow is replaced by a statistical description in which bulk transport of oil is at velocity $V_o(z)\hat{z}$ while the random components of velocity add the stochastic element. The spatial diffusion coefficients $D_{zz}(z)$ and $D_{ij}(z)$ then describe the relative root mean square displacement of an element of oil. Following along the well-developed lines of the classical Markov analysis (Frisch, 1968), in the representation where a diffusion picture is appropriate, we can use the spatial diffusion coefficients given by Eqs. (A73) and (A74) in the appendix at the end of this chapter to write the probability, P, of finding an element of oil at spatial coordinates z, x, and y at time t ($>t_0$) in the form

$$\frac{\partial P}{\partial t} = \frac{\partial}{\partial z}\left[|V_o(z)|D_{zz}(z)\,\frac{\partial P}{\partial z} + V_o(z)P\right] + \frac{\partial}{\partial x_i}\left[|V_o(z)|D_{ij}(z)\,\frac{\partial P}{\partial x_j}\right] \tag{12.35}$$

where $i = x$ or y, $j = x$ or y, and the Einstein double summation convention is understood over repeated indices in Eq. (12.35). The initial conditions on P are that, on $t = t_0$, we have

$$P = \delta(x - x_0)\,\delta(y - y_0)\,\delta(z - z_0) \tag{12.36}$$

so that we are guaranteed to find the oil at the source point. The probability is normalized as usual by

$$\int dx\,dy\,dz\,P(x, y, z, t) = 1 \tag{12.37}$$

so that the element of oil is guaranteed to be somewhere in the coordinate space.

From Eqs. (A73) and (A74), together with Eqs. (12.33b) and (12.33c), it follows that we can write the diffusion coefficients D_{ij} and D_{zz} in the form

$$D_{ij} = \delta_o \exp(\delta_o^2/2)\left(1 + m + \frac{1}{2}\delta_o^2\right)^{-1}[k_o(z)/k_o(0)]^{(1-m)/2}\,h_{ij} \tag{12.38a}$$

and

$$D_{zz} = \delta_o \exp(\delta_o^2/2)\left(1 + m + \frac{1}{2}\delta_o^2\right)^{-1}[k_o(z)/k_o(0)]^{(1-m)/2} h_{zz} \quad (12.38b)$$

where the geometry factors are

$$h_{ij} = (2\pi)^{1/2} \int d^3k\, R(\mathbf{k}) k_z^2 k^{-4} k_i k_j (k_\alpha k_\beta s_{\alpha\beta})^{-1/2} \qquad (\alpha \neq z \neq \beta;\ i \neq z \neq j) \quad (12.39a)$$

and

$$h_{zz} = (2\pi)^{1/2} \int d^3k\, R(\mathbf{k}) |k_z|\, (k^2 - k_z^2)\, k^{-4} \quad (12.39b)$$

and have the dimensions of length, so that $V_o D$ has the dimensions of (length)^2/time—representing temporal diffusion. With Eq. (12.30b) for the mean oil velocity, the diffusion equation (12.35) can then be written

$$\frac{\partial P}{\partial t} = \frac{\partial}{\partial z}\left[|V_{o*}|\delta_o \exp\left(\frac{\delta_o^2}{2}\right)\left(\frac{k_o(z)}{k_o(0)}\right)^{1+m/2} h_{zz}\frac{\partial P}{\partial z} - |V_{o*}|\left(1 + m + \frac{1}{2}\delta_o^2\right)\left(\frac{k_o(z)}{k_o(0)}\right)^{m} P\right] + \delta_o|V_{o*}| \exp\left(\frac{\delta_o^2}{2}\right) h_{ij}\left(\frac{k_o(z)}{k_o(0)}\right)^{1+m/2} \frac{\partial^2 P}{\partial x_i \partial x_j} \quad (12.40)$$

Note that with z measured from the sediment–water interface into the sediments then $V_{o*} = -|V_{o*}|$. Note also that in the absence of any fluctuations ($\delta_o = 0 = m$) the diffusion equation (12.40) reduces to

$$\frac{\partial P}{\partial t} = -|V_{o*}|\frac{\partial P}{\partial z} \quad (12.41)$$

representing the usual kinematic transport at the oil flow speed $|V_{o*}|$.

Again we see that the behavior of oil particle transport depends sensitively upon both the magnitude of the spatial fluctuations in log permeability and the statistical orientation of correlation through the geometry scale length factors h_{zz} and h_{ij}.

Without further knowledge of the detailed statistical properties of δ_o^2, δ_w^2, and the normalized spatial power spectra $\nu(\mathbf{k})$ and $R(\mathbf{k})$, it is difficult to proceed further in general.

The next section of the chapter explores the behavior of flow properties under two different assumptions for the spatial power spectra of logarithmic permeability fluctuations.

12.3. Model Behaviors of Fluctuations

12.3.1. Orientational Aspects

12.3.1.1. Isotropic Fluctuations

By far the simplest expressions are available when the power spectra $r(\mathbf{k})$ and $R(\mathbf{k})$ are isotropic, independent of the orientation of the spatial wave number vector $\mathbf{k}$, and functions only of $|\mathbf{k}|$. In that case we have

$$n = \frac{1}{3}\delta_w^2; \qquad m = \frac{1}{3}\delta_o^2$$

$$s_{xx} = 1/15 = s_{yy}; \qquad s_{ij} = 0, \text{ for } i \neq j; \qquad s_{zz} = 1/5; \qquad q_{ij} = s_{ij} \tag{12.42a}$$

together with

$$h_{xx} = \frac{\pi}{32}(30\pi)^{1/2}\ell \tag{12.42b}$$

$$h_{yy} = h_{xx} \tag{12.42c}$$

$$h_{zz} = \frac{1}{4}(2\pi)^{1/2}\ell \tag{12.42d}$$

and $h_{ij} = 0$ for $i \neq j$, where the spatial scale length ℓ is given exactly by

$$\ell = \frac{\int_0^\infty kR(k)\, dk}{\int_0^\infty k^2R(k)\, dk} \tag{12.42e}$$

12.3.1.2. Bedding Plane Fluctuations Dominant

It can be argued that while isotropic power spectra are useful as a pedagogical device, they are less representative of the effect of fluctuations in real geologic situations where strata tend to be laid down in roughly parallel beds. Thus, sand bodies tend to be laid down in roughly random fashion in surrounding shale, but the forms of the bodies are roughly lenticular, being considerably wider than they are thick. In the Gulf of Mexico, for instance, the sands are those of barrier and distributary bars and stream channels, with lateral scale sizes of the order of miles and vertical thicknesses of the order of tens to hundreds of feet. Thus, the spatial power spectra are likely to be dominated by fluctuations in the vertical direction rather than horizontally. To model this situation simply, we write

$$R(\mathbf{k}) = \varepsilon R(k_\perp)(2\pi^2)^{-1}(1 + \varepsilon^2 k_z^2/k_\perp^2)^{-1} \tag{12.43}$$

with ε (>0) measuring the root mean square vertical to horizontal scale size. We take ε to be a small parameter ($\varepsilon << 1$). Here $\mathbf{k}_\perp = (k^2 - k_z^2)^{1/2}$. The normalization condition is then

$$\int d^3k\, R(\mathbf{k}) = 1 = \int_0^\infty k_\perp^2\, dk_\perp \int_{-\infty}^{\infty} dk_z\, \varepsilon(2\pi^2)^{-1}(1 + \varepsilon^2 k_z^2/k_\perp^2)^{-1} R(k_\perp) \tag{12.44}$$

In this case we have

$$n = \delta_w^2/(1 + \varepsilon); \qquad m = \delta_o^2/(1 + \varepsilon) \tag{12.45}$$

where we have taken $r(\mathbf{k})$ to be identical in structural form to $R(\mathbf{k})$. We can also write

$$q_{xx} = q_{yy} = \frac{1}{4}\varepsilon(1 + \varepsilon)^{-2} = s_{xx} = s_{yy}; \qquad q_{ij} = 0, \qquad i \neq j \tag{12.46a}$$

together with

$$h_{xx} = h_{yy} = \ell(2\pi)^{1/2}\, \varepsilon^{3/2}(1 + \varepsilon/2)/(1 + \varepsilon) \tag{12.46b}$$

$$h_{zz} = \ell(2/\pi)^{1/2}\, \varepsilon(1 - \varepsilon^2)^{-2}(1 - \varepsilon^2 + 2\varepsilon^2\ell\pi\varepsilon) \tag{12.46c}$$

and $h_{ij} = 0$, $i \neq j$; the correlation scale length ℓ is given by

$$\ell = \frac{\int_0^\infty k_\perp r(k_\perp)\, dk_\perp}{\int_0^\infty k_\perp^2 r(k_\perp)\, dk_\perp} \tag{12.46d}$$

Note that for $\varepsilon << 1$ we have $h_{xx}/h_{zz} \cong \pi\varepsilon^{1/2}$, so that for $\varepsilon \leqslant \pi^{-2}$, diffusion perpendicular to the mean flow direction is less dominant than diffusion parallel to the mean flow; that is, for a typical thickness-to-width ratio of a sand body of less than 10%, perpendicular diffusion is a small statistical correction, whereas for a "chunkier" ratio perpendicular diffusion dominates over parallel diffusion.

12.3.2. Magnitude Aspects

While the orientation of the spatial power spectra is of concern in determining the overall systematic pattern of mean and fluctuating behaviors, the modification to mean values and the absolute size of fluctuations in physical parameters are determined by the magnitudes of the variances in the logarithmic fluctuations of permeabilities for oil and water. Two major factors control the relative fluctuations in permeability: lithology and wetness.

For 100% water-wet rock fabrics, the dominant control on permeability fluctuations is the degree of sandiness. The point here is that the contrast in sand/shale permeabilities for water flow is extremely large (sands typically have permeabilities around 1000–100 mD, whereas shale permeabilities are typically around micro- to nanodarcies, leading to contrasts characterized by values of about 10^4–10^6 : 1 for the relative ratio). Thus, sand stringers in a dominantly shale matrix provide high-permeability pathways for flow; conversely, shale stringers in a predominantly sand matrix provide low-permeability barriers, forcing flow around the shale stringers. In either event, there is a marked degree of fluctuation in permeability with sand percentage. On the other hand, at 0% and 100% sand the variations in permeability due to lithology differentiation are zero. Instead, all that remains is variations due to, say, coarse-grained and fine-grained sand bodies being interbraided, that is, an effect on permeability due more to grain size variations within a given lithology. However, by and large, the fluctuations in permeability produced by such intraformational grain size variations are much less than the fluctuations produced by interformational differences.

In addition, as subsurface burial increases, there is a systematic lowering of permeability owing to increased overburden being compensated for by increased grain-to-grain contact pressure, thereby lowering the available interconnected pore space and so also the permeability to flow. Although the detailed manner in which this loss of permeability takes place is complex, and dependent upon the precise evolutionary character of each individual basin, nevertheless there is a general trend character available that suffices to illustrate the main points of this discussion. Porosity, $\phi(z)$, tends to decrease roughly exponentially with increasing depth (Sclater and Christie, 1980). In addition, there is a rough rule of thumb connecting porosity to permeability in a power law fashion: $\kappa \propto \phi^A$ with A different for different lithologies. When combined (i.e., $\phi \propto \exp(-z/z_*)$ and $\kappa \propto \phi^A$), the two rules of thumb imply an approximate behavior of

$$k \propto \exp(-z/L)$$

where L is the scale depth for permeability decrease, related to z_* by $L = z_*/A$.

To handle both the lithology and the compactional dependences at the same time, we take the approximate representation (Lewis and Rose, 1970)

$$k(z) = k_{sh}^{1-S}\, k_s^S \tag{12.47a}$$

where S is the fractional sandiness, and k_{sh} and k_s are, respectively, the shale and the sand permeability at depth z, given by

$$k_{sh}(z) = k_{sh}(0)\exp(-z/L_{sh}) \tag{12.47b}$$

$$k_s(z) = k_s(0)\exp(-z/L_s) \tag{12.47c}$$

where L_{sh} and L_s are the lithology-dependent scaling depths for permeability decrease.

On the basis of this representation, it is the fluctuations in sandiness which control the fluctuations in permeability. Thus, with $\Lambda(z) = \ln(k_s/k_{sh})$ we have

$$\langle \ln k \rangle = \ln k_{sh} + \langle S \rangle \Lambda(z) \tag{12.48a}$$

$$\langle (\delta \ln k)^2 \rangle = \Lambda^2 \langle \delta S^2 \rangle \tag{12.48b}$$

where $\langle S \rangle$ is the mean sandiness, and $\langle \delta S^2 \rangle$ is the variance in sandiness.

On the other hand, the permeability of a formation to oil or water flow depends on the fractional wetness, W, of the grains. As Hubbert (1953) noted, for a 100% water-wet rock, capillary pressure forces oil from fine-grained to coarse-grained regions, whereas for a 100% oil-wet rock fabric, oil is forced into the fine-grained regions and water into the coarse-grained domains. Thus, the permeability to oil flow relative to that to water flow in a complex stratigraphic situation must be dependent on the fractional wetness of rock strata. In order to model such situations, we suppose that the relative permeability of oil with respect to that of water, k_o/k_w, is related to the fractional wetness, W, through

$$q^{1-W}r^W = k_o/k_w \tag{12.49}$$

For a 100% water-wet lithology ($W = 1$) the oil permeability is then given by $k_o = k_w r$, whereas for a 100% oil-wet lithology ($W = 0$) we have $k_o = k_w q$.

On the basis of this representation, the fluctuations in sandiness *and* the fluctuations in water wetness control the fluctuations in oil permeability. Thus, with

$$\ln k_o = \ln k_w + q + W \ln(r/q)$$

and with the fluctuations in sandiness and water wetness being taken to be independent, we have

$$\langle \ln k_o \rangle = \langle \ln k_w \rangle + q + \langle W \rangle \ln(r/q) \tag{12.50a}$$

$$\langle (\delta \ln k_o)^2 \rangle = \Lambda^2 \langle \delta S^2 \rangle + \langle \delta W^2 \rangle \lambda^2, \text{ with } \lambda = \ln(r/q) \tag{12.50b}$$

The fluctuations in sandiness, $\langle \delta S^2 \rangle$, are bounded. For a medium with either 100% ($\langle S \rangle = 1$) or 0% ($\langle S \rangle = 0$) mean sandiness, there can be no fluctuations in sandiness around the mean. For instance, if the probability, $p(S)$, of finding a fractional sandiness in the range S to $S + dS$ is given by a lognormal distribution in $0 \leqslant S \leqslant 1$, that is,

$$p(S) \propto \exp(a \ln S)/S$$

then $a = \langle S \rangle/(1 - \langle S \rangle)$ and $(\delta S)^2 = 0$ on $S = 0$ or $S = 1$ as required. The same sort of limit behavior on $\langle \delta S^2 \rangle$ is obtained with any other form chosen for $p(S)$.

Equally, fluctuations in water wetness, $\langle \delta W^2 \rangle$, are bounded. For a 100% ($\langle W \rangle = 1$) water-wet or 100% oil-wet ($\langle W \rangle = 0$) rock, there can be no fluctuations in water wetness.

In order to illustrate the magnitude of the effects of fluctuations in both sandiness and water wetness on the mean and fluctuating components of physical flow characteristics for oil and water, throughout the remainder of this section we treat the situation in which

$$\langle (\delta S)^2 \rangle = \alpha \langle S \rangle^2 (1 - \langle S \rangle)^2 \tag{12.51a}$$

and

$$\langle(\delta W)^2\rangle = \beta\langle W\rangle^2(1 - \langle W\rangle)^2 \tag{12.51b}$$

with α and β numerical factors, of order unity, whose precise values depend on the statistical distribution chosen for sandiness and wetness variations. When a numerical value is required, we choose $\alpha = \beta = 1/4$. There are, undoubtedly, other functional forms that can be chosen for the variation of permeability with both sandiness and wetness and for the dependence of statistical fluctuations in sandiness and wetness on the mean values. However, the forms chosen encompass the dominant physical characteristics and provide approximate representations of the modifications induced in mean flow properties by spatial variations in permeability.

12.3.3. Numerical Estimates of Mean and Fluctuating Quantities

12.3.3.1. Mean Water Flow Properties

With $\delta_w^2 = C(\ln k_s/k_{sh})^2\langle S\rangle^2(1 - \langle S\rangle)^2$, $C \approx 1/4$, and $\ln k_s/k_{sh} \approx 5 \ln_e 10 \approx 11$, we can write the mean vertical water flow speed as

$$V_w/V_{w*} \cong (1 + a\delta_w^2)\exp(-z/L_{\text{eff}}) \tag{12.52a}$$

where the effective scale length, L_{eff}, is related to the scale lengths L_s and L_{sh} for porosity decrease with depth of sand (L_s) and shale (L_{sh}), respectively, through the harmonic addition formula

$$L_{\text{eff}}^{-1} = n[(1 - \langle S\rangle)L_{sh}^{-1} + \langle S\rangle L_s^{-1}] \tag{12.52b}$$

Note that, depending on the statistical distribution shape of the regions fluctuating in permeability, n is given by

$$n = \gamma\delta_w^2 = \gamma C(1 - \langle S\rangle)^2\langle S\rangle^2(\ln k_s/k_{sh})^2$$

where $\gamma = 1$ for isotropic fluctuations, and $\gamma = (1 + \varepsilon)^{-1}$ for a spatial power spectrum of fluctuations oriented dominantly horizontally.

With the illustrative numerical choices $\gamma \cong 1$, $C \cong 1/4$, and $(\ln k_s/k_{sh}) \approx 11$, it follows that $n \approx 30\langle S\rangle^2(1 - \langle S\rangle)^2$. For a spatial scale of porosity decrease roughly the same in shales as in sands (i.e., $L_{sh} \approx L_s$), it then follows from Eq. (12.52b) that

$$L_{\mathrm{eff}} \cong L_s/[30\langle S\rangle^2(1 - \langle S\rangle)^2] \tag{12.52c}$$

Unless the mean sandiness, $\langle S\rangle$, is less than about 20% or greater than about 80%, it follows that $L_{\mathrm{eff}} << L_s$ so that the mean flow velocity decreases in the fluctuating medium with increasing depth relative to the value in the nonfluctuating medium over a scale depth L_{eff} ($<<L_s$) measured from the sediment–water interface down. Thus, the random fluctuations are responsible for an increased mean fluid flow at shallow subsurface depths but are equally responsible for a decreased mean flow at depths roughly in excess of L_{eff}, which depth scale is much less than the depth scale L_s over which the mean porosity decreases.

For mean sandiness values greater than 80% or less than 20%, $L_{\mathrm{eff}} >> L_s$, and the mean flow is enhanced by spatial fluctuations to greater depths than in the absence of fluctuations by the preexponential factor $1 + a\delta_w^2$ in Eq. (12.52a), which has an approximate value of $30\langle S\rangle^2\langle 1 - \langle S\rangle)^2 \leqslant 2$.

Equally, the mean vertical excess fluid pressure gradient can be expressed as

$$f(z)/f_*(z) = \exp(z/L_{\mathrm{eff}}) \tag{12.53a}$$

so that, relative to the excess fluid pressure gradient, $f_*(z)$, in the absence of spatial fluctuations, the relative mean pressure gradient is increased by the fluctuations over a scale L_{eff}. The *absolute* spatial variation of mean excess pressure gradient in the presence of spatial fluctuations increases with increasing depth as long as $1 > n$, that is, when

$$1 > \gamma\delta_w^2 \cong \gamma C\langle S\rangle^2(1 - \langle S\rangle^2)(\ln k_s/k_{\mathrm{sh}})^2 \tag{15.53b}$$

With the numerical values given, it follows that the absolute excess fluid pressure gradient will increase with increasing depth for mean sandinesses in excess of about 80% or less than about 20%, whereas in the intermediate regime of $20\% \leqslant \langle S\rangle \leqslant 80\%$ the mean excess pressure gradient will decrease—representing the fact that the connection provided by Darcy's law between mean flow speed and excess pressure gradient will require a lower excess pressure gradient for a lower mean flow speed.

Average vertical flow speed divided by average vertical pressure gradient defines the effective vertical permeability, whereas effective lateral permeability is defined by dividing average vertical pressure gradient into the root mean square (r.m.s.) lateral flow because random flow variations laterally average out in the mean. Details of how to calculate relevant quantities are provided in the appendix, where we also show in detail how both the mean steady-state flow pattern and spatially varying fluctuations in flow pattern are related to the average degree of sandiness.

Model r.m.s. fluctuations in sandiness, ΔS, around the mean sandiness, S_0, by

$$(\Delta S)^2 = mS_0^2(1 - S_0)^2$$

when m is a numerical constant of order unity. For an alternately layered sand–shale sequence of equal thickness beds, $m = ½$, corresponding to 42% fluctuations in sandiness.

The physical basis for selecting a value of m depends on the precise statistical distribution chosen to model sand variations (Gaussian, Kolmogorovian, etc.). Because this detailed information is not generally available, the more realistic method of letting m vary from about 0.5 to 1 is followed here to illustrate the rugged nature of an increase of effective permeability.

The main results derived from calculations presented in the Appendix are:

(i) For a mean sandiness, S_0, somewhat less than 20%, little decrease from the fluid-flow pattern that would prevail in a 100% shale medium is seen.

(ii) For a mean sandiness somewhat in excess of about 80%, little difference from the fluid-flow pattern that would prevail in a 100% sand medium occurs.

(iii) However, in the intermediate regime of mean sandiness (20% < S_0 < 80%), the effective vertical permeability depends both upon the degree of mean sandiness and upon fluctuations (ΔS) in sandiness about the mean value. An approximate formula for effective vertical permeability is

$$K_{\text{eff}}\ (\text{vertical}) = K_{\text{sh}}(K_s/K_{\text{sh}})^S \exp\left[\frac{1}{6}(\ln K_s/K_{\text{sh}})^2(\Delta S)^2\right]$$

where K_s is the permeability of sand and K_{sh} the permeability of shale.

(iv) In the intermediate regime of mean sandiness, an effective lateral permeability, K_{eff} (lateral), exists such that r.m.s. lateral flow speed responds to a vertical pressure gradient. An approximate formula for effective lateral permeability is

$$K_{\text{eff}}\ (\text{lateral}) = \frac{1}{4} K_{\text{sh}}(K_s/K_{\text{sh}})^S(\Delta S \ln (K_s/K_{\text{sh}})) \exp\left[\frac{1}{2}(\ln (K_s/K_{\text{sh}})^2 (\Delta S)^2\right]$$

Physical assumptions that have gone into the determination of these estimates of effective lateral and vertical permeability are that (a) on average, the system is laterally and vertically homogeneous so that average vertical pressure gradient must be constant and (b) flow is represented by a steady-state model. Mathematical assumptions that are involved in making the assessments are that (a) the logarithm of permeability at any given location is proportional to the sand ratio at that point, (b) variations in sand ratio are lognormally distributed around the mean sand ratio, and (c) the mean-field renormalization technique is adequate because one is interested in effective system permeabilities (lateral and vertical) on a scale that is large compared with typical sand-body size. Similar assumptions to (a) and (b) were made in the previously cited investigations into one-dimensional fluid flow, and equivalent assumptions are made here so that a body of work performed under the same, or similar, constraints is available. Within the context of mean-field renormalization, assumption (c) is the canonical condition under which mean-field renormalization has been developed as an accurate and powerful tool in many areas of physics [see, e.g., the review by Frisch (1968), and copious references therein].

The degree of concentration of fluid flowing through a "typical" sand body is measured by the ratio of r.m.s. lateral velocity to vertical velocity, that is, by the ratio of effective lateral permeability to vertical permeability as follows.

In the absence of any lateral flow, the concentration factor, C, is unity by definition. In the presence of lateral flow, the concentration factor increases because flow is directed toward more permeable sand beds (see Fig. 12.1). Because lateral flow is randomly oriented, define an r.m.s. angle of influence, θ_c, from a small source of fluid to be given by

$$\tan \theta_c = (\langle V_\perp^2 \rangle / \langle V_z^2 \rangle)^{1/2} = 2^{1/2}\, K_{\text{eff}}\,(\text{lateral})/K_{\text{eff}}\,(\text{vertical})$$

The concentration factor for a typical sand body is then estimated as

$$C = 1 + \tan \theta_c$$

where $\tan \theta_c$ measures bending and focusing of fluid stream lines. The concentration factor measures capability of a sand body to "attract" fluid flow from the surrounding medium.

Total volume of influence of a sand body is simply the volume of the sand body multiplied by the concentration factor C. Hence, a typical scaled radial distance from which a sand body can attract fluid is $C^{1/3}$.

Effective permeability of the composite shale–sand system, both for flow in the direction of net pressure gradient and for flow in the direction

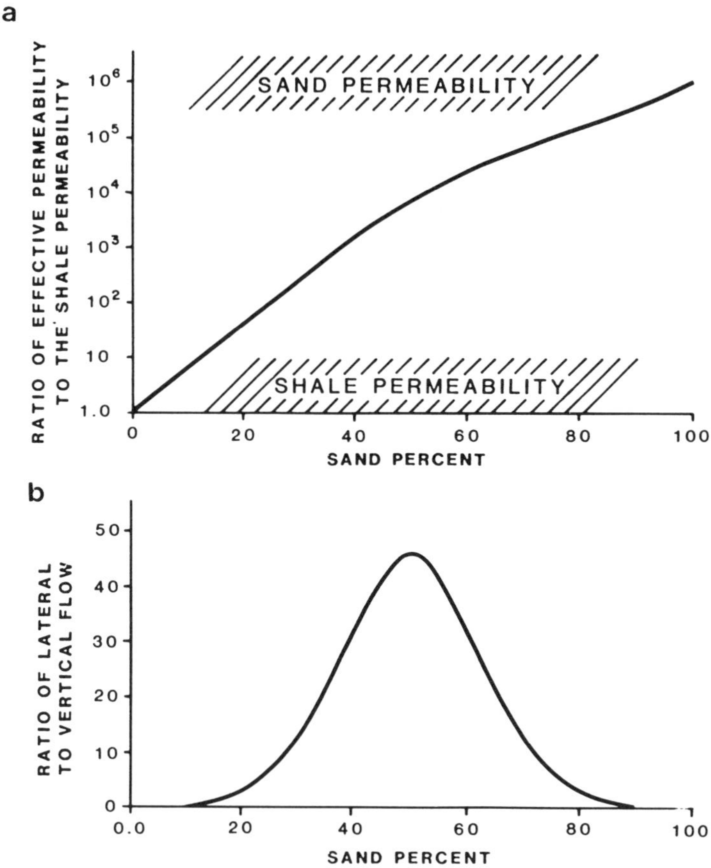

Figure 12.1. Migration characteristics for random sands in a shale matrix in response to a vertical pressure gradient. (a) Effective permeability as a function of S_0, mean sand percent. Sand permeability = $10^6 \times$ shale permeability, "m" = 1.0. Note that for a sand percent of 30%, effective permeability is 10^3 times that for a pure shale ($S_0 = 0$). (b) Effective concentration volume of a sand body as a function of S_0, mean sand percent. Note that in a pure shale ($S_0 = 0$) or a pure sand ($S_0 = 1$), no concentration occurs. For the parameters used in the text, the maximum concentration radius of about 7 occurs at 50% sandiness.

perpendicular to net pressure gradient (Fig. 12.2), shows that in a range of about 10–90% sandiness, K_{eff} (lateral) exceeds K_{eff} (vertical), implying that lateral flow between sand bodies exceeds flow in the direction of regional fluid pressure gradient. Calculations have been performed for values of the ratio of sand permeability to shale permeability from 10^5 to 10^7 and for values of m from 0.5 to 1. We think that this suite of values

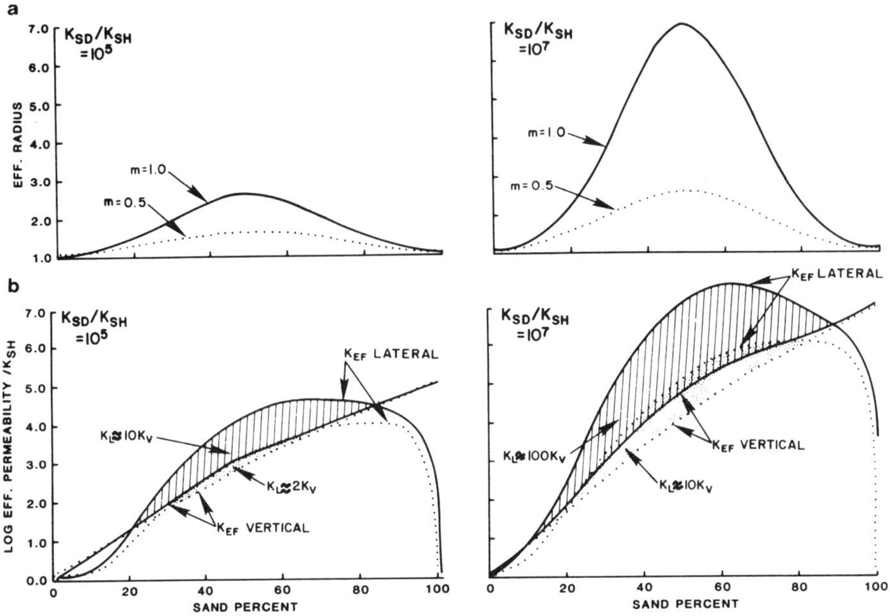

Figure 12.2. (a) Effective concentration radius as a function of sand percent for K_s/K_{sh} in the range 10^5–10^7 and for values of m in the range 0.5–1.0. (b) Logarithm of effective lateral (K_L) and vertical (K_V) permeabilities as a function of S_0, mean sand percent, for K_s/K_{sh} in the range 10^5–10^7 and for values of m in the range 0.5–1.0. Note that K_L exceeds K_V by a factor of 10 for sandiness in the range of about 25–75%. The shaded area delineates the region in which lateral permeability exceeds vertical permeability.

more than encompasses the dynamic range of estimates suggested for conditions prevailing in Gulf Coast sediments (Jones, 1975).

The reason why K_{eff} (lateral) drops to 0 at 100% and 0% sandiness is that flow then follows precisely the direction of applied pressure gradient because no regions of increased permeability exist through which flow can detour.

If one were to choose an effective permeability beyond which flow is so unhindered that the composite system cannot trap hydrocarbons, one might use a value a few times the geometric mean of sand and shale permeabilities as an estimate, that is, a few times $3 \times 10^3 K_{sh}$. We might then consider $10^4 K_{sh}$ as an appropriate cutoff. We see (Fig. 12.2) that this cutoff occurs for K_{eff} (vertical) at a mean sandiness of some 45% and for K_{eff} (lateral) at a mean sandiness of some 28% (for $K_s/K_{sh} \cong 10^7$ and $m = 1$).

Thus, much lateral migration occurs for a mean sandiness in excess of about 20–30%, whereas hydrocarbons would not be expected to be trapped for a mean sandiness in excess of 40–50%.

The effective spatial range (in units of equivalent radius of a typical sand body) that is influenced by a "typical" sand body is estimated by the concentration factor, $C^{1/3}$, which is largest at a sandiness of 50% and decreases on either side of 50% sandiness (Fig. 12.2) for the range $K_s/K_{sh} = 10^5$–10^7 and $m = 0.5$–1.0. Crudely, a sand body "attracts" fluid out to about 2–7 radii at 50% sandiness, with the effective spatial range decreasing roughly linearly to 1 radius at 0% and 100% sandiness. Note that $m = 1$ corresponds to 100% r.m.s. fluctuations in sandiness, whereas $m = 0.5$ corresponds to 40% r.m.s. fluctuations in sandiness.

As an illustration, consider the number of pore volumes flushed per unit time per unit area for a sandiness of 30%, $K_s/K_{sh} = 10^7$, and $m = 1$. Effective vertical permeability (Fig. 12.2) is about $10^3 K_{sh}$, and effective lateral permeability is about $5 \times 10^4 K_{sh}$. For a vertical pressure gradient of 0.1 psi/ft and $K_{sh} = 10^{-15}$, r.m.s. lateral flow is about 100 ft/1000 yr, whereas vertical flow is only about 2 ft/1000 yr. Thus, lateral flow flushes about 50 pore volumes per unit time per unit area for every pore volume flushed by vertical flow.

If no fluctuation in sand ratio existed, vertical permeability would have been only $K_{sh}(K_s/K_{sh})^{0.3} \cong 10^2 K_{sh}$. Thus, for $m = 1$, corresponding to 100% r.m.s. fluctuations in sandiness, vertical flow is 10 times that expected for a rigorously homogeneous system with the same average sandiness. The number of pore volumes flushed by r.m.s. lateral flow is about 500 more than in a truly homogeneous system.

12.3.3.2. Mean Oil Flow Properties

From Eq. (12.30b) we can write the mean vertical oil flow relative to the value obtaining in the absence of fluctuations as

$$V_o(z)/V_{o*} = \left(1 + m + \frac{1}{2}\delta_o^2\right) \exp(-z/l_{\text{eff}}) \qquad (12.54a)$$

where the effective scale length, l_{eff}, for oil flow variation with depth is given by

$$l_{\text{eff}}^{-1} = \frac{m}{n} L_{\text{eff}}^{-1} \qquad (12.54b)$$

with L_{eff} the scale length for water flow. Under the conditions given, we have

$$\frac{m}{n} = 1 + \frac{(\ln r/q^2)\beta\langle W\rangle^2(1 - \langle W\rangle)^2}{\alpha\langle S\rangle^2(1 - \langle S\rangle)^2(\ln k_s/k_{sh})^2} \tag{12.55a}$$

For $\alpha = \beta$, $\ln r/q = 0(1)$, and $\ln k_s/k_{sh} \approx 11$, we have

$$\frac{m}{n} \cong 1 + \frac{1}{120}\left(\frac{\langle W\rangle^2(1 - \langle W\rangle)^2}{\langle S\rangle^2(1 - \langle S\rangle)^2}\right) \tag{12.55b}$$

Thus, unless the mean sandiness is less than 20% or greater than 80%, the relative contrast factor $m/n = 0(1)$. For mean sandiness fractions outside of the range $20\% < \langle S\rangle < 80\%$, m/n exceeds unity provided that the mean water wetness of rocks is not too small or too large, in the sense that $m/n >> 1$ when

$$\langle W\rangle(1 - \langle W\rangle) \geqslant 11\langle S\rangle(1 - \langle S\rangle) \tag{12.56}$$

For $m >> n$, it follows that $l_{\text{eff}} << L_{\text{eff}}$ so that the mean oil flow speed is more reduced by spatial fluctuations in permeability than is the mean water flow, provided only that the water wetness is not too close to 100% or 0%. For $m \approx n$ [which occurs either at high (or low) water wetness $\langle W\rangle \to 1$ ($\langle W\rangle \to 0$) or at sandiness fractions in the range $20\% \leqslant \langle S\rangle \leqslant 80\%$], then $l_{\text{eff}} \approx L_{\text{eff}}$ so that the mean vertical oil flow is diminished in approximately the same proportion as is the mean vertical water flow by the presence of spatial fluctuations in permeability.

For the mean total vertical pressure gradient we have, from Eq. (12.30a), that

$$G(z)/G_*(z) = \exp(-\,z/l_{\text{eff}}) \tag{12.57a}$$

so that, in like manner to the mean water properties, the relative mean total pressure gradient drops over the scale l_{eff} in the presence of spatial fluctuations in permeability, whereas the absolute value of the total pressure gradient increases according as $1 \gtrless m$, i.e.,

$$1 \gtrless \left(\frac{m}{n}\right) n \tag{12.57b}$$

With $m/n \approx 1$, we see that the physical behavior of the mean total pressure gradient then follows the corresponding behavior for water (i.e., as $l \geqslant n$ or $l \leqslant n$, respectively), while in the region where $m/n >> 1$, the pattern of behavior for mean total pressure evolution is contrary to that for water (unless m is itself much less than unity).

12.3.3.3. Fluctuating Water Properties

Apart from changes in the mean vertical flow and the mean vertical pressure brought about by the statistical variations in spatial permeability, there are also randomly fluctuating components. Perhaps a concern of some importance is to estimate the magnitude and spatial scale of such fluctuations.

To handle this estimation problem in a simple manner, concentrate on the mean square pressure gradient fluctuations, for which, from Eq. (12.31a), we have the ratios related to the square of the mean pressure gradient given by

$$\langle \delta f_z^2 \rangle / \langle f(z) \rangle^2 = \delta_w^2 \, q_{zz} \tag{12.58a}$$

and

$$\langle \delta f_x^2 \rangle / \langle f(z) \rangle^2 = \langle \delta f_y^2 \rangle / \langle f(z) \rangle^2 = \delta_w^2 \, q_{xx} \tag{12.58b}$$

For an isotropic statistical distribution of fluctuations, we have $q_{zz} \cong 1/5$ and $q_{xx} \cong 1/15$, whereas for the highly anisotropic "layered" distribution, $q_{xx} \approx \varepsilon/4$ and $q_{zz} \cong 1$.

Thus, the r.m.s. excess pressure gradient fluctuations, $\nabla \delta p$, are related to the mean pressure gradient by

$$\frac{\partial \delta p}{\partial z} \frac{1}{f(z)} \approx \pm \delta_w (q_{zz})^{1/2}$$

$$\frac{\partial \delta p}{\partial x} \frac{1}{f(z)} \approx \frac{\partial \delta p}{\partial y} \frac{1}{f(z)} \approx \pm \delta_w (q_{xx})^{1/2}$$

For an isotropic statistical distribution, we then have

$$\frac{\partial \delta p}{\partial z} \frac{1}{f(z)} \approx \pm\, 11 \langle S \rangle (1 - \langle S \rangle)(10)^{-1/2} \approx 3 \langle S \rangle (1 - \langle S \rangle) \tag{12.59}$$

so that all components of fluctuations in pressure gradient are comparable to the mean pressure gradient for sandiness fractions greater than about 20% and less than about 80%. For highly anisotropically distributed fluctuations in permeability, we have $q_{xx}/q_{zz} \approx \varepsilon/4$, so that horizontal pressure gradient fluctuations are reduced relative to vertical pressure fluctuations by the ratio $\varepsilon^{1/2}/2$; that is, the vertical pressure gradient fluctuations are comparable to the mean vertical pressure gradient, but the horizontal components are reduced by order $\varepsilon^{1/2}/2$—leading to roughly uniform "layers" of horizontal pressure fluctuations.

The spatial compartmentalization scale of the pressure fluctuations is related to the correlation length, l, of the statistical fluctuations in spatial permeability. For isotropic fluctuations, the spatial pressure scales are h_{xx}, h_{yy}, and h_{zz} and, apart from factors of order unity, are all roughly equal to l, suggesting that fluctuations in excess fluid pressure occur on the same scale as spatial fluctuations in permeability. For a highly anisotropic distribution of permeability fluctuations, the relative scales are all much smaller than the permeability fluctuation scale with

$$h_{zz} \approx \varepsilon \ell; \qquad h_{xx} \cong h_{yy} \approx \varepsilon^{3/2} \ell$$

so that the horizontal scale, h_{xx}, of pressure compartmentalization is smaller than the vertical scale, h_{zz}, by a factor of order $\varepsilon^{1/2}$, and the vertical scale of fluctuations is itself smaller than the permeability correlation scale by a factor ε; thus, the pressure variations occur quickly and are of large magnitude on a statistical scale that is small compared to mean fluctuations.

12.3.3.4. Fluctuating Oil Properties

The fluctuations in vertical oil flow speed, $\langle \delta v_z^2 \rangle$, relative to the corresponding fluctuations in vertical water flow, $\langle \delta w_z^2 \rangle$, are given in approximate magnitude by the ratio R_{ow}, which, from Eq. (12.32d), is

$$R_{ow} \approx (\delta_o/\delta_w)^2 \exp(\delta_o^2 - \delta_w^2)(V_{o*}/V_{w*})^2[k_o(z)/k_o(0)]^{m+1}[k_w(z)/k_w(0)]^{-(n+1)} \tag{12.60}$$

Inserting the functional forms for numerical estimates given earlier in this section, we have

$$R_{ow} \approx \left(\frac{V_{o*}}{V_{w*}}\right)^2 \exp\left[-z\left(\frac{1}{l_{\text{eff}}} - \frac{1}{L_{\text{eff}}}\right)\right] \times \exp\left[\left(\ln\frac{r}{q}\right)^2 \beta \langle W\rangle^2 (1 - \langle W\rangle)^2\right] \left[1 + \frac{\beta}{\alpha}\left(\ln\frac{r}{q}\right)^2 \frac{\langle W\rangle^2(1 - \langle W\rangle)^2}{\langle S\rangle^2(1 - \langle S\rangle)^2}\right] \tag{12.61}$$

With $\ln(r/q) \approx 1$, $\beta \approx \alpha \approx 1/4$, and $l_{\text{eff}}^{-1} = (m/n) \quad L_{\text{eff}}^{-1}$, we have

$$R_{ow} \cong \left(\frac{V_{o*}}{V_{w*}}\right)^2 \exp\left[-zL_{\text{eff}}^{-1}\left(\frac{m}{n} - 1\right) \times \exp\left[\frac{1}{4}\langle W\rangle^2(1 - \langle W\rangle)^2\right] \left[1 + \frac{\langle W\rangle^2(1 - \langle W\rangle)^2}{\langle S\rangle^2(1 - \langle S\rangle)^2}\right] \tag{12.62}$$

Thus, for $m \geqslant n$, the fluctuations in vertical oil velocity relative to the fluctuations in vertical water velocity decrease with increasing depth. However, at a given depth, the fluctuations are large compared to the mean flow speeds, V_{o*} and V_{w*}, provided the mean water wetness is in the rough range $20\% \leqslant \langle W \rangle \leqslant 80\%$. For values outside this range, the relative magnitudes of the oil and water velocity fluctuations are controlled by the exponential factor, varying with increasing depth.

Thus, oil flow tends to be more vertically oriented with less fluctuations than does water flow for water-wet ($\langle W \rangle = 100\%$) or oil-wet ($\langle W \rangle = 0$) rocks. As depth increases, the fluctuations in vertical oil flow speed "iron out" relative to water fluctuations, leading to a more vertically oriented laminar type of flow for oil.

12.4. Convection and Diffusion of Oil

The spatial variation of stratigraphically controlled fluctuations in permeability impresses a random component onto the pathways available for oil migration. In its simplest form, this random motion is best handled as a random walk type of behavior for the secondary migration of oil with an ordered component of motion superimposed. As shown in Eq. (12.40), for a vertically oriented mean flow, the probability, P, of finding an element of oil in the coordinate range z to $z + dz$, x to $x + dx$, y to $y + dy$ at time t, given that the element was at coordinates z_0, x_0, y_0 at time $t = t_0$, is governed by

$$\frac{\partial P}{\partial t} = |V_{o*}| \left\{ \frac{\partial}{\partial z} \left[a \exp\left(-\frac{z}{\ell_z}\right) \frac{\partial P}{\partial z} - b \exp\left(-\frac{z}{l_{\text{eff}}}\right) P \right] + a \exp\left(-\frac{z}{l_z}\right) \left(\frac{h_{xx}}{h_{zz}}\right) \left(\frac{\partial^2 P}{\partial x^2} + \frac{\partial^2 P}{\partial y^2}\right) \right\} \tag{12.63}$$

where

$$a = h_{zz}\delta_o \exp(\delta_o^2/2) > 0;$$

$$b = \left(1 + m + \frac{1}{2}\delta_o^2\right) > 0; \qquad h_{xx}/h_{zz} > 0 \tag{12.64}$$

$$l_z^{-1} = l_{\text{eff}}^{-1}(1 + m/2)/m > 0$$

Note that $l_z^{-1} < l_{\text{eff}}^{-1}$ for $m > 1$. We recognize Eq. (12.63) as being of the classical convection–diffusion form. The relative strength of vertical diffusion with respect to vertical convection is governed by the size of $a \exp(-z/l_z)\partial P/\partial z$ relative to $b \exp(-z/l_{\text{eff}})P$. For $l_z^{-1} < l_{\text{eff}}^{-1}$ it follows that as depth, z, below the subsurface increases, the effects of vertical diffusion of oil, controlled by the term $a \exp(-z/l_z)\partial P/\partial z$, become more important than the mean bulk vertical convection, controlled by the term $b \exp(-z/l_{\text{eff}})P$. The differences between vertical and horizontal diffusion are then encapsulated in the geometry factor ratio h_{xx}/h_{zz}, which provides the effect of orientational dominance of the statistical distribution of the spatial fluctuations in permeability. The effects of water wetness and mean sandiness are picked up in the dependence of the coefficients a, b, l_z^{-1}, and l_{eff}^{-1} on δ_o and m, both of which, in turn, are beholden to $\langle S \rangle$ and $\langle W \rangle$ as we have shown in detail in the previous section.

While the general solution to Eq. (12.63) is complex, a simple situation sheds light on the evolution of oil elements. Suppose first that we consider a steady-state situation ($\partial/\partial t \equiv 0$) such that the horizontal to vertical diffusion is small, that is, $h_{xx} << h_{zz}$, corresponding to the highly anisotropic statistical distribution of permeability fluctuations described in the previous section. Then Eq. (12.63) reduces to

$$\frac{\partial}{\partial z}\left[a \exp\left(-\frac{z}{l_z}\right)\frac{\partial P}{\partial z} - b \exp\left(-\frac{z}{l_{\text{eff}}}\right)P\right] = 0 \qquad (12.65)$$

so that the divergence of the vertical convective flux of hydrocarbons is balanced by the vertical diffusion. A first integral of Eq. (12.65) provides

$$a \exp\left(-\frac{z}{l_z}\right)\frac{\partial P}{\partial z} - b \exp\left(-\frac{z}{l_{\text{eff}}}\right)P = \text{constant} \qquad (12.66)$$

representing the conservation of flux. At great depths where $z >> l_z$, it would then follow that $P \propto \exp(z/l_z)$ if the constant on the right-hand side of Eq. (12.66) is not zero. However, P is the differential probability of finding the oil element and must have unit normalization when integrated over all z. Thus, the constant must be zero so that

$$\frac{\partial P}{\partial z} - \frac{b}{a}\exp\left[-z l_{\text{eff}}^{-1}\left(\frac{1}{2} - \frac{1}{m}\right)\right]P = 0 \qquad (12.67)$$

The solution of Eq. (12.67) yields

$$P \propto \exp\left[-\frac{b}{ak}\exp(-kz)\right] \qquad (12.68)$$

where $k = l_{\text{eff}}^{-1}\,(\tfrac{1}{2} - 1/m)$. Equation (12.68) illustrates the point that the relative probability of finding an oil element at a given depth is dependent on the relative strengths of convection and diffusion, measured through b/a, and upon the effective scale over which the statistical fluctuations impact the spatial dependence of the diffusion term relative to the convective term, measured through the spatial scale k^{-1}.

Thus, the vertical diffusion has a significant role to play in modifying the evolution of oil elements. Notice from Eq. (12.67) that $\partial P/\partial z > 0$ everywhere, representing the gradient flow of oil stringers from a deep source toward the sediment–water interface at $z = 0$.

The lateral (x and y) spatial diffusion of oil elements, represented through the term $a \exp(-z/l_z)(h_{xx}/h_{zz})(\partial^2 P/\partial x^2 + \partial^2 P/\partial y^2)$ on the right-hand side of Eq. (12.63), modifies the purely vertical nature of the previous steady-state calculation, whereas the variation with respect to time, which allows for intrinsic source evolution of hydrocarbon release, is picked up through the factor $\partial P/\partial t$ in Eq. (12.63).

12.5. Discussion

One of the central problems in assessing compaction and hydrocarbon migration pathways in the subsurface is related to the behavior of permeability. The difficulties in assessing the pressure and flow speeds are related to three or four major factors: the first is the high contrast in intrinsic permeabilities of sands and shales; the second is the variation in lithology both laterally and vertically in juxtaposed permeability barriers, leading to focusing and defocusing of flow and so to diffusion; the third is the fact that in water-wet rocks, the oil experiences a capillary pressure, driving it from fine to coarse-grained sediments, so that the spatial structuring of oil diffusion in complex stratigraphic sequences is quite different from that for water; and the fourth is the fact that oil is buoyant relative to water so that the vertical force on oil is greater than on water, leading to the more restricted capability for lateral flow and diffusion of oil relative to that of water, all other conditions being identical.

What we have shown here is that these differences and commonalities in behavior for oil and water can be treated in a relatively straightforward (albeit tedious) manner using the highly successful techniques of mean-field theory. In particular, both the mean flow properties and their variation with depth can be addressed at the same time as the spatially fluctuating components of the problem are being allowed for. The influence of the spatial fluctuations on the mean flow behavior is specifically incorporated.

In a sense, then, a considerable increase in understanding of the behavior of statistically steady-state flow for both oil and water can be achieved using the mean-field procedure, as we have shown. Yet all is not well, even with this improvement in visualizations of oil and water flow in complex stratigraphic sequences. First, we note that the characterization of flow has been developed only within the steady-state framework so that an implicit assumption is made that flow reaches a steady-state behavior on a time scale that is short compared to dynamic evolution of the basin. However, we know [see many references in Lerche (1990a,b)] that at least the loss of water from a sedimentary basin is tightly coupled to the deposition and compactional evolution of the basin. However, the compaction is tied directly to the rate at which fluid can be lost from a basin, implying that a self-consistent behavior of compaction and permeability is required.

Second, we note that the effects of diagenesis (both cementation and dissolution) are not incorporated in the behavior recorded here. Thus, the time-evolving properties of intrinsic rock fabrics need to be incorporated in describing both the mean flow and the fluctuating components of flow in relation to time-evolving, spatially varying permeability.

Third, we note that no direct allowance for rock fracturing, faulting, or meteoric recharge has been made, despite the recognition that such factors often play dominant roles during basin evolution and surely have impacts on the loss of fluid from the sedimentary pile.

The point is that, despite the advance in understanding brought about by using mean-field theory techniques to provide a representation of mean and fluctuating flow properties in complex stratigraphic terrains, nevertheless some considerable problems remain to be investigated before the interaction and evolution of both sediments and fluids in the subsurface is completely understood.

The mean-field results provide a representation of effective flow and pressure variations, both vertically and horizontally, in the presence of spatial variability in lithologic units. The converse problem, that is, the representation of evolution of spatially variable lithologic units in the presence of complex flow, and the self-consistent interaction of fluids and lithology remain outstanding concerns.

The significance of our results with regard to hydrocarbon exploration is that concentration of flow by random sand bodies allows large numbers of pore volumes to be flushed both laterally and vertically, providing a much greater chance that commercial oil accumulations can be found in sands than otherwise would be the case. The results also demonstrate that lateral flow is a dominant process with a good degree of fluid connectivity between sand bodies.

More specific predictions of our calculations are that permeable, randomly placed sand bodies in a shale matrix produce the following effects:

1. Bulk permeability is greatly increased over that of pure shale with addition of just a small percentage (~20–30%) of sand.
2. System permeability becomes anisotropic with a large component in the direction perpendicular to the applied pressure gradient and with lateral fluid connectivity between sand bodies being more effective than the increase in vertical permeability for sandinesses in the range ~10–90%.
3. The effective region of influence over which a sand body attracts fluid varies roughly from about two to seven times its typical linear size at about 50% sandiness down to its characteristic size at 0% and 100% sandiness.
4. Good lateral fluid connectivity between sand bodies is expected for sandiness greater than about 20–30%; hydrocarbon trapping should not be expected for sandiness in excess of about 50–60%.

Appendix

A.1. General Considerations

The driving force acting on hydrocarbons is (Hubbert, 1953) the combined pressure field

$$M = p + Z\,\Delta\rho g\hat{z} + p_c \tag{A1}$$

where p is the fluid pressure in excess of hydrostatic, $\Delta\rho$ ($\equiv \rho_w - \rho_o$) is the density difference between water (density ρ_w) and oil (density ρ_o), g is the acceleration due to gravity, $\hat{z}$ is a unit vector in the vertical direction, and p_c is the capillary pressure as a consequence of the physical difference in surface tension, τ, between oil and water and dependent also on the changes in rock properties. Hubbert (1953) has written

$$p_c = c\tau \cos\alpha \left(\frac{1}{r_{\text{coarse}}} - \frac{1}{r_{\text{fine}}}\right) \tag{A2}$$

where r_{coarse} and r_{fine} are the rock pore sizes in a fine- grained and a coarse-grained rock, respectively, c is a constant of order 15 (Hubbert, 1953), and α is the wetting angle. If permeability, κ, is taken to be a function of pore size in the form of a power law (Hubbert, 1953) with

$$\kappa \propto r^{1/a} \tag{A3}$$

then

$$p_c \propto \left(\frac{1}{\kappa_{\text{coarse}}^a} - \frac{1}{\kappa_{\text{fine}}^a}\right) \tag{A4}$$

For $\kappa_{\text{fine}} = \kappa_{\text{coarse}} + \delta\kappa$, we can write

$$p_c \propto \frac{\delta\kappa}{\kappa^{a+1}} \propto \frac{1}{\kappa^a}\delta \ln \kappa \tag{A5}$$

The flow speed, $\mathbf{v}_o$, of oil is then connected to the gradient of the combined pressure field M through Darcy's law:

$$\mathbf{v}_o = -\frac{\kappa_o}{\mu_o}\nabla M \tag{A6}$$

where κ_o is the local rock permeability to oil, ρ_o is the oil density, and μ_o is oil viscosity. The vector displacement, ξ_o, of oil from some initial position ξ_0, occurring at time $t = 0$, is then given through

$$\frac{d\xi_0}{dt} = -\kappa_o \nabla M \tag{A7}$$

with $\xi_o = \xi_0$ on $t = 0$.

As the permeability of the subsurface lithologic units changes, so too does the capillary pressure, the excess fluid pressure, and, accordingly, the velocity and spatial displacement of the oil. Under incompressible flow, the condition relating changes in combined pressure to changes in permeability is described through $\nabla \cdot \mathbf{v}_o = 0$ so that, using Eq. (A6), we can write

$$\nabla^2 M + \nabla \ln k_o \cdot \nabla M = 0 \tag{A8}$$

where $k_o = \kappa_o/\mu_o$.

A.2. Mean-Field Results for Oil Flow

The essence of the random functional methodology is to recognize that actual spatial variations in $\ln k_o$ occur randomly and often. Instead of attempting to solve Eq. (A8) for a precise spatial representation of $\ln k_o$, a more relevant solution in terms of generating an understanding of controls on fluid-flow pattern of hydrocarbons is to search for statistically average properties of Eq. (A8).

To this end, $\ln k_o$ is split into its ensemble average part, $\langle \ln k_o \rangle$, plus fluctuations around the average value; that is,

$$\ln k_o = \langle \ln k_o \rangle + \delta \ln k_o \tag{A9}$$

where angular brackets denote an ensemble averaging procedure in the manner of Frisch (1968) and Lumley (1970), about which we will have more to say later in this appendix.

Because of spatial fluctuations in permeability, a spatially fluctuating total pressure field is generated, superposed on a mean gradient. Both excess fluid pressure, p, and capillary pressure, p_c, are then split into their ensemble average parts, $\langle p \rangle$ and $\langle p_c \rangle$ respectively, plus fluctuations around the average:

$$p = \langle p \rangle + \delta p; \qquad p_c = \langle p_c \rangle + \delta p_c \tag{A10}$$

The hydrocarbon flow velocity vector then also has an average plus a fluctuating part with

$$\langle \mathbf{v}_o \rangle = - \langle k_0 \rangle \nabla \langle M \rangle - \langle \delta k_0 \nabla \delta M \rangle \tag{A11a}$$

and

$$\delta \mathbf{v}_o + \langle k_o \rangle \nabla \delta M + \delta k_o \nabla \langle M \rangle = - (\delta k_o \nabla \delta M - \langle \delta k_o \nabla \delta M \rangle) \tag{A11b}$$

with

$$\langle M \rangle = \langle p \rangle + \Delta \rho g z + \langle p_c \rangle \tag{A12a}$$

and

$$\delta M = \delta p + \delta p_c \tag{A12b}$$

The mean and fluctuating components of oil velocity impart ordered and random components to the displacement vector with

$$\frac{d \langle \xi_o \rangle}{dt} = \langle \mathbf{v}_o \rangle \tag{A13a}$$

and

$$\frac{d\delta\xi_o}{dt} = \delta\mathbf{v}_o \tag{A13b}$$

and with $\langle\xi_o\rangle = \xi_0$ on $t = 0$, and $\delta\xi_o = 0$ on $t = 0$.

From Eq. (A8) it follows that

$$\nabla^2\langle M\rangle + \nabla\langle\ln k_o\rangle \cdot \nabla\langle M\rangle + \langle\nabla\delta \ln k_o \cdot \nabla\delta M\rangle = 0 \tag{A14a}$$

and

$$\begin{aligned}\nabla^2\delta M + \nabla\langle\ln k_o\rangle \cdot \nabla\delta M + \nabla\delta \ln k_o \cdot \nabla\langle M\rangle \\ = -\left[\nabla\delta \ln k_o \cdot \nabla\delta M - \langle\nabla\delta \ln k_o \cdot \nabla\delta M\rangle\right]\end{aligned} \tag{A14b}$$

The task before us then is to solve Eqs (A14) for $\langle M\rangle$ and δM in terms of $\langle\ln k_o\rangle$ and $\delta \ln k_o$ so that we may relate the mean velocity $\langle\mathbf{v}_o\rangle$ and displacement ($\langle\xi_o\rangle$) and their mean square fluctuations ($\langle\delta v_i\, \delta v_j\rangle$, $\langle\delta\xi_i\, \delta\xi_j\rangle$) to spatially fluctuating properties of the permeability. In that way, we can determine how statistical fluctuations in permeability influence the flow and spatial diffusion of hydrocarbons in stratigraphically complex regions.

The essence of the mean-field method is to recognize that the statistical properties of interest are the mean values and ensemble averages of quadratic random components taken in pairs. These properties can be obtained to a good enough degree of approximation by setting to zero the right-hand sides of Eqs. (A11b) and (A14b). This maneuver is tantamount to neglecting the ensemble-averaged values of randomly fluctuating quantities, producted three at a time. The criteria for such neglect have been well documented in the literature (Frisch, 1968; Lumley, 1970), and we see no need to repeat them here.

On the basis of this assumption, it follows that Eq. (A14b) has the solution at location $\mathbf{x}$ of:

$$\begin{aligned}\delta p + \delta p_c \equiv \delta M(\mathbf{x}) = (2\pi)^{-3} \int (k^2 - i\mathbf{k} \cdot \mathbf{b})^{-2}\, d^3k\, d^3\,\mathbf{x}' \\ \exp[i\mathbf{k} \cdot (\mathbf{x} - \mathbf{x}')]\, \frac{\partial\delta \ln k_o(\mathrm{x}')}{\partial x_i'}\, \frac{\partial\langle M(\mathbf{x}')\rangle}{\partial x_i'}\end{aligned} \tag{A15}$$

with $\mathbf{x}'$ a dummy variable of integration; where the Einstein double summation convention over repeated indices is understood, and where we have taken the gradient $\nabla\langle\ln k_o\rangle$ to be an effectively constant vector $\mathbf{b}$.

Inserting Eq. (A15) into Eq. (A14a) gives the equation satisfied by the mean combined pressure:

$$\nabla^2\langle M\rangle + \mathbf{b}\cdot\nabla\langle M\rangle + i(2\pi)^{-3}\int d^3k\, d^3\mathbf{x}'\,(k^2 - i\mathbf{k}\cdot\mathbf{b})^{-1} \exp[i\mathbf{k}\cdot(\mathbf{x}-\mathbf{x}')]\,\frac{\partial\langle M(\mathbf{x}')\rangle}{\partial x_i'}\,k_j\,\langle\frac{\partial\delta\ln k_o(\mathbf{x})}{\partial x_j}\,\frac{\partial\delta\ln k_o(\mathbf{x}')}{\partial x_i'}\rangle = 0 \quad \text{(A16)}$$

where angular brackets are reserved for ensemble averaging operations.

Equally, from Eq. (A11a), the mean oil velocity is then given by

$$\langle\mathbf{v}_o\rangle = -\langle k_o\rangle\nabla\langle M\rangle - i(2\pi)^{-3}\int d^3k\, d^3\,\mathbf{x}'(k^2 - i\mathbf{k}\cdot\mathbf{b})^{-1} \exp[i\mathbf{k}\cdot(\mathbf{x}-\mathbf{x}')]\,\frac{\partial\langle M(\mathbf{x}')\rangle}{\partial x_i'}\,\mathbf{k}\,\langle\delta k_o(x)\,\frac{\partial\delta\ln k_o(\mathbf{x}')}{\partial x_i'}\rangle \quad \text{(A17)}$$

The correlated behavior of combined pressure fluctuations is then given by

$$\langle\delta M(\mathbf{x})\,\delta M(\mathbf{x}')\rangle = (2\pi)^{-6}\int d_3k\, d^3\ell\, d^3u\, d^3v\,(k^2 - i\mathbf{k}\cdot\mathbf{b})^{-1}(\ell^2 - i\ell\cdot\mathbf{b})^{-1} \exp[i\mathbf{k}\cdot(\mathbf{x}-\mathbf{u}) + i\ell\cdot(\mathbf{x}'-\mathbf{v})] \frac{\partial\langle M(\mathbf{u})\rangle}{\partial u_i}\,\frac{\partial\langle M(\mathbf{v})\rangle}{\partial v_j}\,\langle\frac{\partial\delta\ln k_o(\mathbf{u})}{\partial u_i}\,\frac{\partial\delta\ln k_o(\mathbf{v})}{\partial v_j}\rangle \quad \text{(A18)}$$

whereas the correlated behavior of oil velocity fluctuations is given by

$$\begin{aligned}\langle\delta v_i(\mathbf{x})\,\delta v_j(\mathbf{x}')\rangle &= \langle\delta k_o(\mathbf{x})\,\delta k_o(\mathbf{x}')\rangle\,\frac{\partial\langle M(\mathbf{x})\rangle}{\partial x_i}\,\frac{\partial\langle M(\mathbf{x}')\rangle}{\partial x_j'} \\ &+ \langle k_o(\mathbf{x})\rangle\langle k_o(\mathbf{x}')\rangle\,\frac{\partial^2}{\partial x_i\partial x_j'}\,\langle\delta M(\mathbf{x})\,\delta M(\mathbf{x}')\rangle \\ &+ \langle k_o(\mathbf{x})\rangle\,\frac{\partial\langle M(\mathbf{x}')\rangle}{\partial x_j'}\,\frac{\partial}{\partial x_i}\,\langle\delta M(\mathbf{x})\,\delta k_o(\mathbf{x}')\rangle \\ &+ \langle k_o(\mathbf{x}')\rangle\,\frac{\partial\langle M(\mathbf{x})\rangle}{\partial x_i}\,\frac{\partial}{\partial x_j'}\,\langle\delta M(\mathbf{x}')\,\delta k_o(\mathbf{x})\rangle\end{aligned} \quad \text{(A19)}$$

Following Frisch (1968) and Lumley (1970), the assumption is made that the fluctuations in ln $k_o(\mathbf{x})$ are homogeneously correlated so that

$$\langle \delta \ln k_o(\mathbf{x})\, \delta \ln k_o(\mathbf{x}') \rangle = \delta_o^2 R(\mathbf{x} - \mathbf{x}') \tag{A20}$$

with $R(0) = 1$, so that δ_o^2 measures the variance of fluctuations in ln k_o (and may vary on the slow spatial scale of the mean values).

A.2.1. The Mean Total Pressure Field, Equation (A16)

On the assumption in Eq. (A20), Eq. (A16) for the combined mean pressure can be written

$$\nabla^2\langle M\rangle + \mathbf{b} \cdot \nabla\langle M\rangle - i(2\pi)^{-3}\, \delta_o^2 \int d^3k\, d^3\mathbf{x}' \exp[i\mathbf{k} \cdot (\mathbf{x} - \mathbf{x}')] \\ [k^2 - i\mathbf{k} \cdot \mathbf{b}]^{-1} \frac{\partial\langle M(\mathbf{x}')\rangle}{\partial x_i'} k_j \frac{\partial^2}{\partial x_i \partial x_j} R(\mathbf{x} - \mathbf{x}') = 0 \tag{A21}$$

Introduce the total mean pressure gradient, $\mathbf{G} = \nabla\langle M\rangle$. Then

$$\nabla \cdot \mathbf{G} + \mathbf{b} \cdot \mathbf{G} - i(2\pi)^{-3}\, \delta_o^2 \int d^3k\, d^3y \exp \\ [i\mathbf{k} \cdot (\mathbf{x} - \mathbf{y})]\, G_j(\mathbf{y}) k_r \frac{\partial^2}{\partial y_i \partial y_r} R(\mathbf{x} - \mathbf{y}) = 0 \tag{A22}$$

Introduce $\xi = \mathbf{y} - \mathbf{x}$, and remembering that $G_j(\xi + \mathbf{x}) \approx G_j(\mathbf{x})$, insofar as the correlation functions are taken to vary on a much more rapid scale than the mean-field components, we can then write Eq. (A22) in the form

$$\nabla \cdot \mathbf{G} + \mathbf{b} \cdot \mathbf{G} - i(2\pi)^{-3}\, \delta_o^2\, G_j(\mathbf{x}) \int d^3k\, d^3\xi \\ (k^2 - i\mathbf{k} \cdot \mathbf{b})^{-1} k_r \exp(-i\mathbf{k} \cdot \xi) \frac{\partial^2}{\partial \xi_j \partial \xi_r} R(\xi) = 0 \tag{A23}$$

The integral in the last term of Eq. (A23) can be integrated twice by parts to yield

$$\nabla \cdot \mathbf{G} + \mathbf{b} \cdot \mathbf{G} + i(2\pi)^{-3} G_j(\mathbf{x}) \int d^3k\, d^3\xi \\ (k^2 - i\mathbf{k} \cdot \mathbf{b})^{-1}\, k_r k_r k_j \exp(-i\mathbf{k} \cdot \xi) R(\xi) = 0 \tag{A24}$$

Introduce the spatial power spectrum through

$$R(\mathbf{k}) = (2\pi)^{-3} \int d^3\xi \exp(-i\mathbf{k} \cdot \xi)R(\xi) \qquad \text{(A25a)}$$

and note that $R(\mathbf{k})$ is intrinsically positive, and $R(-\mathbf{k}) = R(\mathbf{k})$. Then from Eq. (A23) we have

$$\nabla \cdot \mathbf{G} + \mathbf{b} \cdot \mathbf{G} + iG_j(\mathbf{x})\, \delta_o^2 \int d^3k\, (k^2 - i\mathbf{k} \cdot \mathbf{b})^{-1}\, k_r^2 k_j R(\mathbf{k}) = 0 \qquad \text{(A25b)}$$

Since $R(-\mathbf{k}) = R(\mathbf{k})$, it then follows that the integral in the last term of Eq. (A25b) can be written

$$\int d^3k\, (k^2 - i\mathbf{k} \cdot \mathbf{b})^{-1}\, k_r^2 k_j R(\mathbf{k}) = ib_j \int d^3k\, k_r^2(\mathbf{k} \cdot \hat{\mathbf{b}})^2\, [k^4 + (\mathbf{k} \cdot \mathbf{b})^2]^{-1}\, R(\mathbf{k}) \qquad \text{(A26)}$$

where $\hat{\mathbf{b}}$ is a unit vector in the direction of $\mathbf{b}$. Then

$$\nabla \cdot \mathbf{G} + \mathbf{b} \cdot \mathbf{G}\, [1 - \delta_o^2 \int d^3k\, k^2(\mathbf{k} \cdot \hat{\mathbf{b}})^2[k^4 + (\mathbf{k} \cdot \mathbf{b})^2]^{-1}\, R(\mathbf{k})] = 0 \qquad \text{(A27)}$$

Remembering that $\mathbf{b} = \nabla\langle \ln k_o \rangle$, it follows to quadratic order of mean-field renormalization that the mean total pressure gradient is vertically (z) directed ($\mathbf{G} = \hat{z}G$) as long as $\langle \ln k_o \rangle$ varies solely with z, and then from Eq. (A27) we have

$$G(z) = G(z = 0) \exp\{\gamma_o[\ln\langle k_o(z = 0)\rangle - \ln \langle \mathrm{k}(z)\rangle]\} \qquad \text{(A28)}$$

with

$$\gamma_o = 1 - \delta_o^2 \int d^3k\, k_z^2\, k^{-2} R(\mathbf{k}) \qquad \text{(A29)}$$

We use Eqs. (A28) and (A29) in the body of the chapter.

A.2.2. The Mean Oil Velocity, Equation (A11a)

Again introduce the vector mean total pressure gradient $\mathbf{G} = \nabla\langle M \rangle$, when, to the same order of approximation as previously, Eq. (A11a) yields

$$\langle \mathbf{v}_o \rangle = -\langle k_o \rangle \mathbf{G} + \mathbf{G}(1 - \gamma_o) \exp\langle \ln k_o \rangle$$
$$= -\mathbf{G} \exp(\langle \ln k_o \rangle)[\exp(\delta_o^2/2) - (1 - \gamma_o)] \quad \text{(A30)}$$

Thus, the mean oil velocity is parallel to the mean total pressure gradient vector. We use Eq. (A30) in the body of the chapter.

A.2.3. Correlated Total Pressure Fluctuations, Equation (A18)

In this case, to the same order of approximation, we have

$$\langle \delta M(\mathbf{x})\, \delta M(\mathbf{x}') \rangle = G^2 \delta_o^2 \int d^3k\, k_o^2 (k^4 + (\mathbf{k} \cdot \mathbf{b})^2)^{-1} R(\mathbf{k}) \exp[i\mathbf{k} \cdot (\mathbf{x} - \mathbf{x}')] \quad \text{(A31)}$$

so that the correlated total pressure fluctuations are also homogeneously distributed (function only of $\mathbf{x} - \mathbf{x}'$). The local mean square total pressure fluctuations are then given by

$$\langle \delta M(\mathbf{x})^2 \rangle = G^2 \delta_o^2 \int d^3k\, k_z^2\, (k^4 + (\mathbf{k} \cdot \mathbf{b}^2)^{-1}\, R(\mathbf{k}) \quad \text{(A32)}$$

We use these results in the main body of the chapter.

A.2.4. Correlated Oil Velocity Fluctuations, Equation (A19)

From Eq. (A19) we have, to the same order of approximation as previously, that

$$\langle \delta v_i(\mathbf{x})\, \delta v_j(\mathbf{x}') \rangle \cong \langle k_o \rangle^2 G^2 \delta_o^2 \int d^3k\, (k^4 + (\mathbf{k} \cdot \mathbf{b})^2)^{-1}\, R(\mathbf{k}) \exp[i\mathbf{k} \cdot (\mathbf{x} - \mathbf{x}')]$$
$$\{k_i k_j k_z^2 - \delta_{iz} k_j k_z k^2 - \delta_{jz} k_i k_z k^2 + \delta_{iz}\delta_{jz}[k^4 + (\mathbf{k} \cdot \mathbf{b})^2]\} \quad \text{(A34)}$$

For vertical oil velocity fluctuations ($i = z = j$), Eq. (A33) yields

$$\langle \delta v_z(\mathbf{x})\, \delta v_z(\mathbf{x}') \rangle = \langle k_o \rangle^2 G^2 \delta_o^2 \int d^3k\, (k^4 + (\mathbf{k} \cdot \mathbf{b})^2)^{-1}\, R(\mathbf{k})$$
$$(k_\perp^4 + (\mathbf{k} \cdot \mathbf{b})^2) \exp[i\mathbf{k} \cdot (\mathbf{x} - \mathbf{x}')] \quad \text{(A34)}$$

where $k_\perp^2 = k^2 - k_z^2$; while for the horizontal fluctuations of oil velocity ($i = \mathrm{x}$ or $y = \mathrm{j}$) we have

$$\langle \delta v_j(\mathbf{x})\ \delta v_j(\mathbf{x}')\rangle = \langle k_o\rangle^2 G^2 \delta^2 \int d^3k \, (k^4 + (\mathbf{k}\cdot\mathbf{b})^2)^{-1} R(\mathbf{k}) k_z^2 k_j^2 \exp[i\mathbf{k}\cdot(\mathbf{x}-\mathbf{x}')] \qquad \text{(A35)}$$

The mean square fluctuations in x- and y-components of oil velocity are given, respectively, by

$$(\delta v_x^2, \delta v_y^2) = \langle k_o\rangle^2 G^2 \delta_o^2 \int d^3k\, [k^4 + (\mathbf{k}\cdot\mathbf{b})^2]^{-1}\, R(\mathbf{k})[k_\perp^4 + (\mathbf{k}\cdot\mathbf{b})^2] \qquad \text{(A36)}$$

A.2.5. Statistical Properties of ln k_o

With ln k_o normally distributed we have (Feller, 1966; Lumley, 1970)

$$\langle k_o\rangle = \exp\left[\langle \ln k_o\rangle + \frac{1}{2}\delta_o^2\right] \qquad \text{(A37a)}$$

$$\langle \delta k_o(\mathbf{x})\ \delta \ln k_o(\mathbf{x}')\rangle = \delta_o^2 R(\mathbf{x}-\mathbf{x}') \exp\left[\langle \ln k_o\rangle + \frac{1}{2}\delta_o^2 R(\mathbf{x}-\mathbf{x}')\right] \qquad \text{(A37b)}$$

and

$$\langle \delta k_o(\mathbf{x})\ \delta \ln k_o(\mathbf{x}')\rangle / \langle k_o(\mathbf{x})\rangle^2 = \exp[\delta_o^2 R(\mathbf{x}-\mathbf{x}')] - 1 \qquad \text{(A37c)}$$

The explicit functional forms of Eqs. (A37) can then be used in Eqs. (A28), (A30), (A31) and (A32), and (A34) and (A35) to provide corresponding expressions for the total mean pressure gradient, the mean oil velocity, correlated total pressure fluctuations, and correlated oil velocity fluctuations, respectively, in terms of statistical properties of the fluctuations in ln k_o.

A.2.6. Reference to a Nonfluctuating Medium

To provide a basis for estimating the effects of fluctuations in controlling or modifying the flow of hydrocarbons in the subsurface, we refer all physical quantities of interest to the values such quantities would have had in the absence ($\delta_o^2 = 0$) of fluctuations in the medium, but with the same value for $\langle \ln k_o\rangle$.

a. Mean Total Pressure Gradient. In the absence of any fluctuations in ln k_o, the mean total pressure gradient would be given by

$$G_*(z) = G(z = 0) \exp[\langle \ln k_o(z = 0)\rangle - \langle \ln k_o(z)\rangle] \tag{A38}$$

so that

$$G(z)/G_*(z) = [G_*(z)/G(z = 0)]^{-m} \tag{A39}$$

where

$$m = \delta_o^2 \int d^3k\, k_z^2\, k^{-2} R(\mathbf{k}) > 0 \tag{A40}$$

Notice that for $\langle \ln k_o(z)\rangle$ decreasing with increasing subsurface depth, it follows that $G_*(z)$ *increases* with increasing depth and that the total mean pressure gradient, $G(z)$, increases more slowly than in the case of no fluctuations—representing the "bleeding off" of total pressure because of the greater connectivity of highly permeable pathways for hydrocarbon flow.

b. Mean Oil Velocity. In the absence of any fluctuations in ln k_o, the mean oil velocity would be

$$\langle \mathbf{v}_{o*}\rangle = -G_*(z)\hat{z} \exp\langle \ln k_o(z)\rangle \equiv \hat{z}\langle V_{o*}\rangle \tag{A41}$$

while, in the presence of fluctuations, with $\langle \mathbf{v}_o\rangle = \langle V_o\rangle \hat{z}$ we have

$$\langle V_o\rangle/\langle V_{o*}\rangle = (G/G_*)[\exp(\delta_o^2/2) + m] \cong [G_*(z)/G(z = 0)]^{-m} \left\{1 + \delta_o^2\left[\frac{1}{2} + \int d^3k\, k_z^2 k^{-2} R(\mathbf{k})\right]\right\} \tag{A42}$$

so that, as $G_*(z)$ *increases* with increasing depth, so $\langle V_o\rangle$ *decreases* relative to its surficial value, and $\langle V_o\rangle$ exceeds $\langle V_{o*}\rangle$ at shallow depths near the sediment–water interface by the factor $\exp(\delta_o^2/2 + m)$.

Notice that in the absence of any fluctuations then

$$G_*(z) \exp\langle \ln k_o(z)\rangle = G(z = 0) \exp[\langle \ln k_o(z = 0)\rangle] = \text{constant} \tag{A43}$$

so that $V_{o*}(z) = V_{o*}(z = 0) =$ constant, representing a steady-state loss of hydrocarbons at a constant vertical speed, V_{o*}.

c. Fluctuations in Total Pressure and Oil Flow. In the absence ($\delta_o^2 = 0$) *of any spatial fluctuations in the medium properties, both* $\langle \delta M \, \delta M \rangle$ and $\langle \delta v_i \, \delta v_j \rangle$ are identically zero. In terms of the mean total pressure gradient, G, we can write the statistical fluctuations in total pressure gradient, $\delta \mathbf{G} = \nabla \delta M$, in the form

$$\langle \delta G_i(\mathbf{x}) \, \delta G_j(\mathbf{x}) \rangle = G^2 \delta_o^2 \, s_{ij} \tag{A44}$$

where the angular (dimensionless) correlation matrix is given approximately by

$$s_{ij} = \int d^3k \; k_z^2 \; k_i k_j k^{-4} R(\mathbf{k}) \tag{A45}$$

The fluctuations in oil velocity are then described through the statistical mean square behavior

$$\langle \delta v_z(\mathbf{x})^2 \rangle = \langle k_o \rangle^2 G^2 \delta_o^2 (s_{xx} + s_{yy}) \tag{A46a}$$

and

$$\langle \delta v_i(\mathbf{x}) \, \delta v_j(\mathbf{x}) \rangle = \langle k_o \rangle^2 G^2 \delta_o^2 \, s_{ij} \qquad i, j = x \text{ or } y \tag{A46b}$$

Let us leave these results to one side for the moment and concentrate now on the equivalent results for *water* flow.

A.3. Mean-Field Results for Water Flow

In the case of water flow, three considerations suggest a different behavior for mean and fluctuating properties than occurs in the case of oil flow. First, as Hubbert (1953) noted, the question of whether capillary pressure has a role to play in impacting flow magnitude and direction is influenced to a dominant degree by whether the lithofabric is water-wet or oil-wet. If a water-wet condition prevails, then there is no capillary pressure resistance to water flow across different lithofabrics or, indeed, through a monolithologic fabric with variable grain size. The capillary pressure force then acts only on the oil, forcing motion of the oil from fine-grained to coarse-grained regions. If an oil-wet condition prevails, then, as Hubbert (1953) noted, the converse situation obtains: water is forced from fine-grained to coarse-grained regions, and there is no capillary pressure impediment to oil flow through different lithofacies. We treat here the water-wet situation, in which case the capillary pressure effect on water flow is absent.

Second, the permeability to flow is different for oil than for water. Thus, the effective hydraulic conductivity $k = \kappa/\rho\mu$ depends on the differences in permeability, density, and viscosity between water and oil. Thus, even if the driving forces on water and oil were to be identical, the flow speeds and directions of the two fluids would differ because of the change in physical properties of the fluids in relation to geometrical considerations of pore throat behavior in the petrofabrics through which the two fluids must flow.

Third, while there is a buoyancy pressure drive acting on the hydrocarbons (because of their lower density relative to that of water), no such force is available to cause water to flow in a similar manner. Indeed, the counter argument can be made. If there is a local *upwardly* directed buoyancy force on oil of $\delta\rho g\hat{z}$, then this force is a buoyancy *relative* to replacement of the oil by water. Thus, the buoyancy force cannot act on the water as well as on the oil.

Thus, insofar as water flow is concerned, the force acting to provide steady-state flow is that due to fluid pressure in excess of hydrostatic. The steady-state flow of water under Darcy's law is then described through

$$\mathbf{V}_w = -\kappa_w(\mu_w)^{-1}\nabla p \tag{A47}$$

while incompressibility considerations ($\nabla \cdot \mathbf{V}_w = 0$) give the equation for excess fluid pressure in the form

$$\nabla^2 p + \nabla \ln k_w \cdot \nabla p = 0 \tag{A48}$$

where κ_w is the permeability to water, ρ_w is the water density, μ_w is the kinematic viscosity of water, and $k_w = \kappa_w/(\mu_w)$.

Direct comparison of Eq. (A48), for water flow driven by excess pressure, with Eq. (A8), for oil flow driven by excess pressure, buoyancy, and capillarity, shows that a virtually identical analysis can be performed. Tracing through the arguments allows us to write down the corresponding steady-state water flow behaviors.

A.3.1. Mean Excess Pressure Gradient

Introduce f_*, the mean excess fluid pressure gradient in the *absence* of any spatial fluctuations in $\ln k_w$,

$$f_*(z) = f_*(z = 0)\exp[\langle \ln k_w(z = 0)\rangle - \langle \ln k_w(z)\rangle] \tag{A49}$$

so that $f(z)$, the mean excess fluid pressure gradient in the *presence* of spatial fluctuations in $\ln k_w$, is then given by

$$f(z)/f_*(z) = [f_*(z)/f_*(z = 0)]^{-n} \tag{A50}$$

where

$$n = \delta_w^2 \int d^3k\, k_z^2\, k^{-2} r(\mathbf{k}) \geq 0 \tag{A51}$$

with $\delta_w^2 = \langle(\delta \ln k_w)^2\rangle$ and with $r(\mathbf{k})$ the power spectral density of spatial fluctuations in $\delta \ln k_w(\mathbf{x})$.

A.3.2. Mean Water Flow Velocity

In the absence of any fluctuations in $\ln k_w$, the mean water flow velocity would be

$$\langle V_{w*}\rangle = \langle \mathrm{V}_{w*}\rangle\hat{z} = f_*(z)\hat{z} \exp\langle \ln k_w(z)\rangle = \text{constant} \tag{A52}$$

while in the presence of fluctuations with $\langle \mathbf{V}_w\rangle = \hat{z}\langle V_w\rangle$ we have

$$\langle V_w\rangle/\langle V_{w*}\rangle = [f_*(z)/f_*(z = 0)]^{-n} \left\{1 + \delta_w^2\left[\frac{1}{2} + \int d^3k\, k_z^{2-2} r(k)\right]\right\} \tag{A53}$$

A.3.3. Fluctuations in Excess Fluid Pressure and Water Flow

In the absence of any spatial fluctuations in the medium properties ($\delta_w^2 = 0$), both $\langle \delta p\, \delta p\rangle$ and $\langle \delta w_i\, \delta w_j\rangle$ are identically zero, where δp is the fluctuation in excess fluid pressure and δw is the fluctuation in water velocity.

In terms of the mean excess fluid pressure gradient, f, we can write the statistical fluctuations in excess fluid pressure gradient, $\delta f = \nabla \delta p$, in the form

$$\langle \delta f_i(\mathbf{x})\, \delta f_j(\mathbf{x})\rangle = f(z)^2\, \delta_w^2\, q_{ij} \tag{A54}$$

where the dimensionless angular correlation matrix is given approximately by

$$q_{ij} = \int d^3k\, k_z^2\, k_i k_j k^{-4} r(\mathbf{k}) \tag{A55}$$

The fluctuations in water velocity are then described statistically through

$$\langle \delta w_z(\mathbf{x})^2 \rangle = \langle k_w \rangle^2 f(z)^2 \delta_w^2 (q_{xx} + q_{yy}) \tag{A56a}$$

and

$$\langle \delta w_i(\mathbf{x}) \, \delta w_j(\mathbf{x}) \rangle = \langle k_w \rangle^2 f(z)^2 \delta_w^2 q_{ij} \qquad i, j = x \text{ or } y \tag{A56b}$$

A.4. Oil Flow Relative to Water Flow

A.4.1. Mean Pressure Fields

For ease in presentation, we introduce several ancillary quantities. We write $k_w(z) = \exp\langle \ln k_w(z) \rangle$ and $k_o(z) = \exp\langle \ln k_o(z) \rangle$. We introduce the excess mean fluid pressure gradient $f_*(0) = \partial\langle p \rangle/\partial z$ and the mean capillary pressure gradient $p'_* = \partial p_c/\partial z$, on $z = 0$, the sediment–water interface.

It then follows that we have the mean excess fluid pressure gradient given by

$$f(z) = f_*(0)[k_w(z)/k_w(0)]^{n-1} \tag{A57a}$$

while the mean *total* pressure gradient, $G(z)$, is provided through

$$G(z) = (\Delta\rho g + f_*(0) + p'_*)[k_o(z)/k_o(0)]^{m-1} \tag{A57b}$$

and the mean capillary pressure gradient is given by

$$\partial\langle p_c \rangle/\partial z = G(z) - f(z) - \Delta\rho g \equiv f_*(0) \{[k_o(z)/k_o(0)]^{m-1} - [k_w(z)/k_w(0)]^{n-1}\}$$
$$+ \Delta\rho g\{[k_o(z)/k_o(0)]^{m-1} - 1\} + p'_*[k_o(z)/k_o(0)]^{m-1} \tag{A57c}$$

A.4.2. Mean Vertical Flow Speed Difference

Using the formulas for V_o and V_w, we can write the difference as

$$\Delta V = V_o - V_w = f_*(0)\{k_o(0)[m + \exp(\delta_o^2/2)][k_o(z)/k_o(0)]^m$$
$$- k_w(0)[n + \exp(\delta_w^2/2)][k_w(z)/k_w(0)]^n\} \tag{A58}$$
$$+ (\Delta\rho g + p'_*)k_o(0)[m + \exp(\delta_o^2/2)][k_o(z)/k_o(0)]^m$$

A.4.3. Fluctuations in Flow Velocity

From the formulas for mean square flow speed fluctuations, we have the relationship between velocity dispersion of oil flow and velocity dispersion of water flow characterized by the ratio

$$R_{ow} = (\delta_o/\delta_w)^2 \exp(\delta_o^2 - \delta_w^2)[k_o(0)/k_w(0)]^2[k_o(z)/k_o(0)]^{m+1} \quad \text{(A59)}$$
$$[k_w(0)/k_w(z)]^{n+1}[\Delta\rho g + f_*(0) + p_*']^2 f_*(0)^{-2}$$

with

$$\langle \delta v_z^2 \rangle = \langle \delta w_z^2 \rangle R_{ow}(s_{xx} + s_{yy})/q_{xx} + q_{yy}) \quad \text{(A60a)}$$

$$\langle \delta v_i \, \delta v_j \rangle = \langle \delta w_i \, \delta w_j \rangle R_{ow} \, (s_{ij}/q_{ij}) \quad \text{(A60b)}$$

and with

$$\langle \delta w_z^2 \rangle = f_*(0)^2 \, \delta_w^2 \exp(\delta_w^2) \; k_w(0)^2[k_w(z)/k_w(0)]^{n+1}(q_{xx} + q_{yy}) \quad \text{(A60c)}$$

$$\langle \delta w_i \, \delta w_j \rangle = \langle \delta w_z^2 \rangle \, q_{ij}/(q_{xx} + q_{yy}) \quad \text{(A60d)}$$

A.5. Oil Displacement

The oil displacement vector, ξ_o, is split into the two components, $\langle \xi_o \rangle$ and $\delta\xi_o$, with

$$\frac{d\langle \xi_o \rangle}{dt} = V_o(z)\hat{z} \quad \text{(A61)}$$

and

$$\frac{d\delta\xi_o}{dt} = \delta \mathbf{v}_o(\mathbf{x}) \quad \text{(A62)}$$

Then note that, with $\langle \xi_o \rangle = \xi_o \hat{z}$, the mean displacement of an element of oil which starts at depth z_0 at time $t = t_0$ is given (with $\xi_o = z_0 - z$ and depth, z, measured increasing from the sediment–water interface) by

$$\frac{dz}{dt} = - V_o(z) \quad \text{(A63)}$$

so that, for $V_o > 0$, z decreases as time increases, representing the rise of the oil toward the sediment surface.

To examine simply the diffusive aspects of oil flow around the mean flow direction, write

$$\frac{dx_i}{dz} = \frac{\delta v_i(\mathbf{x})}{V_o(z)} \qquad i = x \text{ or } y \tag{A64}$$

where $\delta\xi_o = \mathbf{x} = (x\hat{x}, y\hat{y})$ for flow components perpendicular to the mean flow direction.

From Eq. (A64) it follows that

$$x_i(z) = -\int_{z_0}^{z} \delta v_i[\mathbf{x}(z')]V_o(z')^{-1}\, dz' \qquad i = x \text{ or } y \tag{A65}$$

It also follows that

$$\langle x_i(z)\, x_j(z)\rangle = \int_{z_0}^{z} dz_1 \int_{z_0}^{z} dz_2\, V_o(z_1)^{-1}\, V_o(z_2)^{-1}\, \langle \delta v_i[\mathbf{x}(z_1)]\delta v_j[\mathbf{x}(z_2)]\rangle \tag{A66}$$
$$i, j = x \text{ or } y$$

Now the transverse (to z) components of velocity are taken to be random. We are interested in the matrix $\langle x_i x_j\rangle$, which characterizes the rate at which a thin pencil of hydrocarbon flow lines spreads out transversely as evolution proceeds from some point $z = z_o$.

In an obvious notation we can write the two-dimensional Fourier representation

$$\delta v_i(z) \equiv \delta v_i(x_\perp(z), z) = \int d^2\mathbf{k}_\perp\, \delta v_i(\mathbf{k}_\perp, z)\exp[i\mathbf{k}_\perp \cdot \mathbf{x}_\perp(z)] \tag{A67}$$

Defining $z_2 - z_1 = \Delta$ and $z_2 + z_1 = 2S$, we have Eq. (A66) in the form

$$\langle x_i(z)\, x_j(z)\rangle = V_o(z)^{-2} \int ds \int d\Delta\, \langle \delta v_i(0)\, \delta v_j(\Delta)\rangle \tag{A68}$$

where we have assumed that (i) the fluctuations are homogeneously correlated on the scale of interest and so have written $\langle \delta v_i(z_1)\, \delta v_j(z_2)\rangle = \langle \delta v_i(0)\, \delta v_j(z_2 - z_1)\rangle$ and (ii) the correlation scale is short compared to the scale over which $V_o(z)$ changes.

For all physical systems, $\langle \delta v_i(0)\ \delta v_j(\Delta)\rangle \to 0$ as $\Delta \to \infty$, and $\langle \delta v_i(0)\ \delta v_j(\Delta)\rangle$ is even in Δ (Lumley, 1970). For $\Delta\langle \delta v_i(0)\ \delta v_j(\Delta)\rangle \to 0$ as $\Delta \to \infty$, we can consider the case in which $|z - z_0|$ is much larger than the scale length over which $|\langle \delta v_i(0)\ \delta v_j(\Delta)\rangle|$ tends to zero (the correlation length). Then

$$\langle x_i(z)\ x_j(z)\rangle \cong 2(z - z_0)\ V_o(z)^{-2} \int_{0)}^{z-z_0} \langle \delta v_i(0)\ \delta v_j(\Delta)\rangle\ d\Delta \qquad \text{(A69)}$$

The ratio $\langle x_i(z)\ x_j(z)\rangle/(z - z_0)$ then becomes independent of z on the local scale and represents the diffusion coefficient for random walk of the hydrocarbon flow lines around the mean oil flow direction.

The basic argument here is that the mean oil speed, $V_o(z)$, changes slowly on the spatial scale of the fluctuation correlation length and so can be regarded as a constant in working out the statistical properties of the transverse diffusion.

With the transverse (to z) diffusion coefficient for oil displacement defined by

$$D_{ij} = \lim_{z\to\infty} \frac{\langle x_i(z)\ x_j(z)\rangle}{z - z_0}$$

we have

$$D_{ij}(z) \cong 2\ [\langle \delta v_i\ \delta v_j\rangle/s_{ij}]\ V_o(z)^{-2} \int_0^z d\Delta \int d^3k\ R(\mathbf{k}) k_z^2 k^{-4} k_i k_j \\ \langle \exp\{i\mathbf{k}_\perp \cdot [\mathbf{x}(0) - \mathbf{x}(\Delta)]\}\rangle \qquad \text{(A70)}$$

where $\mathbf{k}_\perp = \mathbf{k} - \hat{z}k_z$, and where

$$\mathbf{x}(\Delta) = \mathbf{x}(0) + \int_0^\Delta \frac{\delta\mathbf{v}(\mathbf{x})}{V_o(\Delta)}\ d\Delta \qquad \text{(A71)}$$

It is well known (Frisch, 1968; Lumley, 1970) that if $\delta\mathbf{v}$ is a centered random variable, then the integral in Eq. (A70), being a linear functional of $\delta\mathbf{v}$, is also a centered random variable and that

$$\langle \exp\{i\mathbf{k}_\perp \cdot [\mathbf{x}(0) - \mathbf{x}(\Delta)]\}\rangle = \exp\left[-\frac{1}{2}\, k_{\perp i} k_{\perp j} \int_0^\Delta du_1 \int_0^\Delta du_2\ \langle \delta v_i(u_1) \delta v_j(u_2)\rangle\ V_o^{-2}\right]$$

$$\cong \exp\left[-\frac{1}{2}\, k_{\perp i} k_{\perp j} \langle \delta v_i\ \delta v_j\rangle V_o^{-2}\ \Delta^2\right] \qquad \text{(A72)}$$

where the conventional Markov approximation (Frisch, 1968) has been made. Then use the result (Eq. A72) in Eq. (A70) to obtain the local transverse diffusion coefficient:

$$D_{ij}(z) \cong (2\pi)^{\frac{1}{2}} (|\langle \delta v_\alpha \, \delta v_\beta \rangle^2 \, s_{\alpha\beta}^{-2} \, V_o^{-2}|)^{\frac{1}{2}} \int \frac{d^3k \; R(k) k_z^2 k^4 \; k_i k_j}{(k_{\perp \ell} k_{\perp m} s_{\ell m})^{1/2}} \qquad \text{(A73)}$$

The local diffusion coefficient parallel to the mean flow can be obtained by a similar argument, yielding

$$D_{zz} \cong (2\pi)^{1/2} \, [(s_{xx} + s_{yy})^{-1} \langle \delta v_z^2 \rangle \, V_o^{-2}]^{1/2} \int d^3k \; R(k) |k_z| k_\perp^2 \, k^{-4} \qquad \text{(A74)}$$

13

Research Thrusts and Cross-Discipline Ties

The flow of fluids of similar or heterogeneous properties in permeable media can be conveniently categorized by two major factors. The first is the question of whether the fluid phases have the same or different densities so that buoyancy either is not or is a significant factor to be reckoned with. The second is the question of whether the phases are chemically or physically miscible to provide a new fluid phase with its own specific properties.

The complexity of behavior and transport of fluid phases in permeable media owes much to the above two factors. The presence of differential buoyancy implies a tendency for spatial separation to occur between two fluid phases, if allowed by other conditions in the system, most often with a surface tension separation interface between the two phases. Thus, oil is lighter than water, gas (CH_4) is lighter than water or air, steam is lighter than water, and so on. This basic physical difference is, in itself, responsible for a majority of the fascinating possible behaviors. When density difference is coupled to a direct change from one phase to another (e.g., oil conversion to gas, steam production from water, ice conversion to water under pressure or temperature increase), the two phases often having different physical properties, some surprisingly peculiar effects can arise from what would otherwise appear to be rather mundane systems.

In addition, such direct phase changes are often accompanied by latent heat [exothermically and/or endothermically depending upon which phase is converting to which other phase (steam to water is exothermic, water to steam is endothermic)]. The latent heat then plays a second role in influencing the temperature of the component parts of the system. In turn, because solubilities of different component phases are often highly dependent on temperature, it follows that chemical dissolution and precipitation effects, as well as solubility flux transport of one phase in another (methane in water, carbonate in water, gas in oil), are inextricably intertwined with the physical, chemical, and geometrical conditions that control and define the system under discussion, as is the transport of energy from one location to another. However, the transport of energy, by both convective and conductive transport, then has an impact on the local solubility of one phase in another, which, in turn, has an impact on permeability through either cementation or dissolution of the pore spaces.

The dominance of one phase relative to another means that the simplifying DuPuit approximation is often a very accurate procedure for handling the spatial structuring of the minor fluid phase relative to the dominant one. However, the presence of two dissimilar components requires that there be sources and sinks of the components. Thus, the time-dependent problem of supply and loss is then coupled to the problems of flow and energy transport through the system.

What we lack is a quantitative description in which all of the above factors are combined so that the dominance of individual facets can be explored, as can the combined impact of multiple factors. For example, even the relatively simple problem of the supply of oil to a hydrodynamic aquifer (involving the simultaneous conversion of oil to gas and the ongoing solubility pickup and transport of gas and/or oil by water, thereby eroding the initial hydrocarbon shape and location and shifting the accumulation site elsewhere, as well as the shift in buoyancy, surface tension, and capillarity drives in relation to the hydrodynamic forces, and the change in aquifer properties with time and spatial conditions) is not understood in total, although individual aspects of the problem have been subjected to scrutiny over the years.

In order to improve our capabilities for identifying likely sites of hydrocarbon accumulations prior to expensive drilling, and for delineating hydrodynamically controlled oil and/or gas fields, it is a relatively urgent matter to put together a comprehensive quantitative picture of the ebb and flow of competing and interacting factors influencing the patterns of development. This need is even more urgent as the search for commercial hydrocarbons has entered the last decade of the 20th century with replacement of reserves lagging well behind the rate necessary to maintain the current level of societal dependence on hydrocarbons and their products.

Apart from such fairly clear directions of research thrust in relation to hydrocarbons, there are a host of ancillary fields that could benefit from the availability of a generic quantitative procedure. The point is that once a quantitative protocol is available in general form, it becomes a relatively easy matter to adjust parameters and modify geometric dependences to accommodate a class of applications in a variety of settings. Several such applications have been developed in part or in whole in divers scientific arenas, each with its own particular terminology and style of expression, but all beholden to the flow of fluids through permeable media as the underlying basic support mechanism. From the literature we can immediately identify the problems of seepage through dam faces; basal water flow between an overlying mobile glacier and a low-permeability substrate; the engineering of hydrocarbon reservoir production by replacement with water or steam or by physical proppant or chemical action increasing rock permeability or by fire, CO_2, or nitrogen flooding, which change phase and solubility of the hydrocarbons; hydrothermal "plume" transport of minerals and hot water; steam–water interphasic processes in chemical engineering; sulfur liquefaction production from subsurface salt diapirs; and so on. Good references which handle these problems, ranging from simple overviews to technical treatises of considerable depth and length, are Turcotte and Schubert (1982), Scheiddegger (1960), Muskat (1946), Lister (1980), Kruger and Otte (1973), Goguel (1976), Rybach and Muffler (1981), Fyfe *et al.* (1978), Di Pippo (1980), de Wiest (1969), Combarnous and Bories (1975), Cheng (1978), Bear (1972), and Phillips (1991).

We cannot do justice to the wide variety of problems that invoke Darcy's law as a fundamental linchpin nor, in a book of this size and purpose, would it be appropriate to do so. However, the ever increasing demands placed on the environment by the impact of humanity's endeavors and the problems of hydrogeology and water cleanliness, of nuclear waste disposal impacting subsurface and surface conditions, of waste product disposal of both sewage and man-made products, of acid mine drainage, or indeed of any change artificially made or naturally occurring in the only world in which we live all require an understanding of the consequences of causes and effects on the fragile ecosphere. Such considerations have led to an explosive growth in the past decade in the development of methods and techniques for increasing our understanding and awareness of systems dominated by permeable flow through porous media.

This burgeoning growth of understanding shows no sign of slowing down, and the procedures developed in one arena are often easily transported to another, suggesting that our understanding of hydrodynamic flow of hydrocarbons in subsurface conditions can only increase by application of cross-discipline techniques. Such factors strongly argue for the exciting possibility of improving the chances of finding hydrocarbons in areas that might otherwise have been overlooked or declared barren on the basis of conventional procedures.

It is toward these ends that we hope the investigations presented here contribute in some small way. We will more than have accomplished our purpose if others can build on the patterns of behavior reported here so that the subject of hydrodynamic flow of hydrocarbons in aquifers is converted into a quantitative tool of highly accurate predictive capability in decreasing the exploration risk of finding and producing commercial reserves.

Appendix

The oil stringer code used in the body of the text is described and presented here along with subsidiary routines (e.g., spline fits) used in the main part of the calculations. Also included are graphics routines for a Sun Sparc3 machine, which are X-window based, and so, fairly portable.

The instructions for operation of the oil stringer code are as follows:

1. Set up the geometry control file aq.dat.
2. Set up the permeability control file kp.dat.
3. Set up the simulation control file simu.dat.
4. Run the program module AQUIFER (source code file aq.c).
5. Run the program module PERM (source code file kp.c).
6. Run the program module SPL (source code file spl.c).
7. Run the program module SIMU (source code file oila.c).
8. Run the graphics module PLOT (source code file oilplot.c).

Module AQUIFER

The AQUIFER module reads data from the file aq.dat, which contains the necessary information about the aquifer geometry and the method to be used for setting up the geometry information for the following modules. There are three different ways of defining the geometry:

1. A spline curve is calculated based on digitized *x,y* coordinates.
2. A straight line is calculated between digitized *x,y* coordinates.
3. A trigonometric function calculates the geometry grid from input amplitude, wavelength, and general slope.

Two output files are generated: spl.dat, which is to be used by following modules, and plot.dat, which contains the geometry information for screen graphics.

Functions

dummy( )
input( )
aqcalc ()
spline1( )
spline2( )
line1( )
line2( )

```
SET UP AQUIFER DATA
--------------------------------------------------------
flag  pflag
1     1
limit1   limit2
3.0       55.0
xistep   flowdirection
0.5       1
****Function Aq Parameters**************
ap   a1   a2   w   angle
0.0  12.0 5.0  3.14  30.0
****Spline Aq Parameters***************
num
30
x1   y1   x2   y2
0.0  10.0 0.0  22.0
2.0  13.0 2.0  25.0
4.0  16.0 4.0  28.0
6.0  19.0 6.0  31.0
8.0  22.0 8.0  34.0
10.0 25.0 10.0 37.0
12.0 28.0 12.0 40.0
14.0 31.0 14.0 43.0
16.0 34.0 16.0 46.0
18.0 37.0 18.0 49.0
20.0 40.0 20.0 52.0
22.0 40.0 22.0 52.0
24.0 40.0 24.0 52.0
26.0 40.0 26.0 52.0
28.0 40.0 28.0 52.0
30.0 40.0 30.0 52.0
```

Module AQUIFER

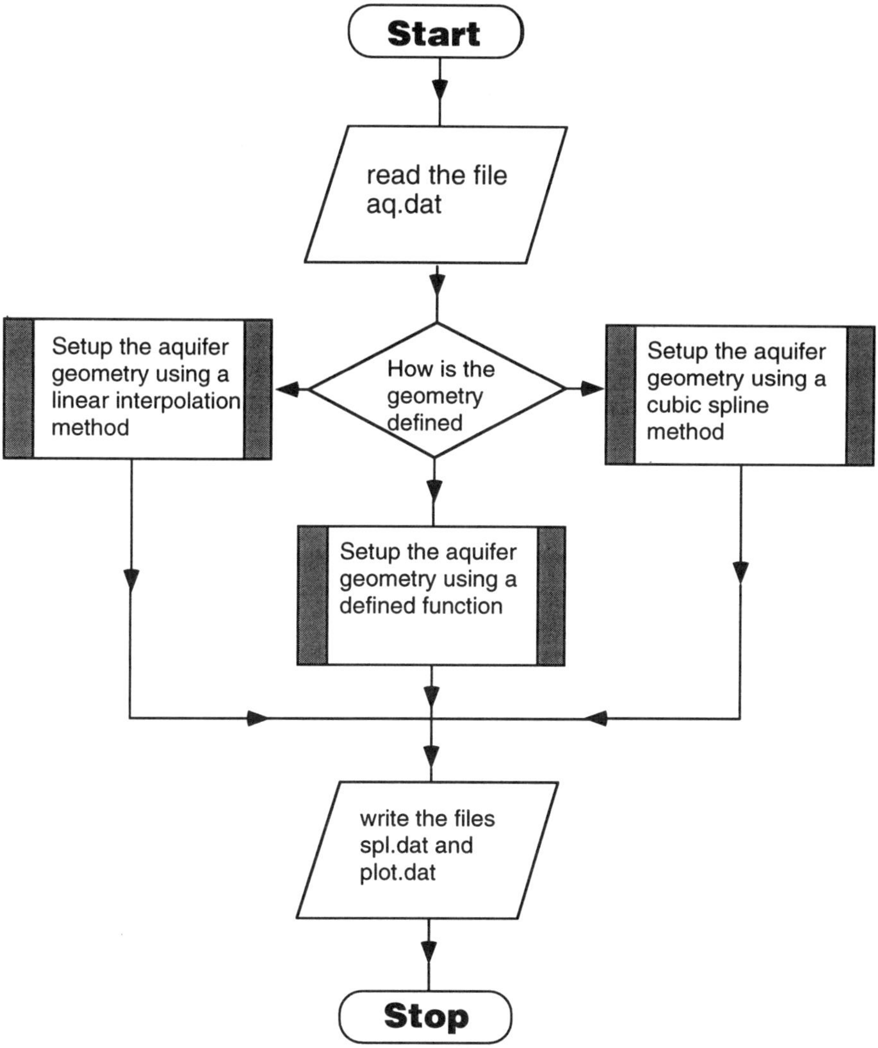

Module AQUIFER

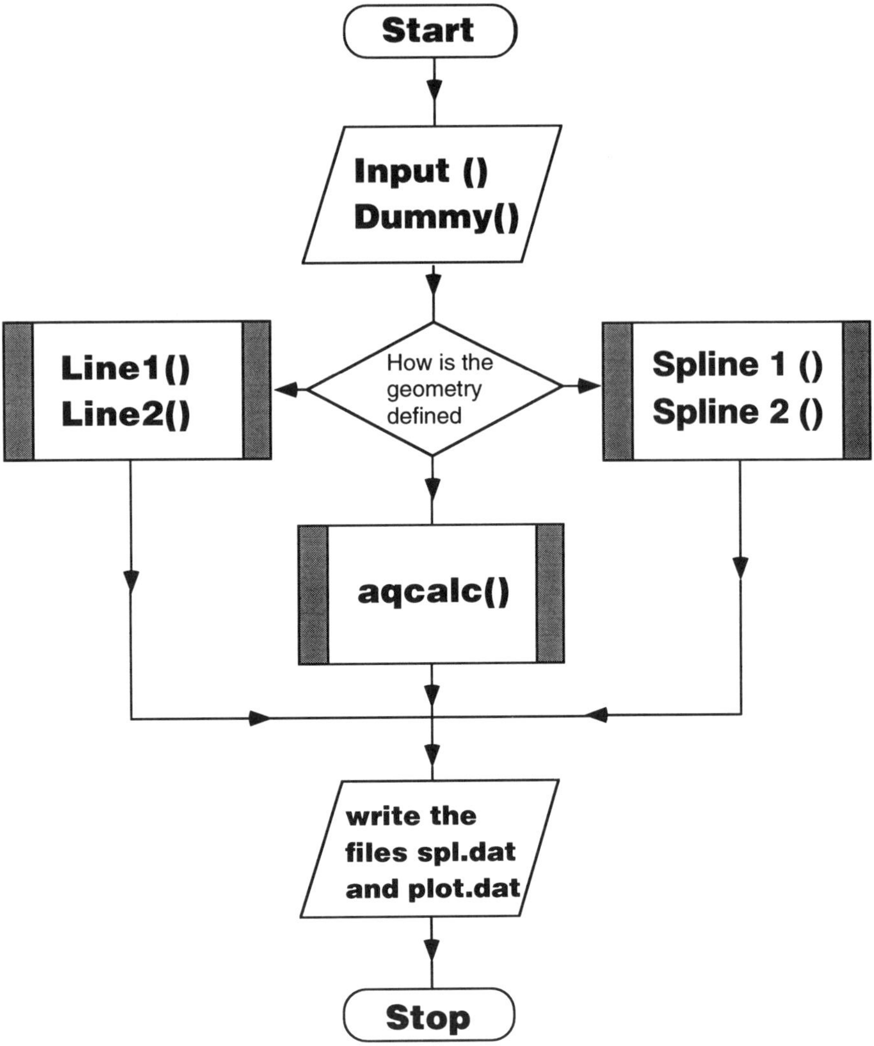

```
32.0 40.0 32.0 52.0
34.0 40.0 34.0 52.0
36.0 40.0 36.0 52.0
38.0 40.0 38.0 52.0
40.0 37.0 40.0 49.0
42.0 34.0 42.0 46.0
44.0 31.0 44.0 43.0
46.0 28.0 46.0 40.0
48.0 25.0 48.0 37.0
50.0 22.0 50.0 34.0
52.0 19.0 52.0 31.0
54.0 16.0 54.0 28.0
56.0 13.0 56.0 25.0
58.0 10.0 58.0 22.0
-------------------------------------------------------
        NOTE
pflag   --- when pflag=0, draw aquifer; when pflag=1, draw
          aquifer and oilstringer
flag    --- flag=0, input data are used to get aquifer shape
          by cubic spline(recommended: use this option
          when the aquifer boundary is relatively smooth);
          --- flag=1, input data are used to get aquifer shape
             by linear interpolation; (recommended: use this option
             when the aquifer boundary has sharp angles( > 40.0 degrees));
        --- flag=2, forward calculation of aquifer shape by
            the functions:
                    y = ap*cos(w*x)+a1          (1)
                    y = x*tan(angle)+a1         (2)
                    y = x*tan(angle)+a2         (3)
x(m)      --- x value of aquifer
y(m)      --- y value of aquifer
ap(m)     --- coefficient of cosine function, if ap is not equal 0.0,
          calculate top of aquifer using (1); if ap=0.0,
          calculate top of aquifer using (2) (default: 0.0)
a1(m)     --- coefficient of straight line for calculating
          top of aquifer (default: 12.0)
a2(m)     --- coefficient of straight line for calculating
          bottom of aquifer (default: 5.0)
w(1/m)    --- wavenumber of cosine function in equation(1) (default: 3.14)
angle(degree)-angle of aquifer when aquifer's angle is constant
          (recommended input data range: 0.0<angle<80.0, default: 30.0)
flowdirection- =1 water flows from right to left
           =-1 water flows from left to right
limit1(m) --- start position(left end of x-axis) of aquifer for plot
limitn(m) --- last position(right end of x-axis) of aquifer for plot
xistep(m) --- interpolation step(along x-axis) for calculating aquifer
            (recommended input data range: 0.001 to 0.6, default: 0.5)
num         --- number of input points for calculating aquifer
            (input data range: 10 <=num <=500, default: see above input data)
x1, y1(m) --- array of input points for bottom of aquifer
                              (default: see above input data)
```

```
x2, y2(m) --- array of input points for bottom of aquifer
                              (default: see above input data)
        (the value of array x1, y1, x2, and y2 always positive)

#include <stdio.h>
#include <math.h>
#include "oilstr.h"

float x[SIZEAQ][3], y[SIZEAQ][3], ya[SIZEAQ], yp1[5], ypn[5], xi, yi;
float limit1, limitn;
int num[5], m, aqflag, pflag, nn, aa, flowflag;
float ap, a1, a2, angle, w, xi0, xistep;
FILE *aqin;
FILE *plo;
FILE *spl;


void dummy(dums)
int dums;
{
  int i;
  char dumchar[100];
  for (i = 0; i < dums; i++)
        fgets (dumchar, 100, aqin);
}


void input()
{
int ii;
float Haq;
dummy(3);
fascanf(aqin, "%d %d", &aqflag, &pflag);
dummy(2);
fscanf(aqin, "%f %f", &limit1, &limitn);
printf("limit1, limitn\n");
printf("%f %f\n", limit1, limitn);
dummy(2);
fscanf(aqin, " %f %d", &xistep, &flowflag);
printf(" xistep\n");
printf(" %f\n", xistep);
aa = 2
nn = (int)((limitn-limit1)/xistep):
if (aqflag ==0 || aqflag == 1)
{
 dummy(6);
 fscanf(aqin, "%d", &num[0]);
 num[1] = num[0];
 printf("num[0] num[1]\n");
 printf("%d aqflag%d\n", num[0], aqflag);
 fprintf(spl, "%d %f %f %f %d\n", num[1], xistep, limit1, limitn, flowflag);
 fprintf(spl, "%d\n", aqflag);
```

```
 fprintf(plo, "%d %d %d %d %f\n", pflag, (int)((limitn-limit1)/xistep), aa,
                                    flowflag, limitn);
 printf("x[ii][0], y[ii][0] x[ii][1] y[ii][1]\n");
 dummy(2);
 if(flowflag == 1)
  {
      for (ii=0;ii<num[0]; ++ii)
       {
        fscanf (aqin, "%f %f %f %f",
&x[ii]{0],&y[ii][0],&x[ii][1],&y[ii][1]);
        printf ("%f %f %f %f\n", x[ii][0],y[ii][0],x[ii][1],y[ii][1]);
        fprintf(spl, "%f %f %f\n", x[ii][1],y[ii][1],y[ii][1]-y[ii][0]);
       }
 }
else
 {
 for (ii=0;ii<num[0]; ++ii)
     {
       fscanf (aqin, "%f %f %f %f", &x[ii][0],&y[num][0]-1-ii]
                                [0],&x[ii][1],&y[num[0]-1-ii][1]);
       }
 for (ii=0;ii<num[0]; ++ii)
       {
        printf ("%f %f %f %f\n", x[ii][0],y[ii][0],x[ii][1],y[ii][1]);
        fprintf (spl, "%f %f %f\n", x[ii][1],y[ii][1],y[ii][1]-y[ii][0]);
       }
 }
 yp1[0] =(y[1][0]-y[0][0])/(x[1][0]-x[0][0]);
 ypn[0] =(y[num[0]-2[0]-y[num[0]-1[0])/(x[num[0]-2][0]-x[num[0]-1][0]);
 yp1[1] =(y[1][1]-y[0][1])/(x[1][1]-x[0][1]);
 ypn[1] =(y[num[1]-2[1]-y[num[1]-1[1])/(x[num[1]-2][1]-x[num[1]-1][1]);
}
 else
{
 dummy(3);
 fscanf(aqin, "%f %f %f %f %f", &ap,&a1,&a2,&w,&angle);
 printf(" ap,a1,a2,w,angle\n");
 printf("%f %f %f %f %f\n", ap,a1,a2,w,angle);
 angle = angle * 3.1416 / 180.0;
}
}


void aqcalc()
{
int i;
float A, Haq, aq_x[SIZESPL], aq_y[SIZESPL], aq_H[SIZESPL];
printf("xi,yi\n");
nn = (int)((limitn-limit1)/xistep);
fprintf(plo, "%d %d %d %d %f\n", pflag, nn, aa, flowflag, limitn);
fprintf(spl, "%d %f %f %f %d %d\n", nn, xistep, limit1, limitn, flowflag,
                                          aqflag);
```

```
xi = 0.0;
A = tan((double)angle);
for (i = 0; i < nn; ++i)
  {
   xi = xi + xistep;
   aq_x[i] = xi;
   if (ap != 0.0)
    aq_y[i] = ap * cos(w*xi) + a1;
   else
    aq_y[i] = xi * A + a1;
   Haq =abs((int)((A*xi-yi+a2)*1000.0));
   Haq = Haq/1000.0/sqrt((double)(A*A+1));
   aq_H[i] = Haq;
  }
if (flowflag == 1)
  {
   for (i = 0; i < nn; ++i)
   {
     fprintf(spl, "%f %f %f\n", aq_x[i], aq_y[i], aq_H[i]);
     printf("%f %f\n",aq_x[i], aq_y[i]);
     fprintf(plo, "%f %f\n",aq_x[i], aq_y[i]);
   }
  }
else
  {
   for (i = 0; i < nn; ++i)
     {
     fprintf(spl, "%f %f %f\n", aq_x[i], aq_y[nn-1-i], aq_H[nn-1-i]);
     printf("%f %f\n",aq_x[i], aq_y[i]);
     fprintf(plo, "%f %f\n",aq_x[i], aq_y[nn-1-i]);
     }
}
printf("xi,yi\n");
xi = 0.0;
for (i = 0; i <= nn; ++i)
{
 xi = xi + xistep;
 aq_x[i] = xi;
 aq_y[i] = xi * A + a2;
 printf ("%f %f\n",aq_x[i],aq_y[i]);
}
if(flowflag == 1)
   {
   for (i = 0; i <= nn; ++i)
     {
     printf("%f %f\n",aq_x[i],aq_y[i]);
     fprintf(plo, "%f %f\n",aq_x[i],aq_y[i]);
     }
  }
else
  {
```

```
  for (i = 0; i <= nn; ++i)
    {
     printf("%f %f\n",aq_x[i],aq_y[i]);
     fprintf(plo, "%f %f\n",aq_x[i], aq_y[nn-i-1]);
    }
  }
}



void spline1()
{
int i,k,ii;
float p,qn,sig,un,u[500],*vector();
if(yp1[m] > 0.99e30)
 ya[1]=u[1]=0.0;
else
 {
  ya[1] = -0.5;
  u[1]=(3.0/(x[2][m]-x[1][m]))*((y[2][m]-y[1][m])/(x[2][m]-x[1] [m]-yp1[m]));
  }
for (i=2[i<=num[m]-1; i++)
  {
  sig=(x[i][m]-x[i-1][m])/(x[i+1][m]-x[i-1][m]);
  p=sig*ya[i-1]+2.0;
  ya[i]=(sig-1.0)/p;
  u[i]=(y[i+1][m]-y[i][m])/(x[i+1][m]-x[i][m]) -
        (y[i][m]-y[i-1][m])/(x[i][m]-x[i-1][m]);
  u[i]=(6.0*u[i]/(x[i+1][m]-x[i-1][m])-sig*u[i-1])/p;
  }
if (ypn[m] > 0.99e30)
 qn=un=0.0;
else
  {
  qn=0.5;
  un=(3.0/(x[num[m]][m]-x[num[m]-1][m]))*(ypn[m]-(y[num[m]][m]-y[num[m]-1]
    [m])/(x[num[m]][m]-x[num[m]-1][m]));
  ya[num[m]]=(un-qn*u[num[m]-1])/(qn*ya[num[m]-1]+1.0);
  for (k=num[m]-1;k>=1;k--) {
  ya[k]=ya[k]*ya[k+1]+u[k];
    }
   }
  }



void spline2()
/* Interpolates by given xi */

 {
 int klo,khi,k;
 float h,b,a;
```

```
klo=1; khi=num[m];

while (khi-klo > 1) {
  k= (khi + klo) >> 1;
  if (x[k][m] > xi) khi=k;
  else klo=k;
  }
  h=x[khi][m]-x[klo][m];
  /* if (h == 0.0) nrerror("Bad XA input to routine SPLINE"); */
  a= (x[khi][m]-xi)/h;
  b= (xi-x[klo][m])/h;
  yi=a*y[klo][m]+b*y[khi][m]+(a*a*a-a)*ya[klo]+(b*b*b-b)*ya[khi]*(h*h)/
6.0;
  if (xi >= limit1 && xi <= limitn && yi >=0.0){
  printf("%f %f\n",xi,yi);
 fprintf(plo, "%f %f\n", xi, yi);
     }
   }


void line1()
 {
 int i, j;

  for(i=0; i<num[m]; i++) {
   ya[i] = (y[i+1][m] - y[i][m])/(x[i+1][m] -x[i][m]);
  printf("%5.2f %5.2f %5.2f\n", x[i][m], x[i+1][m], ya[i]);
   }
  }

void line2()
{
int i;
  for (i=0; i<num[0]; i++) {
  if(xi>=x[i][m] && xi<x[i+1][m]) {
  yi = ya[i]*(xi-x[i][m] + y[i][m];
   printf(" x,y,ya xi yi%5.2f %5.2f %5.2f %5.2f %5.2f\n",x[i][m], y[i][m],ya[i], xi, yi);
     if (xi >= limit1 && xi <= limitn && yi>=0.0)
  fprintf(plo, "%5.2f %5.2f\n", xi, yi);
  }
  }
 }


main()
 {
  int ii,j,i;
  plo = fopen("plot.dat", "w");
  spl = fopen("spl.dat", "w");
  aqin = fopen("aq.dat", "r");

  input();
```

```
if (aqflag == 0) {
printf ("give a choice 0\n");

   for (m = 1; m >= 0; m--)
   {
   xi = limit1;
 spline1 ();

   for (j = 0; j <= nn; ++j)
     {
   xi = xi + xistep;
 spline2 ();
       }
     }
 }
else if (aqflag == 1) {
printf ("give a choice 1\n");
for (m=1; m>=0; m--) {
 xi = limit1;
line1 ();
 for (i=0; i<nn; i++) {
  line2 ();
  xi = xi + xistep;
   }
  }
  }

else {
printf ("give a choice 2\n");
  aqcalc ();
 }

fclose (spl);
}
```

Module PERM

The module PERM reads the permeability control file, kp.dat, for setting up the permeability across the aquifer geometry. There are three ways of defining the permeability:

1. Constant permeability across the aquifer.
2. The permeability can vary linearly, either increasing along the aquifer length or decreasing along the aquifer length.
3. The permeability is defined by a spline curve based on digitized permeability values along the aquifer.

The permeability is added to the file spl.dat.

Functions

dummu( )
input_K( )
K_constant( )
K_linear( )
K_spline( )
manuK( )
permeability( )

```
          SET UP PERMEABILITY DATA
------------------------------------------------------
kflag    choice
0        1
xp1    xpn     xstep
1.0    65.0    1.0
kconstant   kmax     kmin
100.0       1000.0   100.0
nk
21
x(m)     permeability(mD)
0.0      1000.0
1.0      1000.0
5.0      1000.0
10.0     1000.0
15.0     1000.0
20.0     1000.0
25.0     1.0
30.0     1.0
35.0     1.0
40.0     1.0
45.0     1.0
50.0     1.0
55.0     1.0
------------------------------------------------------
```

```
     NOTE
kflag    --- when kflag=0, read parameters for calculating
             constant or linear permeability;
             when kflag=1, read parameters for calculating
             permeability by cubic spline (default: 0)
choice   --- choice=1 constant permeability
             choice=2 calculate permeability using linear function
             choice=3 calculate permeability using cubic spline (default: 1)
xp1(m)   --- start position of calculating permeability (default: 1.0)
xpn(m)   --- last position of calculating permeability (default: 100.0)
```

Module PERM

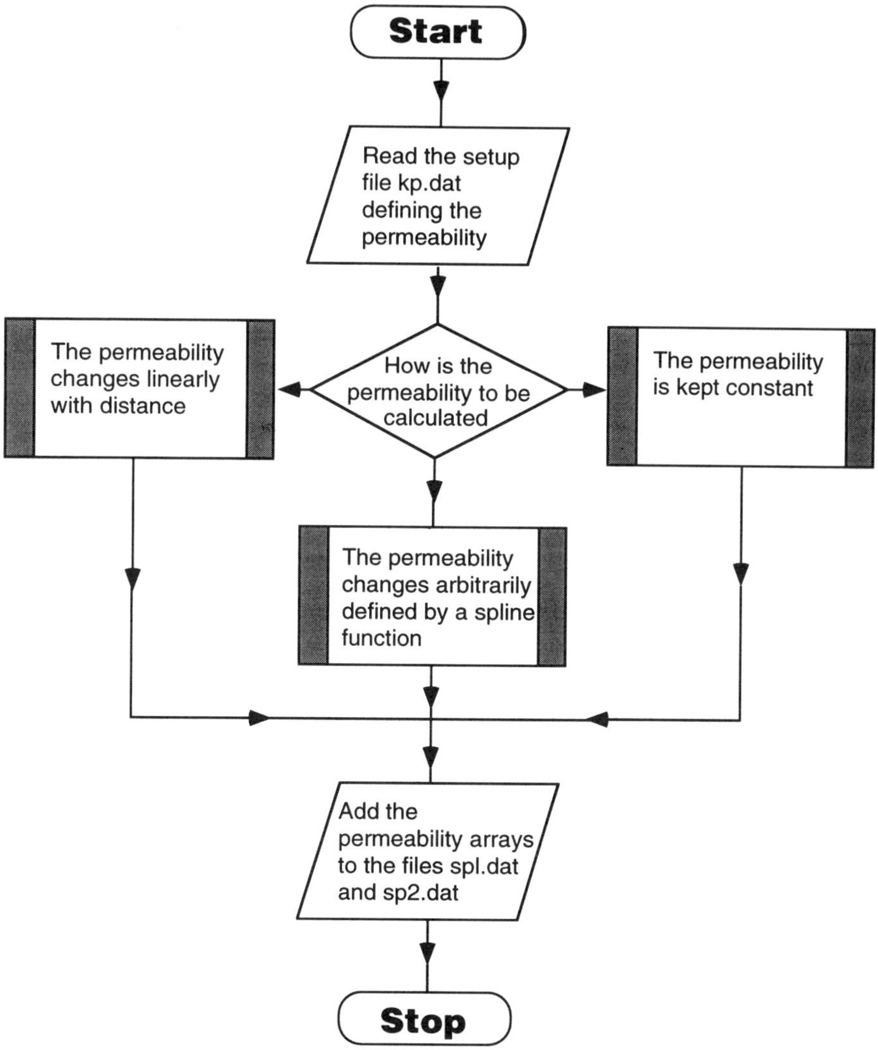
Start
Read the setup file kp.dat defining the permeability
How is the permeability to be calculated
The permeability changes linearly with distance
The permeability is kept constant
The permeability changes arbitrarily defined by a spline function
Add the permeability arrays to the files spl.dat and sp2.dat
Stop

Module PERM

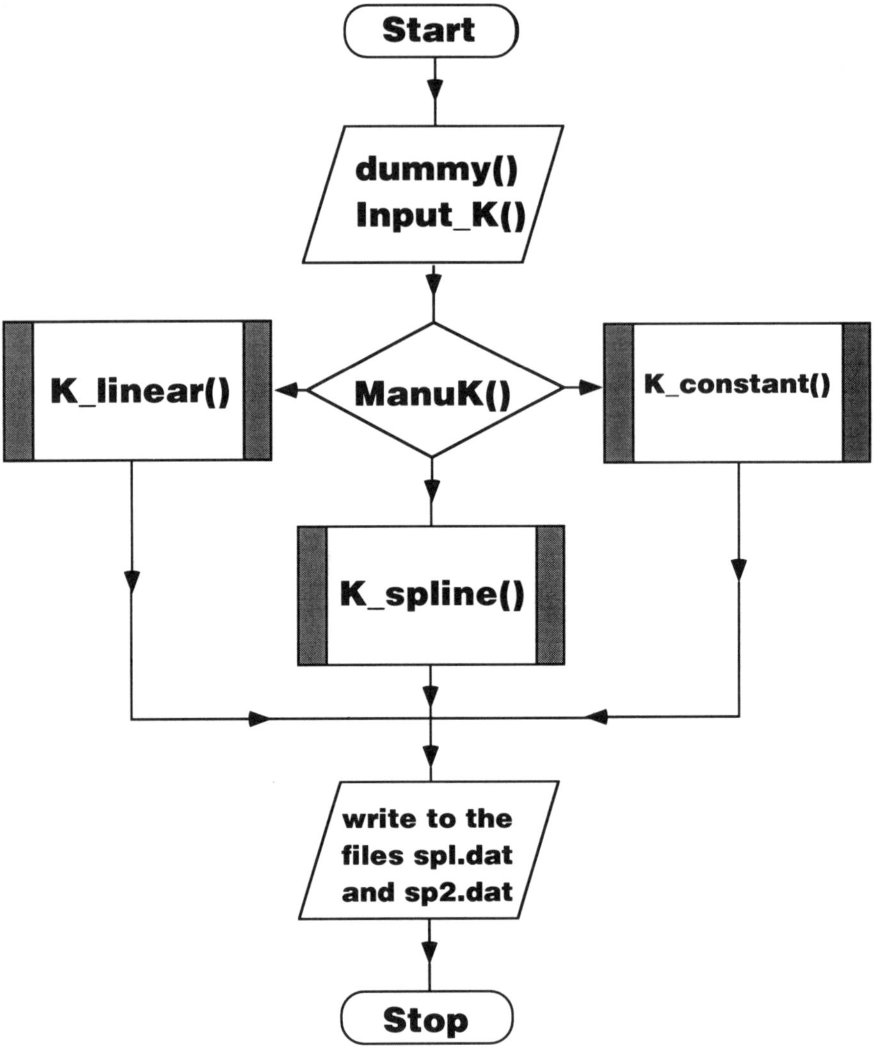
Start
dummy()
Input_K()
K_linear()
ManuK()
K_constant()
K_spline()
write to the
files spl.dat
and sp2.dat
Stop

```
kconstant(mD)-constant permeability (recommended input data range:
            0.0 < kconstant <= 1000.0; default: 100.0);
kmax(mD)  --- maximum permeability (recommended input data range:
           0.0 < kmax <= 1000.0; default: 100.0)
kmin(mD)  --- minimum permeability (recommended input data range:
           0.0 <kmin <= 1000.0; default: 100.0)
nk        --- number of input points of permeability (input data range:
           10 <= nk <= 500; default: see above input data)
xstep(m) --- distance step(x-axis) for calculcating permeability
          (recommended input data range 0.001 to 2.0; default: 1.0)

#include<stdio.h>
#include<math.h>
#include"oilstr.h"
int kflag, ncl, nk, choice;
float kxstep, kxp1, kxpn, ka;
float k_max, k_min;
float ykp1, ykpn;
float k[SIZEAQ], k_xi[SIZEAQ], kx[SIZEAQ], ky[SIZEAQ], kyy[SIZEAQ],
yk2[SIZEAQ];
FILE *kpin;
FILE *spl;

void dummy(dums)
int dums;
{
int i;
char dumchar[100];

for (i = 0; i < dums; i++)
fgets (dumchar, 100, kpin);
}

 void Input_K()
 {
 int i;

 kpin = fopen("kp.dat", "r");

 dummy(3);
 fscanf(kpin, "%d %d", &kflag, &choice);
 printf("kflag, choice\n");
 printf("%d %d\n", kflag, choice);

 dummy(2);
 fscanf(kpin, "%f %f %f", &kxp1, &kxpn, &kxstep);
 printf("kxp1   kxpn   kxstep \n");
 printf("%f %f %f\n", kxp1, kxpn, kxstep);

 ncl = (int) (kxpn/kxstep);
```

```
dummy(2);
fscanf(kpin, "%f %f %f", &ka, &k_max, &k_min);
printf("kconstant, k_max, k_min\n");
printf(" %f %f %f\n", ka, k_max, k_min);
if(kflag == 1){
dummy(2);
fscanf(kpin, "%d", &nk);
/*
printf(" nk\n");
printf("%d\n", nk);
*/

/* printf("kx[i],ky[i]\n"); */
dummy(2);
for(i = 0; i <= nk; i++) {
fscanf(kpin, "%f %f", &kx[i], &ky[i]);
/* printf("%f %f\n", kx[i],ky[i]); */
   }
  }
close(kpin);
}

void K_constant()
{
int i;
float xlen;

fprintf(spl, "%d %d\n", ncl, kflag);
xlen = 0.0;
for(i = 0; i<ncl; i++) {
xlen = xlen + kxstep;
fprintf(spl, "%f %f\n", xlen, ka);
/*
printf("%f %f\n", xlen, ka);
*/
 }

void K_linear()
 {
  int i;
  float a, xlen, ki;
  xlen = kxpn - kxp1;
  a = (k_max - k_min)/xlen;
  xlen = 0.0;
 fprintf(spl, "%d %d\n", ncl, kflag);
  for (i = 0; i<ncl; i++)
   {
 xlen = xlen + kxstep;
  ki = (a * xlen + k_min);
```

```
  if(ki <= 0.5)
  ki = 0.5;
/*
 printf("kx  ky\n");
 printf("%f %f\n", xlen, ki);
*/
 fprintf(spl, "%f %f\n", xlen, ki);

    }
  }


 void K_spline()
 {
 int i, j;

 fprintf(spl, "%d %d\n", nk, kflag);
for (i = 0; i < nk; i++) {
 ky[i] = log((double)ky[i]);
 fprintf(spl, "%f %f\n", kx[i], ky[i]);
/*
 printf("kx  ky\n");
 printf("%f %f\n", kx[i], ky[i]);
*/
   }
 }

void ManuK()
{
/*
printf("\n");
printf("     Permeability menu              \n");
printf("     ******************************* \n");
printf("     * 1---constant Permeability       *\n");
printf("     * 2---linear Permeability         *\n");
printf("     * 3---spline Permeability         *\n");
printf("     ******************************* \n");
printf("\n");
*/
 }

void Permeability()
{

  ManuK();
  printf("put choice please\n");

  switch(choice) {
  case 1:
  K_constant();
```

```
  break;
  case 2 :
  K_linear();
  break;

  case 3 :
  K_spline();
  break;

  default;
  printf("Permeability is not set!\n");

   }
}


main()
{
  spl = fopen("spl.dat", "a");
  Input_K();

  Permeability();
 fclose(spl);

}
```

Module SPL

The module SPL is used for setting up the arrays containing the information about geometry, aquifer width, and aquifer surface angle relative to horizontal and information for using area as an incremental base for the simulation. The arrays are used for spline calculation and linear interpolation. SPL reads the file spl.dat and creates an output file sp2.dat to be used by the simulation module.

Functions

InputSpl( )
Spline1( )
Spline2( )
Line1( )
Line2( )
CalcSpl( )
CalcLine( )

Module SPL

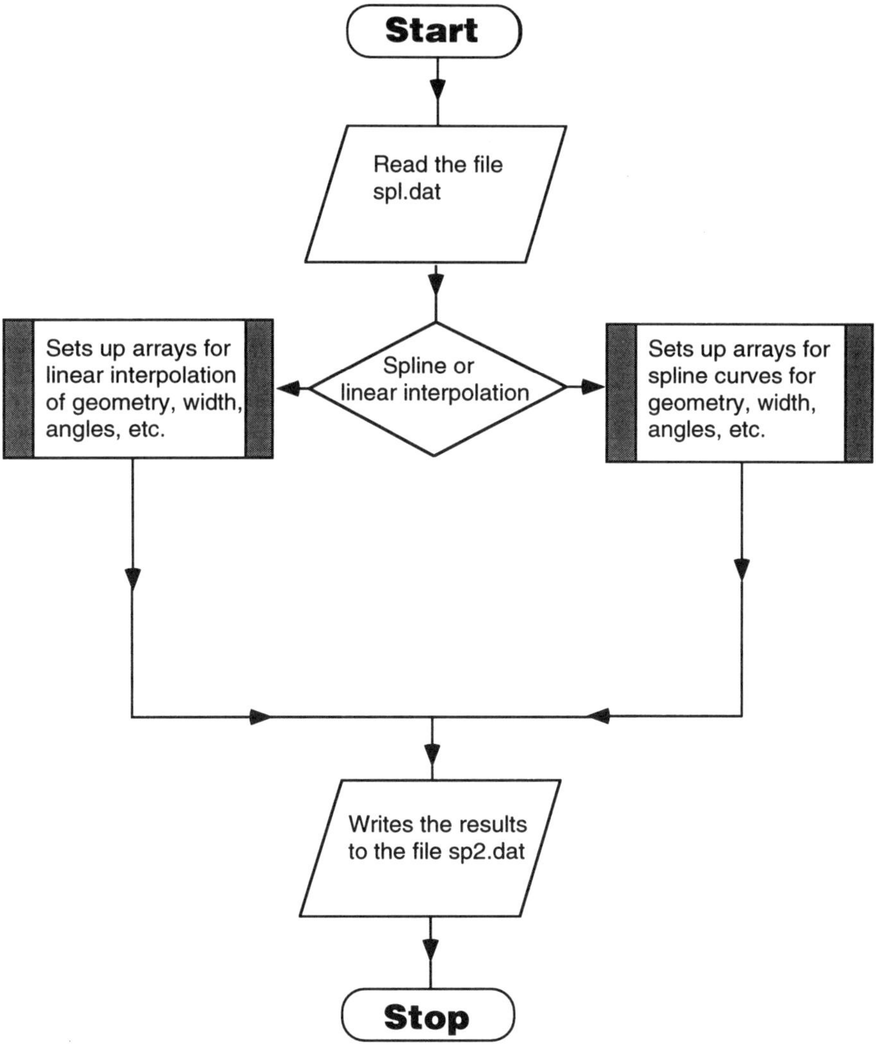
Start
Read the file spl.dat
Sets up arrays for linear interpolation of geometry, width, angles, etc.
Spline or linear interpolation
Sets up arrays for spline curves for geometry, width, angles, etc.
Writes the results to the file sp2.dat
Stop

Module SPL

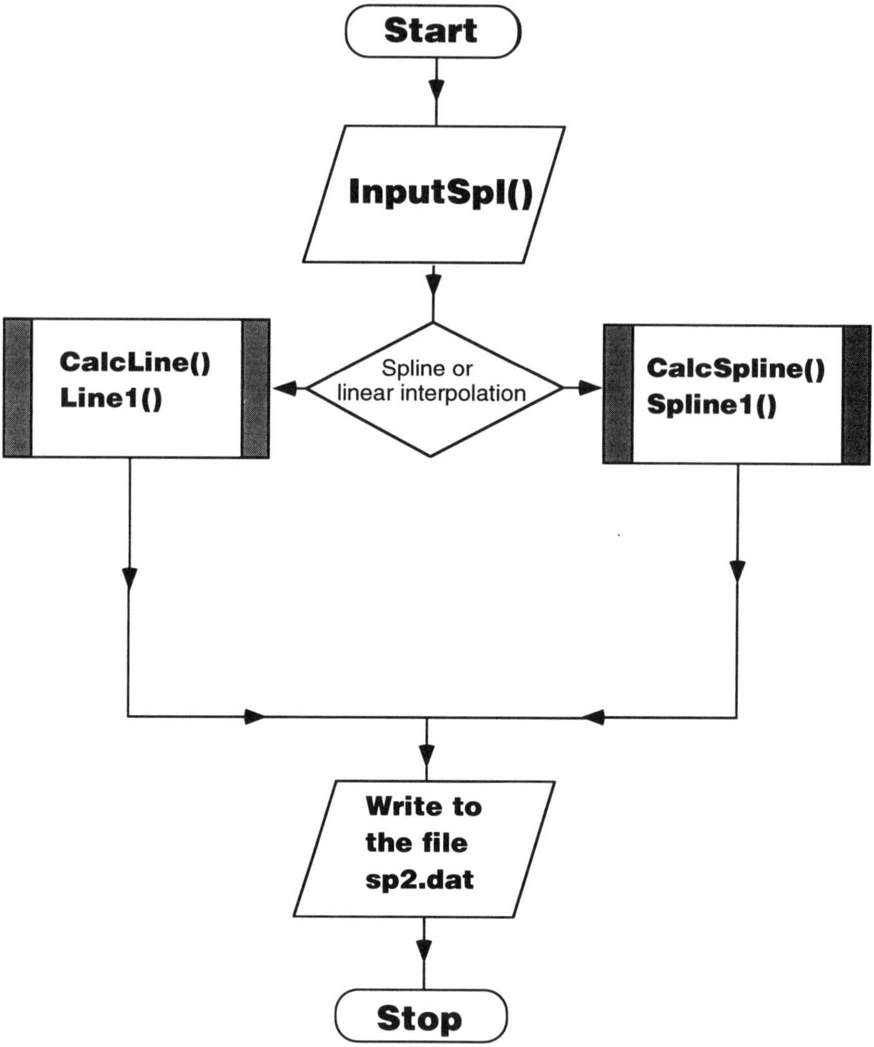
Start
InputSpl()
CalcLine()
Line1()
Spline or
linear interpolation
CalcSpline()
Spline1()
Write to
the file
sp2.dat
Stop

```
/*********************************************
               spl.h
*********************************************/


void InputSpl();
/* Read data */


void CalcSpl();
/* Calculates aquifer spline, angle spline, aquifer width spline and
permeability spline. */


void Spline 1 (float x[100], float y[100], int num, float yp1, float ypn);
/* Set up spline function by input separated points and return ya. */


float Spline2(float x[100], float y[100], float ya[100], int num, float xi);
/* Interpolates by given xi then return yi*/

#include<stdio.h>
#include<math.h>
#include"oilstr.h"

int nu, nuk, logflag, flowflag, aqflag;
float xistep, limit, limitn;

  FILE *sp2;
  FILE *spl;

void InputSpl()
{
  int i;
  float a, b, c, lenth, Hmin;

  spl = fopen("spl.dat", "r");

  fscanf(spl, "%d %f %f %f %d", &nu, &xistep, &limit, &limitn, &flowflag);
  fscanf(spl, "%d", &aqflag);


  printf("nu   xistep  limit flowflag\n");
  printf("%d   %f %f %d %d\n", nu, xistep, limit, flowflag, aqflag);

  printf("Aq1.x  Aq1.y  H1_l.y\n");

  Hmin = 100000.0;
  for (i = 0; i < nu; i++) {
  fscanf(spl, "%f %f %f", &Aq1.x[i], &Aq1.y[i], &H1_l.y[i]);
   if (H1_l.y[i] <Hmin) Hmin = H1_l.y[i];
```

```
printf("%f\n", Hmin);
 }
for(i = 0; i < nu; i++) {
printf("FR: %f %f %f\n", Aq1.x[i], Aq1.y[i], H1_l.y[i]);
 }

lenth = 0.0;
if(aqflag == 0) {
 for(i = 1; i < nu; i++) {
  a = Aq1.x[i] - Aq1.x[i-1];
   b = Aq1.y[i] - Aq1.y[i-1];
    lenth = lenth + (float)(sqrt((double)(a*a + b*b)));
    Ang_l.x[i] = lenth;
    Ang_l.y[i] = atan((double)(b/a));
   H1_l.x[i] = lenth;
   H1_l.y[i] = H1_l.y[i] * cos((double)(Ang_l.y[i]))-0.2;
   if (H1_l.y[i] <=0.0) H1_l.y[i] = H1_l.y[i-1];
  X_l.y[i] = lenth;
  X_l.x[i] = Aq1.x[i];
  L_x.x[i] = lenth;
  L_x.y[i] = Aq1.x[i];
   }
 }
else {
   Ang_l.x[0] = 0.0;
   H1_l.x[0] = 0.0;
  X_l.y[0] = 0.0;
  L_x.x[0] = 0.0;
 for(i = 0; i < nu; i++) {
  a = Aq1.x[i+1] - Aq1.x[i];
  b = Aq.1.y[i+1] - Aq1.y[i];
   lenth = lenth + (float)(sqrt((double)(a*a + b*b)));
   Ang_l.x[i+1] = lenth;
   Ang_l.y[i] = atan((double)(b/a));
   H1_l.x[i+1] = lenth;
   H1_l.y[1]+H1_l.y[i]* cos((double] (Ang_l.y[i]))-0.2;
   if (H1_l.y[i] <=0.0) H1_l.y[i] = H1_l.y[i-1];
  X_l.y[i+1] = lenth;
  X_l.x[i] = Aq1.x[i];
  L_x.x[i+1] = lenth;
  L_x.y[i] = Aq1.x[i];
  }
  }
  for(i = 0; i < nu; i++) {
 printf("XYLHG; %d %f %f %f %f %f\n", i, Aq1.x[i], Aq1.y[i], H1_l.x[i], H1_
l.y[i],
Ang_l.y[i]);
 }

  fscanf(spl, "%d %d", &nuk, &logflag);
  printf("nuk/ logflag\n");
  printf("%d %d\n", nuk, logflag);
```

```
  if(flowflag == 1) {
  for(i = 0; i < nuk; i++) {
  fscanf(spl, "%f %f", &K1_l.x[i], &K1_l.y[i]);
  printf("kk%f %f\n", K1_l.x[i], K1_l.y[i]);
   }
    }
else {
  for (i = 0; i < nuk; i++)
  fscanf(spl, "%f %f", &K_l.x[i], &K_l.y[nuk-1-i]);
      }
  fclose(spl);
   }
void Spline1 (float x[SIZESPL], float y[SIZESPL], int num,
                              float yp1, float ypn)
 {
  int i, k;
  float p, qn, sig, un, u[SIZESPL], ya[SIZESPL], *vector();

  if (yp1 > 0.99e30)
  ya[1]=u[1]=0.0;
  else
  {
  ya[1]=-0.5;
  u[1]=(3.0/(x[2]-x[1]))*((y[2]-y[1])/(x[2]-x[1]-yp1));
  }
  for (i=2; i<=num-1; i++)
  {
  sig=(x[i]-x[i-1])/(x[i+1]-x[i-1]);
  p=sig*ya[i-1]+2.0;
  ya[i]=(sig-1.0)/p;
  u[i]=(y[i+1]-y[i])/(x[i+1]-x[i]) - (y[i]-y[i-1])/(x[i]-x[i-1]);
  u[i]=(6.0*u[i]/(x[i+1]-x[i-1])-sig*u[i-1])/p;
  }
  if (ypn > 0.99e30)
  qn=un=0.0;
  else
  {
  qn=0.5;
  un=(3.0/(x[num]-x[num-1]))*(ypn-(y[num]-y[num-1])/(x[num]-x[num-1]));
  ya[num]=(un-qn*u[num-1])/(qn*ya[num-1]+1.0);
  for (k=num-1; k>=1; k--) {
  ya[k]=ya[k]*ya[k+1]+u[k];
  fprintf(sp2, "%f %f %f\n", x[k], y[k], ya[k]);
    }
   }
 }


float Spline2 (float x[SIZESPL], float y[SIZESPL], float ya [SIZESPL],
                                     int num, float xi)
 {
  float yi;
```

```
int klo,khi,k;
float h,b,a;
klo=1; khi=num;
while (khi-klo > 1) {
k=(khi + klo) >> 1;
if (x[k] > xi) khi=k;
else klo=k;
}
h=x[khi]-x[klo];
a=(x[khi]-xi)/h;
b=(xi-x[klo])/h;
yi=a*y[klo]+b*y[khi]+(a*a*a-a)*ya[klo]+(b*b*b-b)*ya[khi]*(h*h)/6.0;
printf("%f %f\n",xi,yi);

return yi;
}


void line1(int num, float x[SIZESPL], float y[SIZESPL])
  {
  int i, j;
  float ya[SIZESPL];

   for(i=0; i<num; i++) {
    ya[i] = (y[i+1] - y[i])/(x[i+1] -x[i]);
   printf("%5.2f %5.2f %5.2f\n", x[i], x[i+1], ya[i]);
    fprintf(sp2, "%f %f %f\n",x[i],y[i],ya[i]);
  }
}

void line2(int num, float x[SIZESPL], float y[SIZESPL], float ya[SIZESPL])
{
int i;
float xi,yi;

for(i=0; i<num; i++) {
 if(xi>=x[i] && xi<x[i+1]) {
 yi = ya[i]*(xi-x[i]) + y[i];
  printf(" x,y,ya xi yi%5.2f %5.2f %5.2f %5.2f %5.2f\n",x[i], y[i], ya[i]
, xi, yi);
  fprintf(sp2, "%5.2f %5.2f\n", xi, yi);
   }
  }
 }

void CalcSpl()
{
  int i;
  float dfirst, dend;
```

```
  sp2 = fopen("sp2.dat", "w");
  fprintf(sp2, "%d %d %d %f %f %f\n", nu-1, nuk-1, logflag, xistep,
                          limit, limitn);
  fprintf(sp2, "%d %d\n", flowflag, aqflag);

  printf("%d %d %d %f\n", nu, nuk, logflag, xistep);
 dfirst = (Aq1.y[2] - Aq1.y[1])/(Aq1.x[2] - Aq1.x[1]);
 dend = (Aq1.y[nu-1] - Aq1.y[nu-2])/(Aq1.x[nu-1] - Aq1.x[nu-2]);
  Spline1(Aq1.x, Aq1.y, nu, dfirst, dend);
 dfirst = (H1_l.y[2] - H1_l.y[1])/(H1_l.x[2] - H1_l.x[1]);
 dend = (H1_1.y[nu-2] - H1_l.y[nu-1])/(H1_l.x[nu-2] - H1_l.x[nu-1]);
  Spline1(H1_l.x, H1_l.y, nu, dfirst, dend):
 dfirst = (Ang_l.y[2] - Ang_l.y[1])/(Ang_l.x[2] - Ang_l.x[1]);
 dend = (Ang_l.y[nu-2] - Ang_l.y[nu-1])/(Ang_l.x[nu-2] - Ang_l.x[nu-1]);
  Spline1(Ang_l.x, Ang_l.y, nu, dfirst, dend);
 dfirst = (K1_l.y[2] - K1_l.y[1])/(K1_l.x[2] - K1_l.x[1]);
 dend = (K1_l.y[nuk-2] - K1_l.y[nuk-1])/(K1_l.x[nuk-2] - K1_l.x[nuk-1]);
  Spline1(K1_l.x, K1_l.y, nuk, dfirst, dend);
 dfirst = (X_l.y[2] - X_l.y[1])/(X_l.x[2] -X_l.x[1]);
 dend = (X_l.y[nu-2] - X_l.y[nu-1])/(X_l.x[nu-2] - X_l.x[nu-1]);
  Spline1(X_l.x, X_l.y, nu, dfirst, dend);
  for(i = 0; i<nu; i++)
printf("L: %f %f %d %f %f\n", L_x.x[i], L_x.y[i], nu,
                             L_x.y[1], L_x.y[nu-1]);
  dfirst = (L_x.y[2] - L_x.y[1])/(L_x.x[2] = L_x.x[1]);
  dend = (L_x.y[nu-2] - L_x.y[nu-1])/(L_x.x[nu-2] - L_x.x[nu-1]);
  Spline1(L_x.x, L_x.y, nu, dfirst, dend);

}

void CalcLine()
{
  int i;
  float dfirst, dend;

  sp2 = fopen("sp2.dat", "w");

  fprintf(sp2, "%d %d %d %f %f %f\n", nu, nuk, logflag, xistep,
                          limit, limitn);
  fprintf(sp2, "%d %d\n", flowflag, aqflag);
  printf("%d %d %d %f %d\n", nu, nuk, logflag, xistep, aqflag);
 dfirst = (Aq1.y[2] - Aq1.y[1])/(Aq1.x[2] - Aq1.x[1]);
 dend = (Aq1.y[nu-1] - Aq1.y[nu-2])/(Aq1.x[nu-1] - Aq1.x[nu-2]);

  line1(nu, Aq1.x, Aq1.y);

  for(i = 0; i < nu; i++)
  printf("H: %f %f %d\n",H1_l.x[i], H1_l.y[i], nu);
```

```
  dfirst = (H1_l.y[2] - H1_l.y[1])/(H1_l.x[2] - H1_l.x[1]);
  dend = (H1_l.y[nu-2] - H1_l.y[nu-1])/(H1_l.x[nu-2] - H1_l.x[nu-1]);
   line1(nu, H1_l.x, H1_l.y);

  for(i = 0; i<nu; i++)
  printf("AN: %f %f %d\n", Ang_l.x[i], Ang_l.y[i], nu);

  dfirst = (Ang_l.y[2] - Ang_l.y[1])/(Ang_l.x[2] - Ang_l.x[1]);
  dend = (Ang_l.y[nu-2] - Ang_l.y[nu-1])/(Ang_l.x[nu-2] - Ang_l.x[nu-1]);
  line1(nu, Ang_l.x, Ang_l.y);
 dfirst = (K1_l.y[2] - K1_l.y[1])/(K1_l.x[2] - K1_l.x[1]);
 dend = (K1_l.y[nuk-2] - K1_l.y[nuk-1])/(K1_l.x[nuk-2] - K1_l.x[nuk-1]);
  line1(nuk, K1_l.x, K1_l.y);
 dfirst = (X_l.y[2] - X_l.y[1])/(X_l.x[2] - X_l.x[1]);
 dend = (X_l.y[nu-2] - X_l.y[nu-1])/(X_l.x[nu-2] - X_l.x[nu-1]);

  line1(nu, X_l.x, X_l.y);


  for(i = 0; i<nu; i++)
  printf("L: %f %f %d %f %f\n", L_x.x[i], L_x.y[i], nu,
                                L_x.y[1], L_x.y[nu-1]);


  dfirst = (L_x.y[2] - L_x.y[1])/(L_x.x[2] - L_x.x[1]);
  dend = (L_x.y[nu-2] - L_x.y[nu-1])/(L_x.x[nu-2] - L_x.x[nu-1]);
  line1(nu, L_x.x, L_x.y);
  }

main()
{
 float t1,t2,t3,t4;
  InputSpl();
  if(aqflag == 0)
  CalcSpl();
  else
  CalcLine();
}
```

***Module* SIMU**

The SIMU module is the main simulation module. The module performs a search for a steady-state solution to the oil stringer equation. A control file, simu.dat, containing relevant constants and simulation parameters, is read and used. SIMU uses the output files spl.dat and sp2.dat from the previous modules and generates the graphics file draw.dat.

Functions

dummy( )
initial( )
ReadSp2( )
Spline2( )
Line2( )
phi( )
CalcGsum( )
CalcSsum( )
Simu( )
smooth_N( )

```
          SET UP SIMULATION INITIAL VALUES
-----------------------------------------------------
rate      ro        rw          unitscale
50.0      0.8       1.0         100.0
mu         T        pc          philimit
1.0       35.0      0.0         0.1
xstartp    dl        searchflag  phi0
30.0      0.05      1           40.0
mass       dldAflag
0.0       0
plotname
plot
-----------------------------------------------------
     NOTE
rate(cm/yr) --- rate of water flow(recommended input data
          range: 0.0 to 500.0; default: 50.0)
ro(g/cc) ------- density of oil(or gas) (0.0<ro<1.0, recommended input data
          range: for oil: 0.8 to 0.9; default: 0.8;
              for gas: 0.1 to 0.3; default: 0.3)
rw(g/cc) ------- density of water (recommended input data range: 1.0 to 1.3;
         default 1.0)
unitscale ------ a conversion constant (=100.0 from meter to centimeter,
          =100,000.0 from kilometer to centimeter etc.
          default: 100.0 converts meters to centimeters)
mu(cP) --------- viscosity of water (recommended input data range:
          0.8 to 1.5; default: 1.0)
T(dyne/cm) ----- surface tension (recommended input data range: 25.0 to 75.0;
          default: 35.0)
pc(atm) -------- capillary pressure (recommended input data range:
          -10E4 to 10E4; default: 0.0)
xstarp(m) ------ starting point for search of oilstringer in x-axis
```

Module SIMU

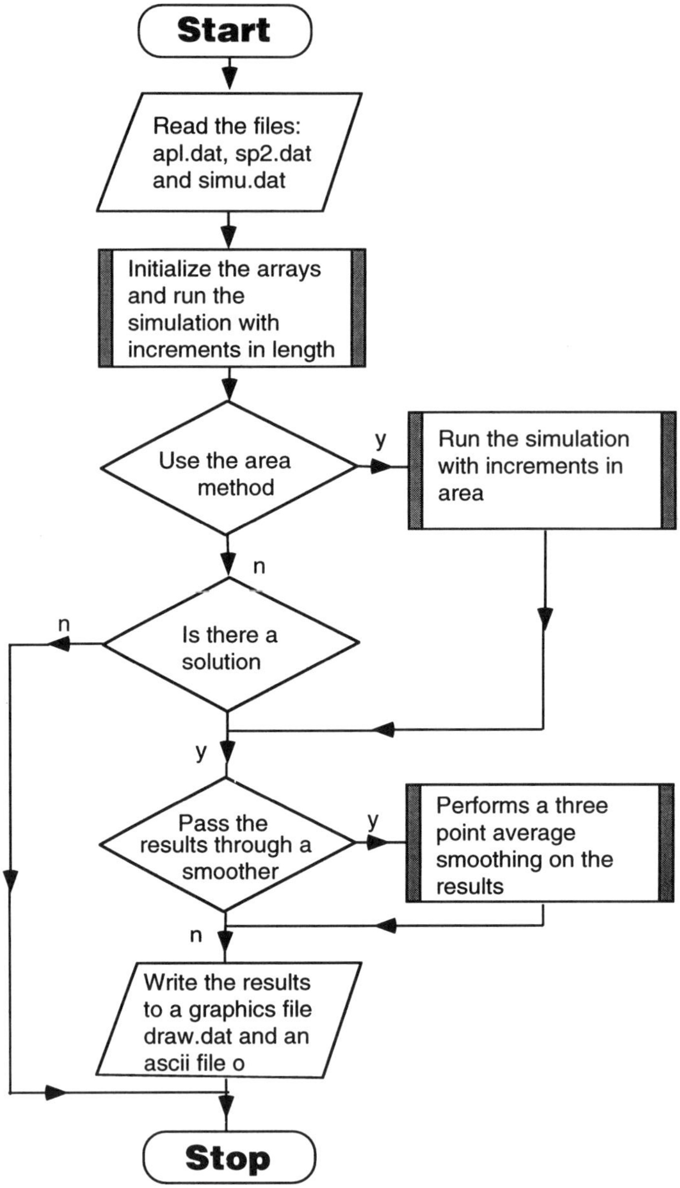
Start
Read the files: apl.dat, sp2.dat and simu.dat
Initialize the arrays and run the simulation with increments in length
Use the area method
y
Run the simulation with increments in area
n
Is there a solution
n
y
Pass the results through a smoother
y
Performs a three point average smoothing on the results
n
Write the results to a graphics file draw.dat and an ascii file o
Stop

Module SIMU

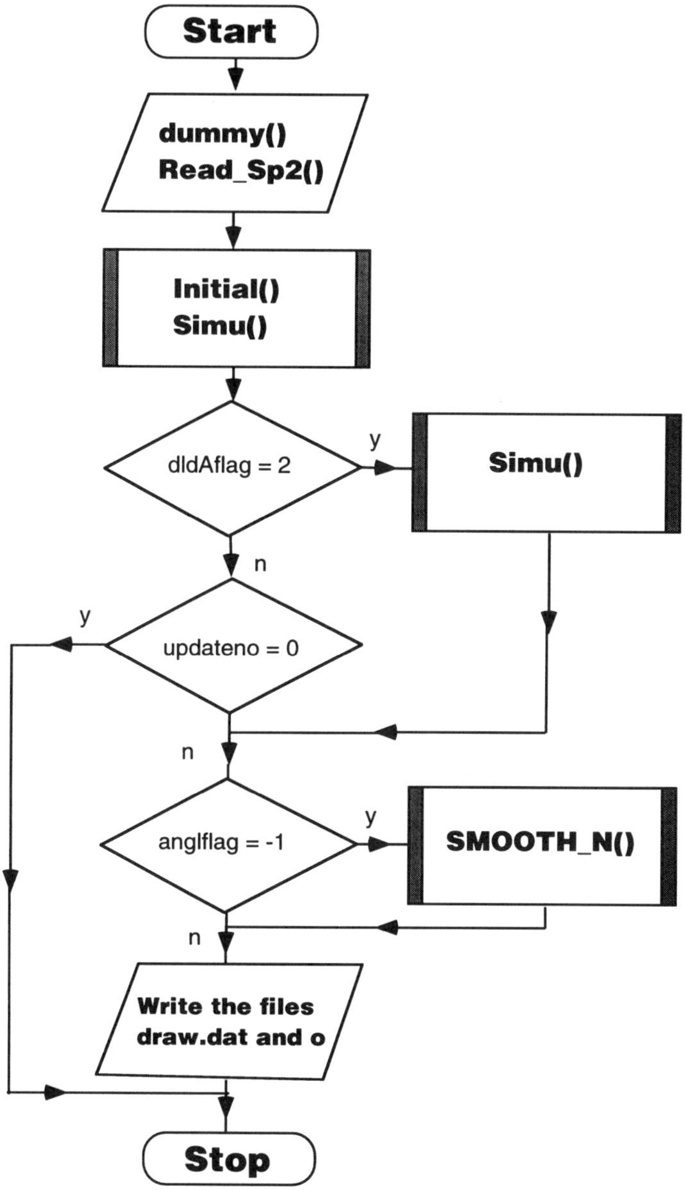
Start
dummy()
Read_Sp2()
Initial()
Simu()
dldAflag = 2
y
Simu()
n
updateno = 0
y
n
anglflag = -1
y
SMOOTH_N()
n
Write the files
draw.dat and o
Stop

```
dl(m) ---------- step length of oilstringer (recommended input data range:
          0.001 to 0.1; default: 0.05)
phi0(degree) --- initial angle of oilstringer with respect to aquifer top
          (recommended input data range: 0.0<phi0<90.0; default: 40.0)
searchflag ----- =1 search maximum initial oilstringer angle;
          =0 fixed initial oilstringer angle
philimit(degree)-limiting search angle of oilstringer (recommended input
          range: 0.001 to 1.0; default: 0.1)
areal mass(m2) - given mass of oil (or gas), if =0.0 mass is not fixed;
          default: 0.0)
dldAflag ------- =0 using length integral method in simulation
          =1 using area integral method in simulation
plotname ------- name of the plot (less than 20 characters, default: plot)

#include<stdio.h>
#include<math.h>
#include"oilstr.h"
float rate, ro, rw, mu, Qw, width;
float T, pc, g, T2, alpha, beta, gamm, unitscale;
float Astar, dl, dh, phi0, length, lengthstart, xstartp;
float Eo, aH, angl, mass, xistep;
double k2, k1;
float updatelength, updatephi, hmax, updatehmax;
int flowdirection, no, p, updateno, nus, nuks, searchflag, logflag;
float pi, phil, phih, phi3, philimit, magic;
float terml, term2, xi, limit, limitn;
float Aterml, Aterm2, dA, dAl[SIZE], updatedAl[SIZE], nostar, total A;
char plotname[20];
int dldAflag, anglflag, flowflag, aqflag;


FILE *fpt;
FILE *plo;

void dummy(dums)
int dums;
{
int i;
char dumchar[100];

for (i = 0; i < dums; i++)
fgets (dumchar, 100, fpt);
 }

void Initial()
 {

   dummy(3);
   fscanf(fpt, "%f %f %f %f", &rate, &ro, &rw, &unitscale);
   dummy(2);
   fscan(fpt, "%f %f %f %f", &mu, &T, &pc, &philimit);
```

```
dummy(2);
fscanf(fpt, "%f %f %d %f", &xstartp, &dl, &searchflag, &phi0);
dummy(2);
fscanf(fpt, "%f %d", &mass, &dldAflag);
dummy(2);
fscanf(fpt, "%s", &plotname);
if(flowflag == -1)
 xstartp = limitn - xstartp;

printf("rate(cm/yr) ro(g/cc)  rw(g/cc)   unitscale\n");
printf("%5.2f %11.2f %11.2f %13.2f\n", rate, ro, rw, unitscale);
printf("\n");
printf("mu(cP)    T(dyne/cm) pc(atm)   philimit\n");
printf("%5.2f %12.2f %10.2f %11.2f\n", mu, T, pc, philimit);
printf("\n");
printf("xstartp  dl(m)  searchflag  phi0\n");
printf("%5.2f %11.2f %10d %12.2f\n", xstartp, dl, searchflag, phi0);
printf("\n");
printf("mass   dldAflag    \n");
printf("%5.2f %10d\n", mass, dldAflag);
printf("\n");
printf("plotname\n");
printf("%s\n", plotname);
printf("\n");

g = 980.0;
pi = 3.1416;
phil = 0.0;
phi3 = phi0*pi/180.0;
phih = phi3;
phi3 = phi3/2.0;
philimit = philimit * pi/180.0;

term1 = 0.0;
term2 = 0.0;
Aterm1 = 0.0;
Aterm 2 = 0.0;

flowdirection = -1;
rate = rate / 3.15582E7;
mu = mu * 0.01;
pc = pc * 1.0E6;
width = H1_1.y[nus-3] * unitscale;
dl = dl * unitscale;
dA = dA * unitscale * unitscale;
if (mass == 0.0) mass = 200.0;
mass = mass * unitscale * unitscale;

rate = rate * (float)flowdirection
Qw = rate * width;
```

```
  T2 = 2.0 * T;
  alpha = 1.0/T2;
  beta = (Qw * mu) / T2;
  gamm = (ro - rw) * g / T2;
  angflag = 0;

}

void Read_sp2()
{
  FILE *sp2;
  int i;

  sp2 = fopen("sp2.dat", "r");

  fscanf(sp2, "%d %d %d %f %f %f", &nus, &nuks, &logflag, &xistep,
           &limit, &limitn);
  fscanf(sp2, "%d %d", &flowflag, &aqflag);
if (aqflag == 0) {
  for (i = nus; i >=1; i--) {
  fscanf(sp2, "%f %f %f", &Aq1.x[i], &Aq1.y[i], &Aq1.spl[i]);
   }

  for (i = nus; i >=1; i--) {
  fscanf(sp2, "%f %f %f", &H1_l.x[i], &H1_l.y[i], &H1_l.spl[i]);
   }

  for (i = nus; i >=1; i--)
  fscanf(sp2, "%f %f %f", &Ang_l.x[i], &Ang_l.y[i], &Ang_l.spl[i]);

  for (i = nuks; i >=1; i--) {
  fscanf(sp2, "%f %f %f", &K1_l.x[i], &K1_l.y[i], &K1_l.spl[i]);
   }

  for (i = nus; i >=1; i--)
  fscanf(sp2, "%f %f %f", &X_l.x[i], &X_l.y[i], &X_l.spl[i]);

  for (i = nus; i >=1; i--) {
  fscanf(sp2, "%f %f %f", &L_x.x[i], &L_x.y[i], &L_x.spl[i]);
     }
   }
else {
  for (i = 0; i< nus; i++) {
  fscanf(sp2, "%f %f %f", &Aq1.x[i], &Aq1.y[i], &Aq1.spl[i]);
  /* printf("X_Y %d %f %f %f\n", i, Aq1.x[i], Aq1.y[i], Aq1.spl[i]); */
   }

  for (i = 0; i< nus; i++) {
  fscanf(sp2, "%f %f %f", &H1_l.x[i], &H1_l.y[i], &H1_l.spl[i]);
           H1_l.spl[i] = 0.0;
   }
```

```
for (i = 0; i< nus; i++) {
fscanf(sp2, "%f %f %f", &Ang_l.x[i], &Ang_l.y[i], & Ang_l.spl[i]);
        Ang_l.spl[i] = 0.0;
   }

for (i = 0; i< nuks; i++) {
fscanf(sp2, "%f %f %f", &K1_l.x[i], &K1_l.y[i], &K1_l.spl[i]);
 }

for (i = 0; i< nus; i++)
fscanf(sp2, "%f %f %f", &X_l.x[i], &X_l.y[i], &X_l.spl[i]);

for (i = 0; i< nus; i++) {
fscanf(sp2, "%f %f %f", &L_x.x[i], &L_x.y[i], &L_x.spl[i]);

   }
 }

fclose(sp2);
 }


floatSpline2(float x[SIZESPL], float y[SIZESPL], float ya[SIZESPL],
                              int num, float xi)
{
  float yi;
  int klo, khi, k;
  float h,b,a;
  klo=1; khi=num;

  while (khi-klo > 1) {
  k=(khi + klo) >> 1;
  if (x[k] > xi) khi=k;
  else klo=k;
  }
  h=x[khi]-x[klo];
  a=(x[khi]-xi)/h;
  b=(xi-x[klo])/h;
  yi=a*y[klo]+b*y[khi]+(a*a*a-a)*ya[klo]+(b*b*b-b)*ya[khi]*(h*h)/6.0;
  return yi;
  }

float line2(float x[SIZESPL], float y[SIZESPL], float ya[SIZESPL], int num, float xi)
{
  int i, inif;
  float yi;

  for(i=0; i< num; i++) {
  inif = 0;
  if(xi>=x[i] && xi<x[i+1]) {
  inif = 1;
```

```
  yi = ya[i]*(xi-x[i]) + y[i];
   }
  if(inif == 1) break;
   }
  return yi;
}


void phi()
{  .
if (h[no-1] <= 0.0) }
phil = phi3;
phi3 =phi3 + (phih - phi3)/2.0;
  }
if (h[no-1] >= aH) {
phih = phi3;
phi3 = phil + (phi3 - phil)/2.0;
  }
 }

float CalcGsum()

 {
   float term3=0.0;
   float term11, term22, dh, gsum, len;
   int i;

    k1 = k2;
    len = length/unitscale;
if (aqflag == 0) {
   aH = Spline2(H1_l.x, H1_l.y, H1_l.spl, nus, len)*unitscale;
    angl = Spline2(Ang_l.x, Ang_l.y, Ang_l.spl, nus, len);
  if (angl <= 0.0)   anglflag = -1;
     k2 = Spline2(K1_l.x, K1_l.y, K1_l.spl, nuks, len);
    }
else {
   aH = line2(H1_l.x, H1_l.y, H1_l.spl, nus, len)*unitscale;
     angl = line2(Ang_l.x, Ang_l.y, Ang_l.spl, nus, len);
   if (angl <= 0.0)   anglflag = -1;
     k2 = line2(K1_l.x, K1_l.y, K1_l.spl, nuks, len);
    }

     if (logflag == 1)
     k2 = exp(k2);
   k2 = k2 * 1.0E-11;

   term1 = term1 + dl/(k2 * (aH - h[no -1]));
   dh = h[no -1] - h[no-2];
   term2 = term2 + (dl * sin(angl) + dh * cos(angl));

   Aterm1 = term1;
   Aterm2 = term2;
```

```
   term11 = beta * term1;
   term22 = gamm * term2;
   term3 = pc * dh / dl;

   gsum = term11 + term22 - term3;
   return gsum/mass;
   }

float CalcSsum()

 {
   float term3=0.0;
   float term11, term22, dh, gsum, len;
   int i;

   len = length/unitscale;
if(aqflag == 0) {
   aH = Spline2(H1_l.x, H1_l.y, H1_l.spl, nus, len)*unitscale;
     angl = Spline2(Ang_l.x, Ang_l.y, Ang_l.spl, nus, len);
   if (angl <= 0.0)   angflag = -1;
     k2 = Spline2(K1_l.x, K1_l.y, K1_l.spl, nuks, len);
   }
else {
   aH = line2(H1_l.x, H1_l.y, H1_l.spl, nus, len)*unitscale;
     angl = line2(Ang_l.x, Ang_l.y, Ang_l.spl, nus, len);
   if (angl <= 0.0)   angflag = -1;
     k2 = line2(K1_l.x, K1_l.y, K1_l.spl, nuks, len);
   }
     if(logflag == 1)
       k2 = exp(k2);

       k2 = k2 * 1.0E-11;

  Aterm1 = Aterm1 + dA/(k2 * (aH - h[no-1]))/h[no-1];
  dh = h[no -1] - h[no-2];
  Aterm2 = Aterm2 + (dA/h[no-1] * sin(angl) + dh * cos(angl));

  term11 = beta * Aterm1;
  term22 = gamm * Aterm2;
  term3 = pc * dh*h[no-1] / dA;

  gsum = term11 + term22 - term3;
  return gsum/mass/h[no-1]/h[no-1];
}


void Simu()
{


  int i;
  float Ao, gsum, xlenth, lengthout, hstar, Astar;
```

```
if(aqflag == 0)
limit =Spline2(X_l.x, X_l.y, X_l.spl, nus-1, limit)*unitscale;
else
limit = line2(X_l.x, X_l.y, X_l.spl, nus-1, limit)*unitscale;
 do {
 printf("\n");
 printf(" Present Oilstringer Initial Angle Is %f\n", phi3*180.0/pi);

 ynn[0] = tan(phi3);
 h[0] = 0.0;
 hstar = 0;
   nostar = 0;

 no = 1;
if(aqflag == 0)
lengthstart = Spline2(X_l.x, X_l.y, X_l.spl, nus-1, xstartp)*unitscale;
else
lengthstart = line2(X_l.x, X_l.y, X_l.spl, nus-1, xstartp)*unitscale;
length = lengthstart;
 Astar = 0.0
hmax = 0.0;
 do {
 if(dldAflag != 2) {
 h[no] = h[no-1] + dl * ynn[no-1];
 ynn[no] =ynn[no-1] + dl * CalcGsum();
 length = length - dl;
  }
 else if (Astar < dA) {
 dAl[no] = dl;
 h[no] = h[no-1] + dl * ynn[no-1];
  ynn[no] =ynn[no-1] + dl * CalcGsum();
  length = length - dl;
    nostar = no;
      hstar = h[no];
Astar = Astar + dl*h[no-1];
  }
else {
  h[no] = h[no-1] + dA/h[no-1] * ynn[no-1];
    ynn[no] = ynn[no-1] + dA/h[no-1] * CalcGsum();
  totalA = total A + dA;

  if (h[no] 0.0 && h[no] > hstar)
     dAl[no] = dA / h[no];

  else
     dA[no] = dAl[no] = dAl[no-1];



      length = length - dAl[no];

  }
```

```
if(aqflag == 0)
xlenth = Spline2(L_x.x, L_x.y, L_x.spl, nus, length/unitscale);
  else
xlenth = line2(L_x.x, L_x.y, L_x.spl, nus, length/unitscale);

no++;

 }   while((h[no-1] >= 0.0) && (h[no-1] < aH) && (length > limit));


  if(length <= limit) {
printf("The oilstringer is too long, please change the parameters.\n");
 exit(1);
 }

  if(h[no-1] <= 0.0) {
   updateno = no;
   updatelength = length;
   updatephi = phi3;

  for (i = 0; i < no; i++) {
   if(h[i] > hmax)   hmax = h[i];
   updatedAl[i] = dAl[i];
  updateh[i] = h[i];
    }
  updatehmax = hmax;
   printf("THE MAXIMUM OILSTR WIDTH IS %f METER\n", hmax/unitscale);
     }

     term1 = 0.0;
      term2 = 0.0;
       Aterm1 = 0.0;
        Aterm2 = 0.0;

if(searchflag ==0) break;
 phi();

   }   while( ((phih-phi3) > philimit) && (phi3 > 0.0));
     dldAflag = dldAflag + 1;
      dA = dl* hmax - dl * hmax / 5.0;
 }



void SMOOTH_N(float aqx[SIZEPLOT], float aqy[SIZEPLOT], float
oilx[SIZEPLOT],
        float datainput[SIZEPLOT], int nn)

/************************************************/
/*                                              */
/* Smoothing procedure that subdivides the initial data     */
/* set into "numb" subsets that each in turn consist of     */
```

```
/* every "numb'th" element, offset by one count from        */
/* the previous subset. The procedure is able to smooth      */
/* data sets with isolated large amplitude spikes as        */
/* well as large wavelength 'noise' without losing any      */
/* of the initial information in the data sets.          */
/*                                                       */
/*****************************************************/

 {

   float dataoutput[SIZEPLOT];
   int nul, numb;
   int ncount, i, j, k;
   float divis_x;
   float help[50][SIZEPLOT];
   FILE *draw;

   nul = 0; numb = 30;
   for(i=0; i< nn; i++) {
    }
nn--;

for (j = 0; j <= numb - 1; j++)
 for (i = 0; i <= nn; i++)
   help[j][i] = 0;

dataoutput[nul] = datainput[nul];
for (j = 0; j <= numb -1; j++)
 {
   i = nul + numb + j; help[j][i -numb] = datainput[i - numb];
   while (i <= nn - numb)
/***************************************************/
/*      3-point averaging of each sub set in turn        */
/***************************************************/
    {
     help[j][i] = (0.25 * datainput[i - numb]) +
              (0.5 * datainput[i]) +
              (0.25 * datainput[i + numb]);
    ncount = numb - 1;
    divis_x = (help[j][i] - help[j][i - numb]) / numb;
    while (ncount >= 1)
/***************************************************/
/* Linear interpolation between the smoothed elements    */
/*           -backwards in the data set              */
/***************************************************/
    {
      help[j][i - ncount] = divis_x * (numb - ncount) +
                   help[j][i - numb];
      ncount--;
    } /* end while */
   i += numb;
```

```
 } /* end while */
ncount = numb - 1; help[j][i] = datainput[i];
divis_x = (help[j][i] - help[j][i - numb]) / numb;
while (ncount >=1)
/**********************************************/
/* Linear interpolation between the smoothed elements */
/*      - last "numb - 1" points in the data set        */
/**********************************************/
  {
   help[j][i - ncount] = divis_x * (numb - ncount) +
                 help[j][i - numb];
   ncount--;
  } /* end while */
 } /* end for */
for (k = 1; k <= numb - 2; k++)
/**********************************************/
/*Arithmetic averaging of the "numb-2" first elements   */
/* (apart from the first) in each data set that have     */
/*              the same index in the array               */
/**********************************************/
 {
  dataoutput[nul + k] = help[0][nul + k] / (k + 1);
  for (j = 1; j<= k; j++)
   dataoutput[nul + k] += help[j][nul + k] / (k + 1);
 }
for (i = nul + numb - 1; i <= nn - (numb - 1); i++)
/**********************************************/
/* Arithmetic averaging of the elements in each sub set */
/*    that have the same index in the array             */
/**********************************************/
 {
  dataoutput[i] = 0;
  for (j = 0; j <= numb - 1; j++)
   dataoutput[i] += help[j][i];
  dataoutput[i] = dataoutput[i] / numb;
 } /* end for */

i = nn - (numb - 2);
dataoutput[nn] = datainput[nn];
k = 1;
while (i < nn)
/**********************************************/
/* Arithmetic averaging of the "numb-2" last elements */
/* (apart from the last) in each data set that have     */
/*              the same index in the array            */
/**********************************************/
 {
  dataoutput[i]=0;
  for (j = 0; j <= numb - 1; j++) {
   dataoutput[i] += help[j][i] / (numb - k);
 }
```

```
  k++; i++;
} /* end while */


 draw = fopen("draw.dat", "w");

fprintf(draw, "%s\n", plotname);
fprintf(draw, "%d\n", nn+1);
 for(i = 0; i<=nn; i++) {
 fprintf(draw, "%f %f %f %f\n", aqx[i], aqy[i], oilx[i], dataoutput[i]);
  }
} /*** end SMOOTH_N ***/



main()
 {

  float aqx[SIZEPLOT], aqy[SIZEPLOT], oilx[SIZEPLOT];
  float datainput[SIZEPLOT];
  int nnn;
float length1, ylen, anglen, Dx, Dy, tx, ty;
float xtemp, ytemp, ytemp1, xtempi, ang;
int nuxstep, istep;
int i, j, nuaq;
FILE *draw;
FILE *spl;
 spl = fopen("spl.dat", "r");
 fpt = fopen("simu.dat", "r");
  draw = fopen("draw.dat", "w");

   Read_sp2();
   Initial();
   Simu();
  if(dldAflag == 2)
   Simu();

if(updateno == 0) {
 printf(''No oilstringer exists for the parameters set up,
                        please change the parameters. \n'');
exit(1);
 }
 else {

 printf("\n");
 printf("$$$$$$$$$$$$ Finds a oilstringer! $$$$$$$$$$$\n");
 printf("\n");
 printf(" Oilstringer Initial Angle %f\n", updatephi*180.0/3.1416);
 printf(" Oilstringer Length  %f meters\n",
                 (lengthstart-updatelength)/unitscale);
 printf("The Maximum Oilstringer Width Is %f meters\n",
```

```
updatehmax/unitscale);
 printf("\n");
 printf("\n");

  fprintf(draw, "%s\n", plotname);
  fprintf(draw, "%d/n", updateno);
  nnn = updateno;
  length1 = updatelength;
   for(i=0; i<updateno; i++) {
   if(dldAflag == 3)
   length1 = length1 + updatedAl[i];
   else
   length1 = length1 + dl;

 if(aqflag == 0) {
  xi = Spline2(L_x.x, L_x.y, L_x.spl, nus, length1/unitscale);
  ylen = Spline2(Aq1.x, Aq1.y, Aq1.spl, nus, xi);
  {
else {
  xi = line2(L_x.x, L_x.y, L_x.spl, nus, length1/unitscale);
  ylen = line2(Aq1.x, Aq1.y, Aq1.spl, nus, xi);
 }

 if(h[i] < 0.0) h[i] = 0.0;
if(aqflag == 0)
  anglen = Spline2(Ang_l.x, Ang_l.y, Ang_l.spl, nus, length1/unitscale);
 else
  anglen = line2(Ang_l.x, Ang_l.y, Ang_l.spl, nus, length1/unitscale);
  h[i] = h[i]/unitscale;
  Dx = h[i] * sin(anglen);
  Dy = h[i] * cos(anglen);
 if(anglflag != -1) {
  tx = xi + Dx; ty = ylen - Dy;
   }
 else {
  tx = xi; ty = ylen - h[i];
   }

  aqx[i] = xi; aqy[i] = ylen;
  oilx[i] = tx; datainput[i] = ty;

  fprintf(draw, %f %f %f %f \n", aqx[i], ylen, oilx[i], ty);



   }
  fclose(draw);
  if(anglflag == -1)
  SMOOTH_N(aqx, aqy, oilx, datainput, nnn);
 }
}
```

Oil Stringer Graphics Routines

Four subsections of the graphics routine are presented here. They are Oilplot.c; Oilplot.h, and OilStr.h, and Oil.Set. Oilplot.c is the main graphics plotting routine. Oilplot.h pulls in graphics subroutines needed by Oilplot.c. OilStr.h sets array size limits. Oil.Set provides a set of screen control commands to ensure the pictures do not overflow screen size limits.

```
/**********************************************/
/*              VARIABLE DECLARATION          */
/**********************************************/
/* global variables declaration */

#include"oilplot.h"
#define PLOTSIZE 10000

int argc;
char **argv;
FILE* info;
int count, rene, fred, t;
int i,ii,jj,abx,aby,gmd,gmt;
int thickline,depth,pp,done,wix,wiy,layersnum,tstepsnum;
int colorflag,plotflag2=1,plotflag1,runflag=0,testflag=0;
int HEIGHT,WIDTH,RATW,RATH;
int wih,wil,size,sizescalex,sizescaley;
flot x1,y1,fdepth,period;
float maxdepth,depthstep,maxlength,lengthstep,ax,ay;
float scalex,scaley,scalet,echx,echy,fgmt,fgmd,fabx,faby,ft,c,Ld,de;
float ftdepthstep,maxftdepth,frene,ffred,time[50];
float f1,f2,f3,f4,f5,fintercept,limitn;
NPOINT npoints[2][PLOTSIZE];
MPOINT mpoints[2][PLOTSIZE],ppoints[PLOTSIZE];
char text[10],dep[8],distance[8],plotname[20];
int oil_flag,no,flowflag;

/**********************************************/
/*                MAIN FUNCTION               */
/**********************************************/

main()
 {

graphics_setup();
open_window();
```

```
done = 0;
wih = HEIGHT*0.65;
wil = WIDTH*3/4;
RATH = HEIGHT/HMAX;
RATW = WIDTH/WMAX;

read_data();

period = time[1]-time[layersnum];
plotflag1 = layersnum;
wix = (WIDTH-wil)/2;
wiy = (HEIGHT-wih)/2+30*RATH;

while(done == 0)
 {
if ( testflag != 0 )
 {
XClearWindow (mydisplay,mywindow);

write_legend();
tic_marks();
drawing_data();

XSetLineAttributes(mydisplay,mygc,size+1,0,0,0);
XSetForeground(mydisplay,mygc,white.pixel);
XDrawRectangle(mydisplay,mywindow,mygc,wix,wiy,wil,wih);

 } /* end testflag */
testflag = 1;

switch_process();

 } /* while (done == 0) */
XFreeGC (mydisplay,mygc);
XDestroyWindow (mydisplay, mywindow);
XCloseDisplay (mydisplay);
free(info);
exit (0);

 }

/***********************************************/
/*              SUBROUTINES LIBRARY            */
/***********************************************/
void graphics_setup()
{
int countion;

if ( (info=fopen("oil.set","r")) == NULL )
   {
```

```
  printf ("Cannot open the file called oil.set!\n");
  exit(1);
   }

/*** FILE INFORMATIONS ***/

read_dummies(4);
fscanf(info, "%f %f\n",&maxlength, &ax);
read_dummies(1);
fscanf(info, "%f\n", &lengthstep);
read_dummies(1);
fscanf(info, "%f\n",&maxdepth, &ay);
read_dummies(1);
fscanf(info, "%f\n",&depthstep);

/***GRAPHICS INFORMATIONS***/

read_dummies(4);
fscanf(info, "%d\n",&size);
read_dummies(2);
fscanf(info, "%d\n",&colorflag);

if ( size == 1 )
   {
   HEIGHT = 400;
   WIDTH = 600;
   }
else if ( size == 2 )
   {
   HEIGHT = 620;
   WIDTH = 900;
   }
else if ( size == 3 )
   {
   HEIGHT = 850;
   WIDTH = 1150;
   }

 }   /* end of graphics_setup */


void open_window()
 {
/* initialization */
mydisplay = XOpenDisplay(" ");
myscreen = DefaultScreen (mydisplay);

/* default pixel values */
mybackground = WhitePixel (mydisplay,myscreen);
myforeground = BlackPixel (mydisplay, myscreen);
```

```
/* default program-specified window position and size */
myhint.x = 0; myhint.y = 0;
myhint.width=WIDTH; myhint.height=HEIGHT;
myhint.flags = USPosition|USSize;

/* window creation */
mywindow = XCreateSimpleWindow (mydisplay,
 DefaultRootWindow (mydisplay),
 myhint.x, myhint.y, myhint.width, myhint.height,
 5, myforeground, mybackground);
XSetStandardProperties (mydisplay, mywindow, hello, hello,
 None, argv, argc, &myhint);

/* GC creation and initialization */
cmap = DefaultColormap(mydisplay,myscreen);
if ( colorflag != 0 )
{
white.red = 65535; white.green = 65535; white.blue = 65535;
}
else
{
white.red = 0; white.green = 0; white.blue = 0;
}
XAllocColor(mydisplay,cmap,&white);
if ( colorflag != 0 )
{
black.red = 0; black.green = 0; black.blue = 0;
}
else
{
black.red = 65535; black.green = 65535; black.blue = 65535;
}
XAllocColor(mydisplay,cmap,&black);
if ( colorflag != 0 )
{
yellow.red = 65535; yellow.green = 65535; yellow.blue = 0;
else
{
yellow.red = 0; yellow.green = 0; yellow.blue = 0;
}
XAllocColor(mydisplay,cmap,&yellow);
if ( colorflag != 0 )
{
peachpuff.red = 61166; peachpuff.green = 52171; peachpuff.blue = 44461;
}
else
{
peachpuff.red = 65535; peachpuff.green = 65535; peachpuff.blue = 65535;
}
XAllocColor(mydisplay,cmap,&peachpuff);
```

```
tradecolor.red = 655353; tradecolor.green = 11780; tradecolor.blue = 52096;
XAllocColor(mydisplay,cmap,&tradecolor);
seacolor.red = 35000; seacolor.green = 52000; seacolor.blue = 65000;
XAllocColor(mydisplay,cmap,&seacolor);
grey.red = 38000; grey.green = 38000; grey.blue = 38000;
XAllocColor(mydisplay,cmap,&grey);
darkbrown.red = 20000; darkbrown.green = 20000; darkbrown.blue = 20000;
XAllocColor(mydisplay,cmap,&darkbrown);
if ( colorflag != 0 )
{
green.red = 0; green.green = 32000; green.blue = 0;
}
else
{
green.red = 0; green.green = 0; green.blue = 0;
}
XAllocColor(mydisplay,cmap,&green);
limegreen.red = 12850; limegreen.green = 52685; limegreen.blue = 12850;
XAllocColor(mydisplay,cmap,&limegreen);
XSetWindowBackground(mydisplay,mywindow,black.pixel);
mygc = XCreateGC (mydisplay, mywindow,0,0);
if ( colorflag == 1 )
{
XSetBackground(mydisplay,mygc, black.pixel);
XSetForeground(mydisplay,mygc,white.pixcl);
}
else
{
XSetBackground(mydisplay,mygc, white.pixel);
XSetForeground(mydisplay,mygc,black.pixel);
}

/* input event selection */
XSelectInput (mydisplay,mywindow,
  ButtonPressMask|KeyPressMask|ExposureMask);

/* window mapping */
XMapRaised (mydisplay, mywindow);
}

void read_data()
{
int count1,count2;
int i;
FILE *plot;
FILE *draw;

plot = fopen("plot.dat", "r");
draw = fopen("draw.dat", "r");
```

```
fscanf (plot, "%d %d %d %d %f", &oil_flag, &tstepsnum, &layersnum,
                        &flowflag, &limitn);

if(flowflag == 1) {
for ( count1 = 0; count1 < layersnum; count1++ )
 {
 for ( count2 = 0; count2 < tstepsnum; count2++ )
  {
  fscanf (plot, "%f %f", &f1, &f2);
  npoints[count1] [count2].px = f1 + ax;
  npoints[count1] [count2].py = f2 + ay;
  }
 }
/* ppoints = (float)malloc((float) (2*tstepsnum+1)); */
 if (oil_flag == 1)
  {
  fscanf (draw, "%s", plotname);
  fscanf (draw, "%d", &no);
/*printf("%d\n", no); */
 for (i = 0; i < no; i++)
  {
  fscanf (draw, "%f %f %f %f", &f1, &f2, &f3, &f4);
  npoints[0] [i].ox = f1 + ax;
  npoints[0] [i].oy = f2 + ay;
  npoints[0] [i].oox = f3 + ax;
  npoints[0] [i].oh = f4 + ay;

/* printf("%f\n", npoints[0] [i].ox); */
   }
  }
 }
else {
for ( count1 = 0; count1 < layersnum; count1++)
 {
 for ( count2 = 0; count2 < tstepsnum; count2++)
  {
  fscanf (plot, "%f %f", &f1, &f2);
  npoints[count1] [count2].px = limitn - f1 + ax;
  npoints[count1] [count2].py = f2 + ay;
  }
 }
 if (oil_flag == 1)
  {
  fscanf (draw, "%s", plotname);
  fscanf (draw, "%d", &no);
/*printf ("%d\n", no); */
 for (i = 0; i < no; i++)
  {
 fscanf (draw, "%f %f %f %f", &f1, &f2, &f3, &f4);
```

```
 npoints[0][i].ox = limitn - f1 + ax;
 npoints[0][i].oy = f2 + ay;
 npoints[0][i].oox = limitn - f3 + ax;
 npoints[0][i].oh = f4 + ay;

/* printf("%f\n", npoints[0][i].ox);*/
    }
   }
  }
 }


void write_legend()
{
sizescalex = 0;
sizescaley = 0;
XSetLineAttributes(mydisplay,mygc,1,0,0,0);
XSetForeground(mydisplay,mygc,peachpuff.pixel);
XFillRectangle(mydisplay,mywindow,mygc,wix,wiy,wil,wih);
XSetForeground(mydisplay,mygc, yellow.pixel);
XDrawRectangle(mydisplay,mywindow,mygc, wix-120*RATW,
     wiy-145*RATH,wil+240*RATW,wih+245*RATH);
XSetForeground(mydisplay, mygc, white.pixel);
XSetLineAttributes(mydisplay,mygc,1,0,0,0);
if(size == 1) {
 sizescaley = -20;
 sizescalex = 120;
 }
XDrawString(mydisplay,mywindow,mygc, (wix-100*RATH)+sizescalex, (wiy-
130*RATH)
              +sizescaley, title0,strlen(title0));
if ( WIDTH >= 1000 )
 {
 XDrawString(mydisplay,mywindow,mygc,wix-80*RATW,wiy+wih/2,
                ord,strlen(ord));
 XDrawString(mydisplay,mywindow,mygc,wix+wil+70*RATW-20,wiy+wih/2,
                ord,strlen(ord));
 XDrawString(mydisplay,mywindow,mygc,wix-80*RATW-
2,wiy+wih/2+20*RATH,
                meter,strlen(meter));
 XDrawString(mydisplay,mywindow,mygc,wix+wil+70*RATW-20-2,
              wiy+wih/2+20*RATH,feet,strlen(feet));
 XDrawString(mydisplay,mywindow,mygc,wix+wil+70*RATW-20-2,
              wiy+wih-10+100*RATH,plotname,strlen(plotname));
XDrawString(mydisplay,mywindow,mygc, (wix+wil/3+40*RATW)+sizescalex,
              (wiy+wih+50*RATH)+sizescaley,abs,
 strlen(abs));
 sizescaley = 160;
 sizescalex = 800;
```

```
if(flowflag == 1) {
XDrawString(mydisplay,mywindow,mygc, (wix-120*RATW)+sizescalex,
          (wiy-140*RATH)+sizescaley, sea,strlen(sea));

  XDrawLine(mydisplay,mywindow,mygc, (wix-120*RATW)+sizescalex, (wiy-
140*RATH)
  +sizescaley+10, (wix-120*RATW)+sizescalex+60, (wiy-
140*RATH)+sizescaley+10);
  XDrawLine(mydisplay,mywindow,mygc, (wix-120*RATW)+sizescalex, (wiy-
140*RATH)
  +sizescaley+10, (wix-120*RATW)+sizescalex+10, (wiy-
140*RATH)+sizescaley+5);
  }
else {
 sizescalex = 140;
XDrawString(mydisplay,mywindow,mygc, (wix-120*RATW)+sizescalex,
          (wiy-140*RATH)+sizescaley, sea, strlen(sea));

 XDrawLine(mydisplay,mywindow,mygc, (wix-120*RATW)+sizescalex, (wiy-
140*RATH)
  +sizescaley+10, (wix-120*RATW)+sizescale+100), (wiy-
140*RATH)+sizescaley+10);
 XDrawLine(mydisplay,mywindow,mygc, (wix-
120*RATW+100)+sizescalex, (wiy-140*RATH)
  +sizescaley+10, (wix-120*RATW)+sizescalex+90, (wiy-
140*RATH)+sizescaley+5);

   }

  }
else
  {
 sizescalex = -45;
 sizescaley = -120;
 XDrawString(mydisplay,mywindow,mygc, (wix-20)+sizescalex, (wiy+wih+10)
          +sizescaley,ord,strlen(ord));
 sizescalex = 35;
 sizescaley = -120;
 XDrawString(mydisplay,mywindow,mygc, (wix+wil)+sizescalex, (wix+wih+10)
                +sizescaley, ord,strlen(ord));
 sizescalex = -40;
 sizescaley = -110;
 XDrawString(mydisplay,mywindow,mygc, (wix-20)+sizescalex, (wiy+wih+20)
                +sizescaley, meter,strlen(meter));
 sizescalex = 38;
 sizescaley = -110;
 XDrawString(mydisplay,mywindow,mygc, (wix+wil)+sizescalex,
              (wiy+wih+20)+sizescaley,feet,strlen(feet));
 sizescalex = 0;
```

```
 sizescaley = 50;
XDrawString(mydisplay,mywindow,mygc, (wix+wil/3+40*RATW)+sizescalex,
          (wiy+wih+50*RATH)+sizescaley,abs,strlen(abs));
 sizescalex = 300;
 sizescaley = 20;
if(flowflag == 1) {
XDrawString(mydisplay,mywindow,mygc, (wix-120*RATW)+sizescalex,
        (wiy-140*RATH)+sizescaley, sea,strlen(sea));
 XDrawLine(mydisplay,mywindow,mygc, (wix-120*RATW)+sizescalex, (wiy-
140*RATH)
 +sizescaley+10, (wix-120*RATW)+sizescalex+60, (wiy-
140*RATH)
 XDrawLine(mydisplay,mywindow,mygc, (wix-120*RATW)+sizescalex, (wiy-
140*RATH)
+sizescaley+10, (wix-120*RATW)+sizescalex+10, (wiy-140*RATH)+sizescaley+5);
   }
else {
 sizescalex = 20;
XDrawString(mydisplay,mywindow,mygc, (wix-120*RATW)+sizescalex,
        (wiy-140*RATH)+sizescaley, sea, strlen(sea));

 XDrawLine(mydisplay,mywindow,mygc, (wix-120*RATW)+sizescalex+10, (wiy-
140*RATH)
 +sizescaley+10, (wix-120*RATW)+sizescale+70, (wiy-
140*RATH)+sizescaley+10);
 XDrawLine(mydisplay,mywindow,mygc, (wix-120*RATW)+sizescalex+60, (wiy-
140*RATH)
 +sizescaley+5, (wix-120*RATW)+sizescalex+70, (wiy-140*RATH)+sizescaley+10);
    }
  }
/*XSetForeground(mydisplay,mygc,tradecolor.pixel); */



/*XDrawString(mydisplay,mywindow,mygc,wix+wil*2/5,wiy-90*RATH,
              title1,strlen(title1));*/
XSetLineAttributes(mydisplay,mygc,2,0,0,0);

/*XSetForeground(mydisplay,mygc,white.pixel);
XDrawRectangle(mydisplay,mywindow,mygc,wix,wiy-
45*RATH,wil,25*RATH);*/
}

void tic_marks()
{
scalex = wil/maxlength;
scaley = wih/maxdepth;
ftdepthstep = 3*depthstep;
maxftdepth = maxdepth/0.3048;
fgmt = maxlength/lengthstep;
```

```
gmt = fgmt;
fabx = scalex*lengthstep;
abx = fabx;
faby = scaley*depthstep;
aby = faby;
fgmd = maxdepth/depthstep;
gmd = fgmd;

tic_lateral();
deep_meters();

/* if using same unit in two sides, common next five statements. */
scaley = wih/maxftdepth;
faby = scaley*ftdepthstep;
aby = faby;
fgmd = maxftdepth/ftdepthstep;
gmd = fgmd;

deep_feet();

scalet = (wil-2)/period;
scalex = wil/maxlength;
scaley = wih/maxdepth;

}

void tic_lateral()
 {
t = wix;
ft = t;
Ld = 0;
for ( count = 0; count <= gmt; ++count)
  {
  sprintf ( distance, "%4.1f",Ld );
  XSetLineAttributes(mydisplay,mygc,1,0,0,0);
  XSetForeground(mydisplay,mygc, white.pixel);
  XDrawString(mydisplay,mywindow,mygc,t-15,wiy+wih+25,distance,
                        strlen(distance));
  XDrawLine(mydisplay,mywindow,mygc,t,wiy+wih,t,wiy+wih+6);
  ft = ft + fabx;
  t = ft;
  Ld = Ld + lengthstep;
  }
 }

void deep_meters()
{
t = wih+wiy;
ft = t;
de = 0;
```

```
for ( count = 0; count <= gmd; count++)
   {
   sprintf ( dep, "%5.0f",de );
   XSetLineAttributes(mydisplay,mygc,1,0,0,0);
   XSetForeground(mydisplay,mygc,white.pixel);
   XDrawString(mydisplay,mywindow,mygc,wix-45,t+4,dep,strlen(dep));
   XDrawLine(mydisplay,mywindow,mygc,wix-2,t,wix-6,t);
   ft = ft - faby;
   t = ft;
   de = de + depthstep;
   }
 }


/* using same units in two side */
/*
void deep_feet()
{
t = wih+wiy;
ft = t;
de = 0;
for ( count = 0; count <= gmd; count++)
   {
   sprintf ( dep, "%5.0f",de );
   XSetLineAttributes(mydisplay,mygc,1,0,0,0);
   XSetForeground(mydisplay,mygc, white.pixel);
if(size == 1)

XDrawString(mydisplay,mywindow,mygc,wix+wil+22*RATW,t+4,dep,strlen(de
p));
else
 XDrawString(mydisplay,mywindow,mygc,wix+wil+22*RATW-
20,t+4,dep,strlen(dep));
 XDrawLine(mydisplay,mywindow,mygc,wix+wil+2,t,wix+wil+6,t);
ft = ft - faby;
t = ft;
de = de + depthstep;
  }
 }
*/
void deep_feet()
{
t = wih+wiy;
ft = t;
de = 0;
for ( count = 0; count <= gmd; count++)
  {
  sprintf ( dep, "%5.0f",de );
  XSetLineAttributes(mydisplay,mygc,1,0,0,0);
  XSetForeground(mydisplay,mygc, white.pixel);
if(size == 1)
```

```
XDrawString(mydisplay,mywindow,mygc,wix+wil+22*RATW,t+4,dep,strlen(de
p));
else
 XDrawString(mydisplay,mywindow,mygc,wix+wil+22*RATW-
20,t+4,dep,strlen(dep));
 XDrawLine(mydisplay,mywindow,mygc,wix+wil+2,t,wix+wil+6,t);
  ft = ft - faby;
  t = ft;
  de = de + ftdepthstep;
  }
 }

void tic_time()
{
for ( ii = 1; ii <= plotflag1; ++i)
  {
  ft = wix + wil - 1 + scalet*(time[layersnum]-time[ii]);
  t = ft;
  sprintf ( dep, "%3.2f",time[ii] );
  XSetForeground(mydisplay,mygc, white.pixel);
  XDrawString(mydisplay,mywindow,mygc,t-6,wiy-65,dep,strlen(dep));
  XDrawLine(mydisplay,mywindow,mygc,t,wiy-45*RATH,t,wiy-51*RATH);
  XSetForeground(mydisplay,mygc, green.pixel);
  XFillRectangle(mydisplay,mywindow,mygc,wix,wiy-45*RATH,t-
wix,25*RATH);
   }   /* end for */
}

void drawing_data()
{
 int i, j;
 for (j=0; j < 2; j++)
   {
 for (i=0; i < tstepsnum; i++)
   {
    mpoints[j][i].pxi = (npoints[j][i].px * scalex) + wix;
    mpoints[j][i].pyi = wih - (npoints[j][i].py * scaley) + wiy;
   }
  }


 for (i = 0; i < tstepsnum; i++)
   {
    ppoints[i].pxi = mpoints[0][i].pxi;
    ppoints[i].pyi = mpoints[0][i].pyi;
   }
 for (i = 0; i < tstepsnum; i++)
    {
    ppoints[2*tstepsnum-i-1].pxi = mpoints[1][i].pxi;
    ppoints[2*tstepsnum-i-1].pyi = mpoints[1][i].pyi;
    }
```

```
    if ( colorflag != 0 )
    XSetForeground(mydisplay,mygc, seacolor.pixel);
    else
    XSetForeground(mydisplay,mygc, black.pixel);

    XFillPolygon(mydisplay,mywindow,mygc,ppoints, (2*tstepsnum),Complex,
            CoordModeOrigin);
    if ( colorflag != 0 )
    XSetForeground(mydisplay,mygc, black.pixel);
    else
    XSetForeground(mydisplay,mygc, white.pixel);

   XSetLineAttributes(mydisplay,mygc,3,0,0,0);
    XDrawLines(mydisplay,mywindow,mygc,mpoints[0], tstepsnum,
        CoordModeOrigin);
    XDrawLines(mydisplay,mywindow,mygc,mpoints[1], tstepsnum,
        CoordModeOrigin);

/* draw oilstrger */

  if(oil_flag == 1)
    {
  for (i = 0; i < no; i++)
    {
    mpoints[0][i].pxi = (npoints[0][i].oox * scalex) + wix;
    mpoints[0][i].pyi = wih - (npoints[0][i].oh * scaley) + wiy;
    mpoints[1][i].pxi = (npoints[0][i].ox * scalex) + wix;
    mpoints[1][i].pyi = wih - (npoints[0][i].oy * scaley) + wiy;
    }

  for (i = 0; i < no; i++)
    {
    ppoints[i].pxi = mpoints[0][i].pxi;
    ppoints[i].pyi = mpoints[0][i].pyi;
    }
  for (i = 0, i < no; i++)
    {
    ppoints[2*no-i-1].pxi = mpoints[1][i].pxi;
    ppoints[2*no-i-1].pyi = mpoints[1][i].pyi;
    }

    XSetForeground(mydisplay,mygc, darkbrown.pixel);
    XFillPolygon(mydisplay,mywindow,mygc,ppoints, (2*no),Complex,
            CoordModeOrigin);
/*
    XSetForeground(mydisplay,mygc, black.pixel);
    XDrawLines(mydisplay,mywindow,mygc,mpoints[0], no,
CoordModeOrigin);
*/
    }
  }
```

```
void switch_process()
{
/* read the next event */
XNextEvent ( mydisplay, &myevent );
switch (myevent.type)
{

/* process mouse-button presses */
case ButtonPress:
if ( plotflag2 == 0 ) plotflag2 = 1;
else plotflag2 = 0;
break;

/* process keyboard input */
case KeyPress:
i = XLookupString ( &myevent, text, 10, &mykey, 0);
 if (i == 1 && text[0] == 'q') done = 1;
 else if (i == 1 && text[0] == 'f' && plotflag1 != layersnum)
  plotflag1 = plotflag1 +1;
 else if (i == 1 && text[0] == 'b' && plotflag1 != 1)
  plotflag1 = plotflag1 - 1;
 else if (i == 1 && text[0] == 'm' && plotflag1 != layersnum)
  plotflag1 = layersnum;
 else if (i == 1 && text[0] == '0' && plotflag1 != 1)
  plotflag1 = 1;
 else testflag=0;
if (plotflag1 >= layersnum+1) plotflag1 = layersnum;
if (plotflag1 <= 1) plotflag1 = 1;
 break;
 case MappingNotify:
  XRefreshKeyboardMapping(&myevent);
 break;

} /* switch (myevent.type) */
}

/***********************************************/
/*                                             */
/*              PERSONAL LIBRARY               */
/*                                             */
/***********************************************/
void read_dummies(dumnum)
int dumnum;
 {
int countion;
char dumchar[100];

for ( countion =0; countion < dumnum; ++countion )
  {
  fgets(dumchar,100,info);
  }
 }
```

```
/* X include files */
#include </usr/include/X11/Xlib.h>
#include </usr/include/X11/Xutil.h>
#include </usr/include/X11/X.h>
#include <stdio.h>
#include <string.h>
#include <malloc.h>

#define FIGURE 5
#define WMAX 1150
#define HMAX 850

typedef struct
    {
    float px, py, ox, oy, oox, oh;
    } NPOINT;

typedef struct
    {
    short int pxi, pyi;
    } MPOINT;

/*************************************************/
/*              SUBROUTINES DECLARATION          */
/*************************************************/
void graphics_setup();
void read_dummies();
void open_window();
void read_data();
void write_legend();
void tic_marks();
void deep_feet();
void deep_meters();
void tic_lateral();
void tic_time();
void drawing_data();
void switch_process();

/* declaration for the window */
Display *mydisplay;
Colormap cmap;
Window mywindow;
XColor black, white, grey, yellow, green, peachpuff, seacolor, limegreen,
                        tradecolor, darkbrown;
GC mygc;
XEvent myevent;
KeySym mykey;
XSizeHints myhint;
Drawable window;
unsigned long myforeground, mybackground;
int myscreen;
```

```
/* declarations */
char hello[] = { "OILSTRINGER" };
char wname[] = { "NAME : " };
char wlocal[] = { "LOCATION : " };
char age[] = { "PRESENT DAY" };
char sea[] = { "DIRECTION OF WATER FLOW" };
char trade[] = { "FM-RT plot " };
char ord[] = { "DEPTH" };
char meter[] = { " (meter) " };
char feet[] = { " (feet) " };
char abs[] = { "LATERAL DISTANCE IN METERS" };
char title0[] = { "SHAPE OF OILSTRINGER IN THE AQUIFER" };
#define SIZE 100000
#define SIZESPL 5000
#define SIZEPLOT 1000
#define SIZEAQ 500

typedef struct {
          float x[SIZESPL], y[SIZESPL], spl[SIZESPL];
        } SPL;

SPL Aq1, Ang_1, K1_1, H1_1, X_1, L_x;
double h[SIZE], updateh[SIZE], ynn[SIZE], hh[SIZE];
```

```
-------------------------------------------------------
            FILE INFORMATIONS
-------------------------------------------------------
1-WHAT IS THE LENGTH OF YOUR SECTION? (M)
          60.0        2.0

2-SPACE BETWEEN TIC MARKS FOR THE LENGTH? (M)
          5.0

3-WHAT IS THE MAXIMUM DEPTH TO BE APPEARING ON THE GRAPHICS? (M)
          60.00       0.0

4-SPACE BETWEEN TIC MARKS FOR THE DEPTH? (M)
          5.0

-------------------------------------------------------
            GRAPHICS INFORMATIONS
-------------------------------------------------------
5-SIZE of the WINDOW (SMALL: 1; MEDIUM: 2; LARGE: 3)
              3

6-COLOR OF THE PICTURE
(0: black and white; 1(2): color with black (white) background)
              1
```

References

Amyx, J. W., Bass, D. M., Jr., and Whiting, R. L. 1960, *Petroleum Reservoir Engineering, Physical Properties,* McGraw-Hill, New York.

Bakr, A. A., Gelhar, L. W., Gutjahr, A. L., and MacMillan, J. R., 1978, Stochastic analysis of spatial variability in subsurface flows, 1. Comparison of one-dimensional flows, *Water Resour. Res.* **14**, 263–271.

Barnes, H. L., 1979, *Geochemistry of Hydrothermal Ore Deposits,* John Wiley & Sons, New York.

Bear, J., 1972, *Dynamics of Fluids in Porous Media,* American Elsevier, New York.

Berg, R. R., 1975, Capillary pressures in stratigraphic traps, *Am. Assoc. Pet. Geol. Bull.* **59**, 939–956.

Braun, R. L., and Burnham, A. K., 1987, Analysis of chemical reaction kinetics using a distribution of activation energies and simpler models, *Energy Fuels* **1**, 153–161.

Brink, A. H., 1974, Petroleum geology of Gabon Basin, *Am. Assoc. Pet. Geol. Bull.* **58**, 216–235.

Cao, S, 1987, Sensitivity analysis of a 1-D dynamical model for basin analysis, Ph.D. Dissertation, University of South Carolina.

Cheng, P., 1978, Heat transfer in geothermal systems, *Adv. Heat Transfer* **14**, 1–105.

Combarnous, M. A., and Bories, S. A., 1975, Hydrothermal convection in saturated porous media, *Adv. Hydrosci.* **10**, 231–307.

Crank, J., 1987, *Free and Moving Boundary Problems,* Oxford Science Publications, Oxford.

Dagan, G., 1979, Models of groundwater flow in statistically homogeneous porous formations, *Water Resour. Res.* **15**, 47–63.

Dahlberg, E. C., 1982, *Applied Hydrodynamics in Petroleum Exploration,* Springer-Verlag, New York.

Darcy, H., 1856, Les fontaines publiques de la ville de Dijon, in *Atlas,* pp. 590–594, Figure 3, Victor Dalmont, Paris.

Davis, R. W., 1987, Analysis of hydrodynamic factors in petroleum migration and entrapment, *Am. Assoc. Pet. Geol. Bull.* **71**, 643–649.

de Wiest, R. J. M. (ed.), 1969, *Flow through Porous Media,* Academic Press, New York.

Di Pippo, R., 1980, *Geothermal Energy as a Source of Electricity,* U.S. Department of Energy, Washington, D.C.

Dutta, N. C., 1987, Fluid flow in low permeable porous media, in *Migration of Hydrocarbons in Sedimentary Basins* (B. Doligez, ed.) pp. 567–596, Editions Technip, Paris.

Feller, W., 1966, *An Introduction to Probability Theory and Its Applications,* John Wiley & Sons, New York.

Freeze, R. A., 1975, A stochastic conceptual analysis of one-dimensional groundwater in nonuniform homogenous media, *Water Resour. Res.* **11**, 725–741.

Frisch, U., 1968, Wave propagation in random media, in *Probabilistic Methods in Applied Mathematics,* Vol. 1 (A. T. Bharucha-Reid, ed.), Pergamon Press, New York.

Fyfe, W. S., Price, N. J., and Thompson, A. B., 1978, *Fluids in the Earth's Crust,* Elsevier, Amsterdam.

Garg, S. K., and Kassoy, D. R., 1981, Convective heat and mass transfer in hydrothermal systems, in *Geothermal Systems: Principles and Case Histories* (L. Rybach and L. J. P. Muffle, eds.), pp. 37–64. John Wiley & Sons, London.

Garrels, R. M., and Christ, C. L., 1965, *Solutions, Minerals and Equilibria,* Freeman, Cooper and Co., San Francisco.

Goguel, J., 1976, *Geothermics,* McGraw-Hill, New York.

Gutjahr, A. L., and Gelhar, L. W., 1981, Stochastic models of subsurface flow: Infinite versus finite domains and stationary flows, *Water Resour. Res.* **17**, 337–350.

Gutjahr, A. L., Gelhar, L. W., Bakr, A. A., and MacMillan, J. R., 1978, Stochastic analysis of spatial variability in subsurface flows, 2. Evaluation and application, *Water Resour. Res.* **14**, 953–959.

Hobson, G. D., 1954, *Some Fundamentals of Petroleum Geology,* Oxford University Press, London.

Hubbert, M. K., 1940, The theory of ground-water motion, *J. Geol.* **48**, 209–217.

Hubbert, M. K., 1953, Entrapment of petroleum under hydrodynamic conditions, *Am. Assoc. Pet. Geol. Bull.* **37**, 1954–2026.

Hunt, J. M., 1979, *Petroleum Geochemistry and Geology,* W. H. Freeman, San Francisco.

Jones, P. H., 1975, Geothermal and hydrocarbon regimes, northern Gulf of Mexico, in *Proceedings of the First Geopressured Geothermal Energy Conference,* U.S. Geological Survey, Austin, Texas, pp. 15–89.

Kruger, P., and Otte, C. (eds.), 1973, *Geothermal Energy, Resources, Production, Stimulation,* Stanford University Press, Stanford, California.

Lamb, H., 1879, *Treatise on the Mathematical Theory of the Motion of Fluids,* Cambridge University Press, Cambridge.

Laplace, P. S., 1789, Mémoire de L'Académie.

Leder, F., and Park, W. C., 1986, Porosity reduction in sandstone by quartz overgrowth, *Am. Assoc. Pet. Geol. Bull.* **70**, 1713–1728.

Lerche, I., 1990a, *Basin Analysis: Quantitative Methods,* Vol. 1, Academic Press, San Diego.

Lerche, I., 1990b, *Basin Analysis: Quantitative Methods,* Vol. 2, Academic Press, San Diego.

Lewis, C. R., and Rose, S. C., 1970, A theory relating high temperatures and overpressures, *J. Pet. Technol.* **22**, 11–16.

Lister, C. R. B., 1980, Heat flow and hydrothermal circulation, *Annu. Rev. Earth Planet. Sci.* **8**, 95–118.

Lumley, D., 1970, *Stochastic Tools in Turbulence,* Academic Press, New York.

Morse, P. M., and Feshbach, H., 1953, *Methods of Theoretical Physics,* McGraw-Hill, Reading, Massachusetts.

Muskat, M., 1946, *The Flow of Homogeneous Fluids through Porous Media,* J. W. Edwards, Ann Arbor, Michigan.

Nind, T. E. W., 1981, *Principles of Oil Well Production,* McGraw-Hill, New York.

Phillips, O. M., 1991, *Flow and Reactions in Permeable Rocks,* Cambridge University Press, Cambridge.

Polubarinova-Kochina, P. Ya, 1962, *Theory of Ground Water Movement,* Princeton University Press, Princeton, New Jersey.

Purcell, W.R ., 1949, Capillary pressures—their measurement using mercury and the calculation of permeability therefrom, *Trans. Am. Inst. Min. Metall. Eng.* **186**, 39–48.

Rybach, L., and Muffler, L. J. P. (eds.), 1981, *Geothermal Systems: Principles and Case Histories,* John Wiley & Sons, London.

Scheiddegger, A. E., 1960, *The Physics of Flow through Porous Media,* University of Toronto Press, Toronto, Canada.

Sclater, J. G., and Christie, P. A. F., 1980, Continental stretching: An explanation of the post-mid-Cretaceous subsidence of the central North Sea basin, *J. Geophys. Res.* **85**, 3711–3739.

Shorthose, D. N., 1936, *Properties of Matter,* W. Heinemann, London.

Smith, L., and Freeze, R. A., 1979, Stochastic analysis of steady-state groundwater flow in a bounded domain, 1. One-dimensional simulations, *Water Resour. Res.* **15**, 521–528.

Tiratsoo, E. N., 1952, *Principles of Petroleum Geology,* McGraw-Hill, Reading, Massachusetts.

Tissot, B., and Welte, D. H., 1978, *Petroleum Formation and Occurrence: A New Approach to Oil and Gas Exploration,* Springer-Verlag, Berlin.

Turcotte, D. L., and Schubert, G., 1982, *Geodynamics: Applications of Continuum Physics to Geological Problems,* John Wiley & Sons, New York.

Warren, J. E., and Price, H. S., 1961, Flow in heterogeneous porous media, *Soc. Pet. Eng. J.* **1**, 153–169.

Waring, G. A., 1965, Thermal springs of the United States and other countries of the world—a summary, U.S. Geological Survey Professional Paper 492.

Yukler, M. A. and Kokesh, F., 1984, A review of models used in petroleum resource estimation and organic geochemistry, *Adv. Pet. Geochem.* **1**, 69–114.

Index